Ecological Studies, Vol. 131

Analysis and Synthesis

Edited by

M.M. Caldwell, Logan, USA
G. Heldmaier, Marburg, Germany
O.L. Lange, Würzburg, Germany
H.A. Mooney, Stanford, USA
E.-D. Schulze, Bayreuth, Germany
U. Sommer, Kiel, Germany

Ecological Studies

Volumes published since 1992 are listed at the end of this book.

Springer

New York
Berlin
Heidelberg
Barcelona
Budapest
Hong Kong
London
Milan
Paris
Santa Clara
Singapore
Tokyo

Erik Jeppesen   Martin Søndergaard
Morten Søndergaard   Kirsten Christoffersen

Editors

The Structuring Role of Submerged Macrophytes in Lakes

With 117 illustrations

Springer

Erik Jeppesen
Department of Lake and Estuarine Ecology
National Environmental Research Institute
DK-8600 Silkeborg
Denmark

Martin Søndergaard
Department of Lake and Estuarine Ecology
National Environmental Research Institute
DK-8600 Silkeborg
Denmark

Morten Søndergaard
Freshwater Biological Laboratory
University of Copenhagen
DK-3400 Hillerød
Denmark

Kirsten Christoffersen
Freshwater Biological Laboratory
University of Copenhagen
DK-3400 Hillerød
Denmark

Cover illustration: As it will appear from this book, submerged macrophytes may have an important impact on the nutrient dynamics, trophic structure, and trophic interactions of shallow lakes. Within certain nutrient limits, submerged macrophytes may via a number of feedback mechanisms maintain a clearwater state despite increased nutrient supply. Drawing by Bjørn Bachmann and Erik Jeppesen.

Library of Congress Cataloging-in-Publication Data
The structuring role of submerged macrophytes in lakes/Erik Jeppesen
      ... [et al.].
         p.   cm.—(Ecological studies; v. 131)
      Includes bibliographical references and index.
      ISBN 0-387-98284-1 (hc: alk. paper)
      1. Lake ecology.   2. Lake plants—Ecology.   I. Jeppesen, Erik.
   II. Series.
   QH541.5.L3S77   1997
   577.63—dc21   97-22884

Printed on acid-free paper.

Production coordinated by Princeton Editorial Associates and managed by Francine McNeill; manufacturing supervised by Jeffrey Taub.
Typeset by Princeton Editorial Associates, Princeton, NJ.
Printed and bound by Maple-Vail Book Manufacturing Group, York, PA.
Printed in the United States of America.

9 8 7 6 5 4 3 2 1

ISBN 0-387-98284-1 Springer-Verlag New York Berlin Heidelberg   SPIN 10632906
ISSN 0070-8356

Preface

Submerged macrophytes have been the object of intensive research, and a large body of literature exists on their growth, reproduction, and physiology. Several studies have focused on the interactions between submerged macrophytes and other autotrophic components and the impact of the plants on the dynamics of nutrients, dissolved organic and inorganic carbon, oxygen, and pH. Comparatively few studies have dealt with the ability of submerged macrophytes to modulate the structure and dynamics of pelagic and benthic food webs. Recently, however, the amount of research into the structuring role of submerged macrophytes in food webs has markedly increased, and the results obtained so far suggest that submerged macrophytes are of significant importance for the food web interactions and environmental quality of lakes, even at relatively low areal plant coverage. For example, plants affect the interactions between predacious, planktivorous, and benthivorous fish and between fish and invertebrates, including key organisms such as large zooplankton and snails. Changes in these interactions in turn may have cascading effects on the entire food web in both the pelagial and the littoral zone.

To provide a forum for discussion of recent results in this growing field of research and to define future research needs, a workshop was held on 16 to 20 June, 1996, at the Freshwater Centre in Silkeborg, Denmark. The present book is a result of the workshop. It is divided into three parts. The first part consists of 10 thematic chapters (Chapters 1 to 10) describing how submerged macrophytes influence various biological and biogeochemical interactions in lakes. These chapters are written by authors having specialized knowledge within the field treated. Cascading effects through the food web as a result of changes in resource and predator/grazer control are given main emphasis in several of these chapters. The authors were given the option of either writing a state-of-the-art review or discussing the subject based on mainly their own investigations. The second part consists of 18 case studies (Chapters 11 to 28) related to the thematic chapters, and the third part (Chapters 29 to 31) summarizes three of the workshop's cross-subject discussions. The authors were here given the option of writing a summary of the discussion or treating the subject more extensively, using the workshop discussions as their starting point.

We thank translator Anne Mette Poulsen of the National Environmental Research Institute for her efficient help in planning and arranging the workshop and in the subsequent editing phase. We are also grateful to assistant editor Janet Slobodien of Springer-Verlag and the staff of Princeton Editorial Associates for fruitful and efficient cooperation. Finally, we thank the Strategical Environmental Research Programme, the Danish Natural Science Research Council, and the National Environmental Research Institute for the financial support that made this workshop possible.

The Editors

Contents

Section 2. Case Studies

Contributors

Gunnar Andersson

County Administration Board, S-205 15
Malmö, Sweden

John W. Barko

Environmental Laboratory, USACE
Waterways Experiment Station,
Vicksburg, MS 39180-6199,
USA

Irmgard Blindow

Department of Limnology, Lund
University, S-223 62 Lund, Sweden

Christer Brönmark

Department of Ecology, Lund University,
S-223 62 Lund, Sweden

Lise Bruun

Department of Lake and Estuarine
Ecology, National Environmental Research
Institute, DK-8600 Silkeborg, Denmark

Stephen R. Carpenter

Center for Limnology, University of
Wisconsin, Madison, WI 53706, USA

Contributors

Kirsten Christoffersen

Freshwater Biological Laboratory,
University of Copenhagen, DK-3400
Hillerød, Denmark

Hugo Coops

Institute for Inland Water Management
and Waste Water Treatment, 8200 AA
Lelystad, The Netherlands

Greg Cronin

Cooperative Institute for Research in
Environmental Sciences (CIRES),
University of Colorado/NOAA, Boulder,
CO 80309-0216, USA

Larry B. Crowder

Marine Laboratory, Nicholas School of
the Environment, Duke University,
Beaufort, NC 28516-9721, USA

Paul Cunningham

Bureau of Fish Management, Department
of Natural Resources, Madison, WI
53703, USA

Sebastian Diehl

Zoologisches Institut,
Ludwig-Maximilians Universität,
D-80333 München, Germany

Bjørn A. Faafeng

Norwegian Institute for Water Research,
Kjelsaas, 0411 Oslo, Norway

Adrienne J. Froelich

Department of Biological Sciences,
University of Notre Dame, Notre Dame
IN 46556, USA

Sarig Gafny

Institute for Nature Conservation Research,
George S. Wise Faculty of Life Sciences,
Tel Aviv University, Tel Aviv 69978,
Israel

Avital Gasith

Institute for Nature Conservation Research,
George S. Wise Faculty of Life Sciences,
Tel Aviv University, Tel Aviv 69978, Israel

Binhe Gu

Division of Environmental Sciences,
St. Johns River Water Management
District, P.O. Box 1429, Palatka, FL
32178-1429, USA

Anders Hargeby

Department of Biology, University College of Karlstad, S-65188 Karlstad, Sweden

Mark V. Hoyer

Department of Fisheries and Aquatic Sciences, University of Florida/Institute of Food and Agricultural Sciences, 7922 NW 71st Street, Gainesville, FL 32653-3071, USA

William F. James

Eau Galle Aquatic Ecology Laboratory, USACE Waterways Experiment Station, Spring Valley, WI 54767, USA

Jens Peder Jensen

Department of Lake and Estuarine Ecology, National Environmental Research Institute, DK-8600 Silkeborg, Denmark

Erik Jeppesen

Department of Lake and Estuarine Ecology, National Environmental Research Institute, DK-8600 Silkeborg, Denmark

John Iwan Jones

Royal Halloway Institute for Environmental Research, Royal Halloway College, University of London, Huntersdale, Callow Hill, Virginia Water, Surrey GU25 4LN, UK

Klaus Jürgens

Max Planck Institute for Limnology, D-24302 Plön, Germany

Timo Kairesalo

Department of Ecological and Environmental Sciences, University of Helsinki, 15210 Lahti, Finland

Eva Kanstrup

Department of Lake and Estuarine Ecology, National Environmental Research Institute, 8600 Silkeborg, Denmark

Ryszard Kornijów

Department of Hydrology and
Ichthyobiology, University of
Agriculture, 20-950 Lublin 1, Poland

Torben L. Lauridsen

Department of Lake and Estuarine
Ecology, National Environmental Research
Institute, DK-8600 Silkeborg, Denmark

David M. Lodge

Department of Biological Sciences,
University of Notre Dame, Notre Dame,
IN 46556, USA

Thomas H. Martin

School of Forest Resources,
Pennsylvania State University, University
Park, PA 16802, USA

Elizabeth W. McCollum

1715 Broadview Lane, Apt. 209, Ann
Arbor, MI 48105, USA

Marie-Louise Meijer

Institute for Inland Water Management
and Waste Water Treatment, 8200 AA
Lelystad, The Netherlands

Stuart F. Mitchell

Department of Zoology, University of
Otago, Dunedin, New Zealand

Marit Mjelde

Norwegian Institute for Water Research,
Kjelsaas, 0411 Oslo, Norway

Robert E. Moeller

Department of Earth and Environmental
Science, 31 Williams Drive, Lehigh
University, Bethlehem, PA 18015, USA

Brian Moss

School of Biological Sciences,
University of Liverpool, Liverpool
L69 3BX, UK

Nathan Nibbelink

Department of Zoology and Physiology,
University of Wyoming, Laramie, WY
82071, USA

Mark Olson

Cornell University Biological Field
Station, Bridgeport, NY 13030, USA

Craig W. Osenberg

Department of Zoology, University of
Florida, Gainesville, FL 32611-8525,
USA

Tom Pellett

Bureau of Integrated Science Services,
Wisconsin Department of Natural
Resources, Madison, WI 53716, USA

Martin R. Perrow

ECON, Biological Sciences, University
of East Anglia, Norwich NR4 7TJ, UK

Lennart Persson

Department of Animal Ecology, Umeå
University, S-901 87 Umeå, Sweden

Birgitte Petersen

Department of Lake and Estuarine
Ecology, National Environmental
Research Institute, DK-8600 Silkeborg,
Denmark

Marten Scheffer

Wageningen Agricultural University,
Department of Water Quality
Management and Aquatic Ecology,
P.O. Box 8080, NL-6700 DD
Wageningen, The Netherlands

Claire L. Schelske

Department of Fisheries and Aquatic
Sciences, University of Florida/Institute
of Food and Agricultural Sciences, 7922
NW 71st Street, Gainesville, FL
32653-3071, USA

Louise Schlüter

The Water Quality Institute, Agern
Allé 11, DK-2970 Hørsholm, Denmark

Jan Simons

Department of Ecology and
Ecotoxicology, Free University, 1081 HV
Amsterdam, The Netherlands

Martin Søndergaard   Department of Lake and Estuarine
                                       Ecology, National Environmental
                                       Research Institute, DK-8600 Silkeborg,
                                       Denmark

Morten Søndergaard   Freshwater Biological Laboratory,
                                       University of Copenhagen, DK-3400
                                       Hillerød, Denmark

Christine Storlie   Bureau of Integrated Science Services,
                                       Wisconsin Department of Natural
                                       Resources, Madison, WI 53716, USA

Jon Theil-Nielsen   Freshwater Biology Laboratory,
                                       University of Copenhagen, DK-3400
                                       Hillerød, Denmark

Anett Trebitz   U.S. Environmental Protection Agency,
                                       Duluth, MN 55804, USA

Marcel S. Van den Berg   Department of Ecology and
                                       Ecotoxicology, Free University, 1081 HV
                                       Amsterdam, The Netherlands

Ellen Van Donk   Netherlands Institute of Ecology, Centre
                                       for Limnology (NIOO-CL),
                                       Rijksstraatweg 6, 3631 AC Nieuwersluis,
                                       The Netherlands

Jan E. Vermaat   International Institute for Infrastructural,
                                       Hydraulic and Environmental
                                       Engineering, 2601 DA Delft,
                                       The Netherlands

Robert T. Wass   Department of Zoology, University of
                                       Otago, Dunedin, New Zealand

Robert G. Wetzel   Department of Biological Sciences,
                                       University of Alabama, Tuscaloosa, AL
                                       35487-0206, USA

Karen Wilson   Center for Limnology, University of
                                    Wisconsin, Madison, WI 53706, USA

Johnstone O. Young   School of Biological Sciences,
                                    University of Liverpool, Liverpool
                                    L69 3BX, UK

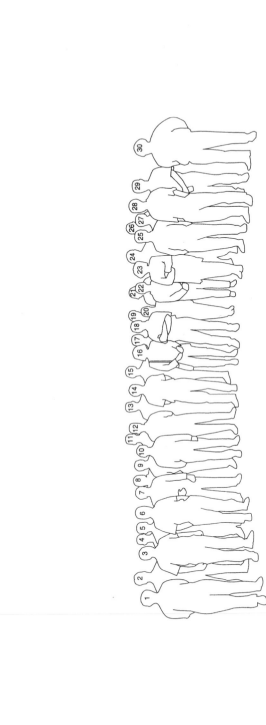

1. Stuart F. Mitchell
2. John Iwan Jones
3. Stephen R. Carpenter
4. Morten Søndergaard
5. Klaus Jürgens
6. Brian Moss
7. Robert G. Wetzel
8. Sebastian Diehl

9. Erik Jeppesen
10. Greg Cronin
11. Christer Brönmark
12. Torben L. Lauridsen
13. Mark V. Hoyer
14. Bjørn Faafeng
15. Jens Peder Jensen
16. Ellen van Donk

17. Irmgard Blindow
18. Jan E. Vermaat
19. David M. Lodge
20. Marie-Louise Meijer
21. Lennart Persson
22. Kirsten Christoffersen
23. Lene Jacobsen
24. Larry B. Crowder

25. Timo Kairesalo
26. Martin Søndergaard
27. Ryszard Kornijów
28. Marten Scheffer
29. Avital Gasith
30. John W. Barko

1. General Themes

1. Fish–Habitat Interactions Mediated via Ontogenetic Niche Shifts

Lennart Persson and Larry B. Crowder

Introduction

A fundamental characteristic of fish is that individuals increase in size by several orders of magnitude over their ontogeny (Werner, 1988). This increase in size generally means that the individual changes its food resource use during development. The change in resource use can take many different routes involving changes between carnivory and herbivory/detritivory (Gerking, 1994). Commonly, an increase in prey size eaten is observed in connection with the increase in consumer size, which potentially involves a change from zooplanktivory, to benthivory, and ultimately, to piscivory (Persson, 1988; Osenberg et al., 1994; Olson et al., 1995). These changes in resource use are, in turn, generally associated with habitat shifts in which complex habitats such as vegetated areas of lakes may function both as a resource base and as a refuge from predation (Heck and Crowder, 1991; Mittelbach and Osenberg, 1993; Persson, 1993; Olson et al., 1995; Persson and Eklöv, 1995).

The focus of this chapter is on how size-dependent processes in fish interact with habitat structure to shape ecological communities. Although we will restrict our treatment to fish–habitat interactions, changes in body size over ontogeny and consequences thereof for ecological interactions are by no means restricted to fish but are rather ubiquitous in aquatic environments (Mittelbach et al., 1988; Persson, 1988; Stein et al., 1988). For example, zooplankton and macroinvertebrate species generally increase substantially in size over their lifetime, an increase that often involves metamorphosis (Werner, 1988). Correspondingly, size-dependent inter-actions have been shown to have substantial effects on population dynamics in *Daphnia* (McCauley and Murdoch, 1990), and major effects of size-structured interactions for overall community dynamics have, for example, been demon-strated in the *Chaoborus* larvae (Neill, 1988).

The impact of vegetation on fish is multifold. Vegetation offers a physical structure that affects both competitive and predatory interactions between dif-ferent species and sizes of fish (Winfield, 1986; Diehl, 1988; Persson, 1991). Vegetation is also associated with high densities of invertebrate prey, which have been shown to affect food consumption and growth of the fish (Crowder and Cooper, 1982; Heck and Crowder, 1991; Diehl, 1993; Persson, 1993; Diehl and Kornijów, this volume, Chapter 2). Macrophytes and associated epiphytic algae may also form a resource for fish (Prejs, 1984; Hansson et al., 1987). Fish have feedback effects on vegetation by their direct consumption of vegetation and indirectly via other trophic components or abiotic routes such as sediment-feeding–induced turbidity. Fish may therefore also affect habitat structure. Several of the indirect effects of fish on vegetation are considered in other chapters of this volume. We therefore largely restrict our treatment to how habitat structure and associated resources affect fish performance, although we include a discussion on the importance of fish for nutrient fluxes between habitats.

Because size plays such a prominent role in interactions among fish species and for fish–habitat interactions, we first consider basic ecological capacities of in-dividual fish in relation to size and discuss how these size-dependent capacities are influenced by habitat structure. The ecological performances of individuals in different habitats are also affected by species-specific characteristics, and we cover some of these characteristics. Based on the individual level characteri-zations, we review how life history phenomena related to ontogenetic constraints affect the performance of different functional groups of fish, specifically pis-civore–prey fish interactions. We also discuss the implications of ontogenetic changes for overall community, ecosystem, and nutrient dynamics. Finally, we point to how our understanding of interactions between fish and habitat structure may be enhanced by the application of stage-structured population models.

Size-Dependent Foraging and Predator Avoidance Abilities and Habitat Structure

Increasing in size imposes a series of constraints on the organism, ranging from physical and mechanical constraints to ecological constraints (Miller et al., 1988; Werner, 1988). An example of the former is that the Reynolds number that a fish larva experiences is quite different from that an adult large fish experiences (Webb, 1978). Examples of the latter are that the foraging and predator avoidance capacities are closely related to the size of the fish. With respect to foraging rate, an increase in size means that two basic components of the individual's competi-tive ability (i.e., its foraging rate and its metabolic demands) change (Peters, 1983;

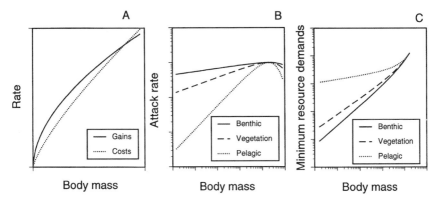

Figure 1.1. (A) General relationship between foraging gains and metabolic costs and body mass. (B) Attack rate as a function of body mass in the benthic, vegetation, and open water habitats based on encounter rates for bluegill sunfish. (Data from Mittelbach, 1981.) The attack rate has been assumed to have a hump-shaped relationship with body mass (see Persson et al., submitted). For the clarity of presentation, the maximum attack rate has been assumed to be the same for all prey types. (C) The minimum resource requirements necessary for maintenance as a function of body mass in three habitats based on the encounter rate function in B. For the clarity of presentation, it has been assumed that the prey weights in the different habitats are the same. Observe the log axes in B and C.

Calder, 1984; Miller et al., 1988; Werner, 1988). Metabolic demands as a function of body weight are generally assumed to be described by a power function with a slope varying between 0.6 and 0.9 (Peters, 1983; Calder, 1984; Werner, 1988) (Fig. 1.1A). The capacity to ingest energy has also been described by a power function of body size. However, for a prey of a specific size, the foraging rate is not expected to increase monotonically with size but to increase to a maximum to thereafter decrease (Tripet and Perrin, 1994; Persson et al., submitted) (Fig. 1.1B). The form of this general function has been substantiated in fish larvae (Bailey and Houde, 1989), and a review of size-dependent interactions in freshwater and marine fish larvae is given in Miller et al. (1988). The decreasing part of the relationship relates, among other things, to the capacity of an individual to discern small prey and make fine-tuned maneuvers, which decreases with body size (Breck and Gitter, 1983; Noakes and Godin, 1988; see also Persson et al., submitted).

The rate by which the foraging capacity increases with body size, generally termed the ontogenetic scaling of foraging rate, varies among taxa (Wilson, 1975; Werner, 1988, 1994; Lundberg and Persson, 1993; Persson et al., submitted). This variation is partly related to differences in foraging methods used by different functional groups of consumers. For example, the ingestion rate of filter feeders is expected to scale to body size with a higher slope than that of particulate feeders. The ontogenetic scaling of foraging rate to body size will also vary within taxa based on habitat-specific constraints on search behavior. The slope of the size

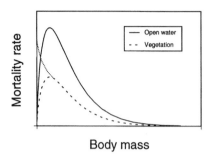

Figure 1.2. General relationship between mortality rate and body mass in the open water (solid line) and vegetation (dashed line) habitats. Due to the presence of predation from other small (albeit larger) fishes that are confined to the vegetation due to their own predation risk and potentially invertebrate predators, the form of the mortality rate function in the vegetation can change so that mortality is highest for the smallest stages (dotted + dashed lines).

scaling of foraging rate is expected to be higher for a fish foraging in a three-dimensional environment such as the pelagic habitat than for a fish foraging in a two-dimensional environment such as the benthic habitat. Correspondingly, it has been found that the slope of the relationship between the attack rate and body weight of the bluegill sunfish (*Lepomis macrochirus*) decreases from pelagic prey to vegetation prey to benthic prey (Mittelbach, 1981) (Fig. 1.1B). Mechanistic explanations for why small and large fish should be differently affected by vegetation are that the encounter rate and swimming speed advantages of larger fish will decrease in structured habitats. The size scaling of foraging rate is thus expected to decrease for fish moving from a pelagic habitat to a vegetation habitat, which, in turn, will affect the competitive abilities of differently sized individuals (Persson et al., submitted) (Fig. 1.1C). This suggested effect of vegetation structure on the size scaling of foraging intake is supported by other studies. Ryer (1987, 1988), for example, showed that the amount of prey encountered and consumed by large pipefish *(Syngnathus fuscus)* decreased in eelgrass (*Zostera marina*), whereas the foraging efficiency of small individuals was unaffected by the amount of structure. Several studies have shown that the growth rate of prey fish is retarded when they are confined to the vegetation due to predation risk (Mittelbach, 1988; Persson, 1993; Diehl and Eklöv, 1995; Persson and Eklöv, 1995). This suggests that, in addition to a decrease in the slope of the size scaling of foraging rate to body size, the maximum foraging rate decreases when the fish shifts from the pelagic to the vegetation habitat (note that the maximum foraging rate is assumed to be habitat independent in Fig. 1.1B and C).

The risk of being consumed by predators is also strongly connected to body size. Until the late 1980s, it was generally thought that the predation mortality rate of larval fish decreased monotonically with increasing prey fish size (Gilliam, 1982; Fuiman and Magurran, 1994) (Fig. 1.2). However, this expectation was based on studies that focused on the capture success (ratio of prey consumed to prey attacked) of predators and neglected other parts of the predation cycle such as the encounter rate (Fuiman, 1994; Fuiman and Magurran, 1994). Once these parts of the predation cycle were considered, it has been predicted and confirmed that the vulnerability of larval fish to raptorial predators increases to a maximum and then decreases as prey fish size increases (Bailey and Houde, 1989; Fuiman, 1989; Litvak and Leggett, 1992; Pepin et al., 1992) (Fig. 1.2). This pattern has been suggested to result from an increase in the encounter rate between predator and prey due to increased swimming speeds and increased pigmentation of the fish prey and a simultaneous decrease in capture success of predators due to better escape responsiveness of the prey fish as they grow and develop (Fuiman and Magurran, 1994).

Complex habitats like vegetated habitats may affect the relationship between body size and predation mortality of prey fish by lowering the overall predation efficiency of piscivores (Savino and Stein, 1982; Heck and Crowder, 1991; Christensen and Persson, 1993; Persson and Eklöv, 1995) (Fig. 1.2). The mechanisms behind the decreased predation efficiency in complex habitats can be both decreased encounter rate between predator and prey and decreased capture success of the predator once the prey has been encountered (Andersson, 1984; Main, 1987; Savino and Stein, 1989a,b; Christensen and Persson, 1993). Complex habitats may also affect the form of the relationship between body size and predation mortality (Fig. 1.2). For example, the size-dependent mortality in the bluegill sunfish can take the form of a monotonically decreasing function rather than a hump-shaped one. The high predation mortality of very small stages of bluegill in the vegetated habitat is a consequence of predation from other small (albeit larger) fishes, which are confined to the vegetation due to their own predation risk, and potential invertebrate predators that prey on the very youngest stages (Werner and Hall, 1988; G. Mittelbach, personal communication) (Fig. 1.2). The monotonically decreasing form of the predation mortality function in the vegetation habitat may be one reason (in addition to differences in resource size distributions between vegation and pelagic habitats) why the smallest stages of bluegills spend the first few weeks after hatching in the pelagic habitat before returning to the vegetation habitat (Werner and Hall, 1988; G. Mittelbach, personal communication).

Predator-Induced Habitat Shifts: Competitive and Predator Avoidance Abilities of Different Fish Species

The presence of piscivores affects the habitat use of small stages of fish and often restricts them to the littoral vegetated habitat (Mittelbach, 1986, 1988; Turner and Mittelbach, 1990; Persson, 1991, 1993; Tonn et al., 1992; Brabrand and Faafeng, 1993; Persson and Eklöv, 1995). This will generally lead to an increased competition intensity among refuging prey fish (Mittelbach, 1988; Turner and Mittelbach, 1990). Predator-mediated habitat use may also release invulnerable size classes from competition from smaller vulnerable size classes (predator-mediated habitat segregation) (Werner et al., 1983; Gilliam and Fraser, 1988; Savino and Stein, 1989a,b; Tonn et al., 1992; Christensen and Persson, 1993; Diehl and Eklöv, 1995; Carpenter et al., this volume, Chapter 11).

How seeking refuge in vegetation habitats affects the competitive abilities of prey fish depends on the species concerned. In their classic work on North American centrarchids, Werner and Hall (1979) showed species-dependent foraging abilities of three sunfish species in the vegetation, which resulted in differences in the timing of niche shifts between the three species. Studies of European species have shown that different fish species are affected differently by the presence of vegetation. Winfield (1986) found that the foraging rate of the cyprinid rudd (*Scardinius erythrophthalmus*) on *Daphnia* was only affected at stem densities greater than 200/m^2 and that the foraging rate of juvenile perch (*Perca fluviatilis*) did not decrease even at the highest stem density used (600 stems/m^2). By contrast, the foraging rate of roach (*Rutilus rutilus*) decreased substantially even at the lowest stem density used. Similar results were obtained by Diehl (1988) in a study of the foraging efficiencies of roach, bream (*Abramis brama*), and perch feeding on chironomids (i.e., the foraging performances of roach and bream decreased strongly in the presence of vegetation whereas the foraging performance of perch was only slightly affected by vegetation).

Because the effects of habitat structure on foraging performance are species- (and size-) specific, predator-induced habitat shifts by prey fish affect and even reverse the outcome of competitive interactions among refuging prey fish. Persson (1991) showed that juvenile perch and roach responded to the presence of piscivorous perch by moving into the vegetation refuge. This resulted in a reversal of the foraging advantage of roach (in the open water habitat) to a foraging advantage of perch (in the vegetation refuge). The shift in relative foraging performance, in turn, resulted in changes in relative growth rates. Although the growth rates of roach were higher than those of perch in the absence of piscivores, this relationship was reversed in the presence of piscivorous perch. The interactions between refuging juvenile roach and perch are thus affected by the structure per se. Because vegetation structure is also associated with vegetation-attached resources (see Diehl and Kornijów, this volume, Chapter 2), the prey communities inhabiting vegetation may have additional effects on the competitive interactions among fish species. Perch have been shown to be superior foragers to roach on macroinvertebrates (Persson, 1988). Correspondingly, Persson (1993; see also Persson and Eklöv, 1995) found that invertebrate resources associated with vegetation structure additionally competitively favored juvenile perch over roach in vegetation refuges as reflected in both diet and growth patterns. Because perch and roach make up most of the total fish biomass in many European lakes, the effects of vegetation structure on the interactions between these two species will have ramifications for overall community and lake ecosystem dynamics. Roach as a competitor with juvenile perch may severely limit the recruitment of perch to large piscivorous stages in the absence of vegetated habitats (see below).

Different species of prey fish do not only vary in their habitat specific foraging capacities but also in their habitat-specific abilities to avoid predation. Christensen and Persson (1993) found that juvenile roach were more efficient in avoiding piscivorous perch than juvenile perch in both open water and simulated vegetation consisting of strings (see also Persson and Eklöv, 1995). However, juvenile perch were more efficient in avoiding predators by using crevices. When simultaneously offered perch and rudd in field enclosures, pike captured more rudd than perch in environments lacking vegetation, whereas the opposite was the case in environments with vegetation (Eklöv and Hamrin, 1989). Interactions between piscivorous predators and prey fish are also affected by the type of predator species present. Pike (*Esox lucius*) have been found to be a more efficient predator than perch and pikeperch (*Stizostedion lucioperca*) in vegetation, whereas pikeperch and perch are more efficient in open water (Eklöv, 1992; Eklöv and Diehl, 1994; Greenberg et al., 1995). The presence of vegetation will also affect the foraging mode of specific piscivorous predators. For example, perch and largemouth bass (*Micropterus salmoides*) change from an active pursuit foraging mode to an ambush sit-and-wait foraging mode with an increase in vegetation density (Savino and Stein, 1982; Eklöv and Diehl, 1994).

Habitat Shifts and Mixed Competition–Predation Interactions and Ontogenetic Constraints

In the previous sections, we pointed out the importance of size when studying interactions among fish populations and that habitat structure affects the size scaling of the performance of the fish in terms of foraging capacity and predator avoidance ability. In this section, we consider how growth in size imposes changes in the nature of ecological interactions and also puts constraints on fish life history. The latter relates to the fact that the most efficient morphology for handling prey varies with prey types used over ontogeny.

As a result of variability in fish individual growth rates and the presence of several size cohorts in fish populations, interactions among fish species are characterized by a mixture of competitive and predatory interactions (Werner et al., 1983; Mittelbach, 1986, 1988; Werner, 1986; Persson, 1988; Persson and Greenberg, 1990). This mixture of competitive and predatory interactions takes place at several temporal and spatial scales. On a short time scale and within a system spatial scale, the behavioral decisions of an individual fish as a function of its size are a result of both competitive and predatory considerations (Gilliam, 1982; Lima and Dill, 1990). Behavioral models have predicted that, given different foraging returns and predation risks in different habitats (i.e., open water versus vegetated habitats), juvenile fish are expected to choose the habitat with the lowest mortality rate/growth rate ratio (assuming equilibrium and no time constraints) (Gilliam, 1982). This prediction (or predictions analogous to this) has been supported in several experimental studies (Gilliam and Fraser, 1988; Turner and Mittelbach, 1990). On a longer time scale, interactions among species may change between competitive interactions and predatory interactions as a result of individual growth (Wilbur, 1988). In fish, these changes between mainly competitive interactions to mainly predatory interactions are often associated with changes in habitat use in which the vegetation habitat plays a crucial role. For example, although juvenile fish of different species often compete for resources when refuging from predators in vegetation habitats, one of the species may start to prey on the other as they increase in size and shift habitat (Mittelbach, 1986; Werner, 1986; Olson et al., 1995; Persson and Eklöv, 1995). The importance of predatory versus competitive interactions may also vary among systems mediated through the size structures of the fish populations (Persson, 1988; Persson and Greenberg, 1990). In this case, the availability of submerged vegetation has been advanced as a major factor influencing the role of competitive and predatory interactions (Persson, 1988; see below).

The presence of size-dependent ontogenetic niche shifts imposes a series of constraints on the organism as a result of ontogenetic covariance (Werner, 1988). Natural selection will operate on morphological and behavioral traits over the whole life cycle of the individual fish and traits that are optimal at one ontogenetic niche are suboptimal in other ontogenetic stages (Werner and Hall, 1979; Werner, 1986, 1988; Persson, 1988). An illustrative example is an adult piscivorous species that as a small planktivore will be burdened with morphologies and behaviors more adapted for piscivory than for planktivory. The morphological traits for a typical particulate feeding planktivore are, for example, a compressed body and a small gape size, which will allow the fish to capture relatively small and non-evasive prey items efficiently at high swimming speed (Werner, 1977; Webb, 1984). By contrast, piscivorous feeding involves large evasive prey, which requires traits such as high attack speed, large gape size, and attacking foraging mode (Webb, 1984). Based on the constraints imposed by ontogenetic covariance, a general hypothesis has been advanced. This hypothesis states that a species undergoing substantial ontogenetic niche shifts during its life will be a less efficient predator on small zooplankton prey compared with a species undergoing less drastic niche shifts such as a planktivore specialist (Werner, 1986; Persson, 1988). Experimental support for this hypothesis has been provided for at least two species constellations—the perch–roach interaction and the largemouth bass–bluegill interaction (Werner, 1977; Persson, 1988). For perch, several aspects of the body morphology and behavior are generally associated with benthivorous feeding in vegetated habitats rather than planktivory or piscivory. These aspects include a relatively low cruising speed, a relatively deep body, laterally inserted pectoral fins, and enlarged dorsal fins (Eklöv and Persson, 1995). We expect that the support for the hypothesis of an ontogenetic trade-off cost in piscivorous species will increase when experimental data for additional species constellations are provided.

As a consequence of ontogenetic trade-offs in piscivores, Persson (1988) suggested that the interactions between piscivores and planktivores are characterized by a high degree of asymmetry. By definition, piscivores have a predatory advantage, which is counteracted by a competitive advantage on juvenile resources for planktivores due to ontogenetic trade-off costs in the piscivore. This type of asymmetric interaction, and particularly changes in the relative strength of the predatory versus competitive advantage has been suggested to have major consequences for overall community and ecosystem dynamics (Persson et al., 1991, 1992). In moderately productive systems, the proportion of piscivores (piscivorous perch making up most of piscivore biomass) of total fish biomass is high, composing up to 80% of total fish biomass (Fig. 1.3). By contrast, the proportion of piscivores of total fish biomass in highly productive systems is low ($\leq$20%) as a result of a severe bottleneck in the recruitment of juvenile perch to piscivorus

Figure 1.3. Changes in percentage pelagic piscivorous perch biomass of total pelagic fish biomass and phytoplankton biomass (chlorophyll *a* in μg/L) along a phosphorus-loading gradient in Swedish lakes. (Data from Persson et al., 1991; Carpenter et al., 1996.)

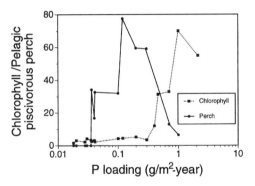

stages caused by planktivorous and benthivorous cyprinids (Persson, 1988; Persson and Greenberg, 1990). These changes in the importance of piscivores also have feedback effects on other trophic levels including zooplankton and phytoplankton. For example, a tenfold increase in phosphorus loading from 0.03 to 0.3 g/m^2 a year only led to minor increase in phytoplankton biomass (Carpenter et al., 1996), which suggests that an increased piscivore biomass may prevent an increased phosphorus loading from being expressed as an increase in phytoplankton biomass within this range of phosphorus loadings (Fig. 1.3). By contrast, in highly productive systems with a low proportion of piscivores, phytoplankton biomass increases steadily with phosphorus loading.

Studies of fish communities show that a shift in the species numerically dominating the fish community takes place along the productivity gradient. This shift involves a change in dominance of percids (mainly perch) in medium-productive lakes to a dominance of cyprinids in highly productive lakes, which is also reflected in changes in size structures of the populations (Persson, 1988). Although correlated to productivity, this major change in fish community structure, which involves feedback effects on lower trophic levels, has been hypothesized to be also affected by changes in the availability of submerged vegetation with increasing productivity (Persson et al., 1992; Persson, 1993; Persson and Eklöv, 1995). This hypothesis is related to the observation that the importance of submerged vegetation generally is at a maximum in moderately productive lakes (Wetzel, 1979), where also piscivore biomass has a maximum. Mechanistic explanations for why vegetation structure should affect piscivore–planktivore/benthivore fish interactions are that the performance of the juveniles of the major piscivore, perch, in relation to competing planktivores is strongly related to the presence of vegetation structure (see above; Diehl and Kornijów, this volume, Chapter 2).

Habitat Structure and Stage-Structured Interactions in Lakes: Two Examples

In many lakes throughout the north central United States, the Centrarchid bluegill sunfish make up most of total fish biomass (Osenberg et al. 1988, 1994; Mittelbach and Osenberg, 1993). This species hatches in the littoral and moves to open water for a few weeks before moving back to the sheltered vegetation habitat. As an adult, it feeds on zooplankton in open water, and the body size at the shift to this habitat depends on predation risk from largemouth bass (Mittelbach and Chesson, 1987; Werner and Hall, 1988). Mittelbach and Osenberg (1993; see also Osenberg et al., 1994) have suggested that the limnetic productivity of zooplankton sets the limit to the production (including fecundity) of adult bluegill sunfish, which, in turn, determines the intensity of competition in the littoral vegetation habitat through juvenile bluegill recruitment. A negative effect of juvenile bluegill sunfish density on the growth of its competitors, including the major adult piscivorous predator largemouth bass, has been demonstrated in cross-lake comparisons as well as in pond/enclosure experiments (Mittelbach, 1988; Osenberg et al., 1994; Olson et al., 1995). The strong competitive effect of juvenile bluegills on other refuging littoral fish can partly be related to the fact that juvenile bluegill outnumber the other species. For the interaction between largemouth bass and bluegill, it has also been experimentally demonstrated that the per capita effect of juvenile bluegill on young of the year (YOY) bass is larger than the reverse (Olson et al., 1995). This was the case despite a substantial resource partitioning between juvenile bluegill and bass, because bluegill caused changes in the size structure of major invertebrate prey (bluegill fed on smaller shared prey than bass and prevented these resources from growing to the larger sizes used by bass).

The interaction between largemouth bass and bluegill is a typical example of a mixture of competitive and predatory interactions that also involves habitat shifts. The effect of bluegill density (both adult and juvenile) on YOY largemouth bass growth has been found to be negative (competitive interactions), whereas the effect of YOY bluegill density on the growth of large largemouth bass is positive (predator–prey interaction) (Olson et al., 1995). The effects of bluegill density on adult bass density is also positive, and as a higher growth rate of adult bass leads to higher per capita fecundity, also YOY largemouth bass density is positively related to juvenile bluegill density (Fig. 1.4) (see also below).

In many Scandinavian lakes, perch and roach are the two dominating species. Roach are efficient zooplanktivores competing with juvenile perch but may also feed on macroinvertebrates and nonanimal food items. Perch are ontogenetic omnivores and start to feed on zooplankton, to thereafter shift to macroinvertebrates to finally become piscivorous (Persson, 1988). Vegetation has, as was considered above, been shown to affect both competitive interactions between roach and perch and predator–prey interactions between piscivorous perch and small roach and perch. The macroinvertebrate feeding stage has been identified to be an important bottleneck in the recruitment of perch to piscivorous stages (Persson, 1986, 1988), and increased availability of habitats with submerged vegetation is likely to decrease the limitations set by this recruitment bottleneck (Diehl and Kornijów, this volume, Chapter 2).

In contrast to the largemouth bass–bluegill density relationships, the density relationship between perch and roach can be both positive and negative, which has been suggested to be related to the availability of submerged vegetation (see

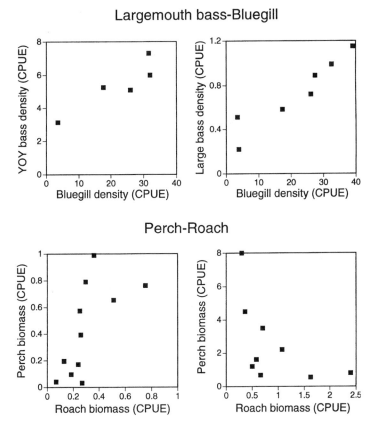

Figure 1.4. (Top) Density catch per unit effect [CPUE] relationships between YOY largemouth bass and bluegill (left) and between large largemouth bass and bluegill (right) in Michigan lakes. (Data from Olson et al., 1995). (Bottom) Biomass (kg CPUE) relationships between roach and perch in low to medium productive lakes (left) and medium to highly productive systems (right). (Data from Persson et al., 1991.)

above) (Fig. 1.4). This difference in density relationship between perch and roach versus bass and bluegill can be related to differences in the life histories of the two piscivores (Olson et al., 1995). Bass eventually become piscivorous in their first year, whereas it may take several years for perch to reach piscivorous stages. The potential for competing prey to affect perch recruitment to piscivorous stages is thus higher.

Littoral-Pelagic Coupling in Lakes: Effects of Fish on Nutrient Fluxes

It has long been recognized that littoral habitats are linked biogeochemically to the open waters of lakes (Wetzel, 1979; Barko and James, this volume, Chapter 10).

Table 1.1. Summary of Potential Mechanisms by Which Fish May Affect Macrophyte
Abundance, Littoral-Pelagic Couplings, and Overall Lake Dynamics

---

I. Routes for effects on submerged vegetation (habitat structure) by fish feeding
   activities in macrophyte habitats
   - Feeding on macrophytes
   - Feeding on epiphytes
   - Feeding on macroinvertebrates
   - Feeding-induced uprooting of plants

II. Routes for effects on submerged vegetation (habitat structure) by feeding activities in
    open water
   - Zooplankton predation induced changes in phytoplankton biomass (transparency)
   - Sediment feeding induced fluxes of nutrients to open water affecting
     phytoplankton biomass (transparency)

III. Other fish induced coupling of littoral and pelagic habitats
   - Transport of nutrients/organic matter to pelagic due to littoral feeding and pelagic
     excretion/egestion
   - Transport of nutrients/organic matter to littoral due to pelagic feeding at night by
     in daytime refuging juvenile fish
   - Recruitment of juvenile fish to the vegetation habitat from pelagic larval stages
     over ontogeny
   - Recruitment of adult fish to the pelagic habitat from the vegetation habitat over
     ontogeny

Littoral zones are extremely productive; in addition to macrophyte production,
epiphyte production is now known to be substantial (Wetzel, 1990; Wetzel and
Søndergaard, this volume, Chapter 7). Biomass, production, and diversity of
littoral invertebrates often exceed that of invertebrates in open water areas. Unfortunately, most studies of food web interactions in lakes have focused on either the
littoral or the pelagic habitat (Lodge et al., 1988). Linking littoral zone into the
whole lake requires us to consider its role as a refuge, habitat, and nutrient source
or sink. The role of food web links in the transformation or translocation of
nutrients between the littoral and pelagic habitats is not well known (Lodge et al.,
1988). We do know that fish often shift from the littoral to the pelagic habitat as
they increase in body size through ontogeny. Life history omnivory in itself thus
means that there will be a coupling between different habitats (Table 1.1). For
example, in bluegill sunfish ontogenetic habitat shifts will couple the dynamics of
the invertebrate prey communities in the littoral zone with the zooplankton prey
community in the open water habitat (Osenberg et al., 1994). Furthermore, fish
move to and from the littoral habitat on a diel basis (Hall et al., 1979; Naud and
Magnan, 1988). Do these movements have implications for nutrients in pelagic
food webs?

Consumers may play a large role in lake nutrient budgets (Table 1.1). Although
limnologists initially focused on external loading of nutrients, for many systems
internal nutrient recycling contributes substantially to the nutrient budget. Nutrient
recycling and transformation by zooplankton are now widely recognized as impor-

tant sources of nutrients for phytoplankton production (Vanni, 1996), but fish effects on nutrient translocation and transformation and subsequent effects on algal productivity and species composition have been overlooked until recently. Only a few experimental studies document direct nutrient recycling by fish affecting phytoplankton community structure (Reinertsen et al., 1986; Vanni and Findlay, 1990; Schindler, 1992; Vanni and Layne, 1997). In an experimental study, Vanni et al. (1997) separate the community-level effects of nutrient recycling by fish from those due to zooplankton and document that in lakes dominated by planktivores, fish effects can exceed nutrient recycling effects of zooplankton on algal community composition. Fish effects may be even more significant if one considers that a substantial portion of fish diets is consumed in the littoral habitat (Schindler et al., 1993). In northern Wisconsin lakes, planktonic prey accounted for less that 30% of diet biomass in planktivorous fishes; they mostly ate benthic insects and periphyton. Even piscivorous fish consumed primarily benthic (65%) or terrestrial (15%) prey; fish only accounted for 16% of the diet biomass of piscivores (He and Kitchell, 1990; He and Wright, 1992; see also Brabrand et al., 1990). But fish may move offshore and release nutrients (Schindler et al., 1996; Vanni, 1996). Excretion by consumers of nutrients derived from the littoral provides new nutrients for pelagic producers (Table 1.1).

Recent biogeochemical analyses of phosphorus cycles in the pelagic zones of lakes suggest that observed levels of primary productivity are too high to be supported by pelagic recycling alone. Caraco et al. (1992) estimated that more than one-third of the pelagic primary production of Mirror Lake (NH) must be supported by "new" phosphorus. Could the source of these nutrients be translocated by nutrients from the littoral? In recent reviews, both Vanni (1996) and Schindler et al. (1996) argued strongly that pelagic nutrient budgets may only be balanced by considering biologically driven phosphorus from the littoral. Our current understanding suggests that the littoral zone is very likely a source of nutrients for pelagic food webs in lakes. Both biogeochemical processes and animal movements suggest that the pelagic habitat is basically a sink for littoral productivity.

The relative importance of biological processes in nutrient translocation will vary among lakes. Translocation of nutrients from the littoral or deep benthic habitat (in stratified lakes) to the pelagic will vary with fish species composition as well as the ontogenetic stage of the fish species. Species that feed exclusively in the pelagic will contribute little to linkages from the littoral. Omnivorous fishes, especially detritivores, will transport nutrients from the sediment and detritus to the pelagic (Brabrand et al., 1990; Mather et al., 1995; Vanni, 1996). Benthic feeding fishes transport nutrients from the bottom substrate back into the water column, fueling pelagic production. Fish that migrate onshore-offshore (or among habitat patches in shallow lakes) may increase nutrient translocation above that accounted for by biogeochemical processes alone (Schindler et al., 1996). Lake geomorphology will also contribute. Large lakes or those with limited shoreline development and relatively small littoral zones (Gasith, 1991) may not foster substantial littoral production or transfer from the littoral to the pelagic. Still, although Carpenter and Kitchell (1993) selected lakes for their experimental manipulations of pelagic food webs that had minimal littoral habitat, the phosphorus budgets suggested strong inputs of phosphorus to the pelagic food webs originating from the littoral habitats (Schindler et al., 1996).

Size-Structured Interactions, Habitat, and Population Dynamics

Given that fish movements on diel, seasonal, and ontogenetic scales can play a large role in the translocation of nutrients, what methods do we have to predict these behaviors? Early efforts used optimal foraging theory to forecast habitat choice among individual juvenile fish (Mittelbach, 1981). Foraging strategies of individuals, of course, also depend on the behaviors of other foragers. Game theoretical approaches led to the prediction of an ideal free distribution among competitors (Fretwell and Lucas, 1970; Milinski, 1979). Distributions of foragers are also modified by the presence of predators (Werner et al., 1983; Werner and Gilliam, 1984; Abrahams and Dill, 1989), and new models of individual foraging behaviors include the simultaneous optimization of several different objectives. These models are fitness-based, and one of the most recent ones assumes that prey fish choose habitats to maximize the ratio of net energy intake to probability of death per unit time (Gilliam and Fraser, 1988). Werner and Hall (1988) used such a model to predict the ontogenetic shifts in bluegills from the littoral to the pelagic habitat as a function of predation risk in different lakes. These models assume that risk of predation is static (i.e., the predators do not move). Recently, Hugie and Dill (1994; see also Sih, in press) expanded the game theoretical approach to predict habitat choice of both prey and predators in a two-habitat model with a sedentary resource of the prey. One very interesting outcome of their model is that, without behavioral interference among predators (e.g., when one would expect predators to distribute themselves according to an ideal free distribution), the density of prey in a habitat is determined only by the inherent riskiness of the habitat and at the behavioral equilibrium is not influenced by habitat productivity (i.e., productivity of the sedentary resource) per se. In other words, prey density is only influenced by habitat features that influence risk of predation such as habitat structure, turbidity, or light levels. In contrast to the prey, the habitat use of the predator is affected by both habitat riskiness of the prey and productivity (see also Diehl and Kornijów, this volume, Chapter 2). When predators do interfere with each other in the model, habitat productivity also plays a role in the expected distribution of prey. The model of Hugie and Dill does not address dynamics on the diel time scale, although one might expect diel behaviors to respond similarly as light levels influence the risk of predation. Because availability of food resources and riskiness of the habitat scale with habitat structure, submerged macrophytes should play a large role in determining both the distribution of prey and predators in littoral habitats.

Ideal free distribution models predict predator and prey habitat distribution on a limited time scale and do not address long-term population dynamics, although the model of Hugie and Dill can be expanded to an equilibrium situation with no net migration between habitats (see Oksanen et al., 1995). To handle population dynamics, we need to derive models that include all relevant vital rates (growth, birth, mortality, migration between habitats). To do this for size-structured populations such as fish is not an easy task. The least complex way to introduce size structure is to use age-based demography. Mittelbach and Chesson (1987) argued that for the Centrarchidae system, the dynamics of bluegill populations and interactions between bluegill and pumpkinseed (*Lepomis gibbosus*) sunfish may be adequately characterized by an age-based model using two stages (juveniles and adults). It has been argued that an approach based on life history stages is sufficient for the description of populations if vital rates are similar within stages but different between stages (Osenberg et al., 1994).

The two life-stage interaction in bluegill is set up by piscivorous largemouth bass in the open water that force the vulnerable size classes of bluegill to stay in the littoral, vegetated area. The size at which bluegill move out to the open water area varies among lakes depending on predation risk (i.e., largemouth bass density) (Werner and Hall, 1988). This flexible and stage-structured behavior of bluegill leads to complex indirect effects between populations at several different trophic levels that do not share the same habitat. In the two life-stage model developed by Mittelbach and Chesson (1987), an increase in open water productivity is expected to lead to an increase in per capita adult fecundity. This will, in turn, lead to an increased number of juveniles, which depletes the resource in the vegetation refuge. As a result, per capita juvenile survival will decrease. The total juvenile survival will still increase, causing an increase in adult numbers. Although adult density thus will increase with increased adult resource productivity, density-dependent juvenile survival will prevent adult density from fully responding to the increase in their resource productivity. As a result of these stage-structured interactions mediated via habitat shifts, there will be a positive relationship between the densities of adults and adult resources and a negative relationship between the densities of juveniles and juvenile resources. These patterns contrast to predictions of standard predator–prey theory. The predictions regarding how different stages of bluegill respond to an increase in productivity of the adult resource as advanced by Mittelbach and Chesson (1987) have subsequently been supported by comparative field studies (Mittelbach and Osenberg, 1993).

Mittelbach and Chesson (1987) extended their stage-based model to include a competitor of bluegill, the pumpkinseed sunfish. They showed that although adult bluegill and adult pumpkinseed sunfish did not share resources (adult bluegill feed on zooplankton and adult pumpkinseed feed on gastropods), an indirect negative effect was present between these stages, mediated via interactions between juveniles of both species in the vegetation refuge.

In many organisms, the ecological capacities in terms of foraging efficiency, metabolic demands, capacity to avoid predators, and fecundity are much more closely related to the size of the organism than to its age (Ebenman and Persson, 1988). Population models having size rather than age as their basic state variable may therefore be more appropriate in many situations when analyzing the dynamics of stage-structured populations such as fish. Physiologically structured models that have been developed during the past decade may be a useful tool here (Metz et al., 1988; DeAngelis and Gross, 1992; De Roos et al., 1992). The analysis of population dynamics in relation to habitat structure by using this modeling approach has not yet been carried out. In a recent contribution, Persson et al. (submitted) showed that in planktivore-zooplankton systems, the population dynamics for biologically realistic parameter values is characterized by recruiter-driven dynamics (regular cycles or quasiperiodic fluctuations). A shift from zooplankton feeding in the open water to feeding in the vegetation will decrease the slope of the size scaling of the attack rate (Fig. 1.1B) and increase the relative competitive ability of small individuals versus large individuals (Fig. 1.1C). This will actually reinforce the tendency for recruiter-driven dynamics. Because population cycles are not commonly observed in fish populations except for obligate planktivores (Hamrin and Persson, 1986), there must be some mechanism that prevents large-amplitude population fluctuations. Vegetated habitats (refuges) in combination with predator-induced restriction of habitat use of small vulnerable size classes may be one way by which the population dynamics is stabilized as larger-size classes are released from competition from recruits. To analyze this potential stabilizing effect of vegetation on population and overall community dynamics is a challenging task for future modeling research.

Acknowledgments. The research on which this review is partly based has been supported by The Swedish Natural Science Research Council and the Swedish Council for Forestry and Agricultural Sciences (to L. Persson) and by the U.S. National Science Foundation and the University of North Carolina Sea Grant for research on species interactions in submerged macrophytes and seagrass habitats. (to L. Crowder). Valuable comments on the chapter were given by S. Diehl and an anonymous reviewer, who are gratefully acknowledged.

References

Abrahams, M.V.; Dill, L.M. A determination of the energetic equivalence of the risk of predation. Ecology 70:999–1007; 1989.

Anderson, O. Optimal foraging by largemouth bass in structured environments. Ecology 65:851–861; 1984.

Bailey, K.M.; Houde, E.D. Predation on eggs and larvae of marine fishes and the recruitment problem. Adv. Mar. Biol. 25:1–83; 1989.

Brabrand, A.; Faafeng, B.A. Habitat shift in roach (*Rutilus rutilus*) induced by pikeperch (*Stizostedion lucioperca*) introduction: predation risk versus pelagic behaviour. Oecologia 95:38–56; 1993.Oecologia 95:38–46; 1993.

Brabrand, A.; Faafeng, B.A.; Nilssen, J.P. Relative importance of phosphorus supply to phytoplankton production: fish excretion verus external loading. Can. J. Fish. Aquat. Sci. 47:364–372; 1990.

Breck, J.E.; Gitter, M.J. Effect of fish size on the reactive distance of bluegill (*Lepomis macrochirus*) sunfish. Can. J. Fish. Aquat. Sci. 40:162–167; 1983.

Calder, W.A., III. Size, function and life history. Cambridge, MA: Harvard University Press; 1984.

Caraco, N.F.; Cole, J.J.; Likens, G.E. New and recycled primary production in an oligotrophic lakes: insights for summer phosphorus dynamics. Limnol. Oceanogr. 37:590–602; 1992.

Carpenter, S.R.; Kitchell, J.F. The trophic cascade in lakes. Cambridge: Cambridge University Press; 1993.

Carpenter, S.R.; Frost, T.M.; Persson, L.; Power, M.; Soto, D. Freshwater ecosystems: linkages of complexity and processes. In: Mooney, H. et al., eds. Biodiversity and ecosystem functions: a global perspective. New York: John Wiley and Sons; 1996:299–325.

Christensen, B; Persson, L. Species specific antipredator behaviours: effects on prey choice in different habitats. Behav. Ecol. Sociobiol. 32:1–9; 1993.

Crowder, L.B.; Cooper, W.E. Habitat structural complexity and the interactions between bluegill and their prey. Ecology 63:1802–1813; 1982.

DeAngelis, D.L.; Gross, L.J. Individual-based models and approaches in ecology—populations, communities and ecosystems. New York: Chapman & Hall; 1992.

De Roos, A.M.; Metz, J.A.J.; Diekmann, O. Studying the dynamics of structured population models: a versatile technique and its application to *Daphnia*. Am. Nat. 139:123–147; 1992.

Diehl, S. Foraging efficiency of three freshwater fish: effects of structural complexity and light. Oikos 53:207–214; 1988.

Diehl, S. Effects of habitat structure on resource availability, diet and growth of benthivorous perch, *Perca fluviatilis*. Oikos 67: 403–414; 1993.

Diehl, S.; Eklöv, P. Effects of piscivore-mediated habitat use on resources, diet and growth of perch. Ecology 76:1712–1726; 1995.

Ebenman, B.; Persson, L. Dynamics of size-structured populations—an overview. In: Ebenman, B.; Persson, L., eds. Size-structured populations: ecology and evolution. Heidelberg: Springer Verlag; 1988:3–9.

Eklöv, P. Group foraging versus solitary foraging efficiency in piscivorous predators: the perch, *Perca fluviatilis,* and pike, *Esox lucius,* patterns. Anim. Behav. 44:313–326; 1992.

Eklöv, P.; Diehl, S. Piscivore efficiency and refuging prey: the importance of predator search mode. Oecologia 98:344–353; 1994.

Eklöv, P.; Hamrin, S.F. Predator efficiency and prey selection: interactions between pike *Esox lucius,* perch *Perca fluviatilis* and rudd *Scardinius erythrophthalmus*. Oikos 56:149–156; 1989.

Eklöv, P.; Persson, L. Species-specific antipredator capacities and prey refuges: interactions between piscivorous perch (*Perca fluviatilis*) and juvenile perch and roach (*Rutilus rutilus*). Behav. Ecol. Sociobiol. 37:169–178; 1995.

Fretwell, S.D.; Lucas, H.L. On territorial behaviour and other factors influencing habitat distribution in birds. I. Theoretical development. Acta Biotheor. 19:16–36; 1970.

Fuiman, L.A. Vulnerability of Atlantic herring larvae to predation by yearling herring. Mar. Ecol. Prog. Ser. 51:291–299; 1989.

Fuiman, L.A. The interplay of ontogeny and scaling in the interactions of fish larvae and their predators. J. Fish Biol. 4:55–79; 1994.

Fuiman, L.A.; Magurran, A.E. Development of predator defences in fishes. Rev. Fish Biol. Fish. 4:145–183; 1994.

Gasith, A. Can littoral resources influence ecosystem processes in large, deep lakes? Verh. Int. Verein. Theoret. Angew. Limnol. 24:1073–1076; 1991.

Gerking, S.D. Feeding ecology of fish. San Diego, CA: Academic Press; 1994.

Gilliam, J.F. Habitat use and competitive bottlenecks in size-structured populations. Thesis, Michigan State Univ., East Lansing; 1982.

Gilliam, J.F.; Fraser, D.F. Resource depletion and habitat segregation by competitors under predation hazard. In: Ebenman, B.; Persson, L., eds. Size-structured populations: ecology and evolution. Heidelberg: Springer Verlag; 1988:173–184.

Greenberg, L.A.; Paszkowski, C.A.; Tonn, W.M. Effects of prey composition and habitat structure on foraging by two functionally distinct piscivores. Oikos 74:522–532; 1995.

Hall, D.J.; Werner, E.E.; Gilliam, J.F.; Mittelbach, G.G.; Howard, D.; Doner, C.G.; Dickerman, J.A.; Stewart, A.J. Diel foraging behavior and prey selection in golden shiner. J. Fish. Res. Bd. Can. 36:1929–1939; 1979.

Hamrin, S.F.; Persson, L. Asymmetrical competition between age classes as a factor causing population oscillations in an obligate planktivorous fish. Oikos 47:223–232; 1986.

Hansson, L.-A.; Johansson, L.; Persson, L. Effects of fish grazing on nutrient release and succession of primary producers. Limnol. Oceanogr. 32:723–729; 1987.

He, X.; Kitchell, J.F. Direct and indirect effects of predation on a fish community: a whole lake experiment. Trans. Am. Fish. Soc. 119:825–835; 1990.

He, X.; Wright, R.W. Piscivore–planktivore interactions in an experimental lake: population and community responses to manipulation. Can. J. Fish. Aquat. Sci. 49:1176–1183; 1992.

Heck, H.L., Jr.; Crowder, L.B. Habitat structure and predator–prey interactions in vegetated aquatic systems. In: Bell, S.S.; McCoy, E.D.; Mushinsky, H.R., eds. Habitat structure—the physical arrangement of objects in space. London: Chapman & Hall; 1991:281–299.

Hugie, D.M.; Dill, L.M. Fish and game: a game theoretic approach to habitat selection by predators and prey. J. Fish Biol. 45 (Suppl. A):151–169; 1994.

Lima, S.L.; Dill, L.M. Behavioural decisions made under risk of predation: a review and prospectus. Can. J. Zool. 68:619–640; 1990.

Litvak, M.K.; Leggett, W.C. Age and size-selective predation on larval fishes: the bigger-is-better hypothesis revisited. Mar. Ecol. Prog. Ser. 81:13–24; 1992.

Lodge, D.M.; Barko, J.W.; Strayer, D.; Melack, J.M.; Mittelbach, G.G.; Howarth, R.W.; Menge, B.; Titus, J.E. Spatial heterogeneity and habitat interactions in lake communities. In: Carpenter, S.R., ed. Complex interactions in lake communities. New York: Springer-Verlag; 1988:181–208.

Lundberg, S.; Persson, L. Optimal body size and resource density. J. Theor. Biol. 164:163–180; 1993.

Main, K.L. Predator avoidance in seagrass meadows: prey behavior, microhabitat selection and cryptic coloration. Ecology 68:170–180; 1987.

Mather, M.E.; Vanni, M.J.; Wissing, T.E.; Davis, S.A.; Schaus, M.H. Regeneration of nitrogen and phosphorus by bluegill and gizzard shad: effect of feeding history. Can. J. Fish. Aquat. Sci. 52:2327–2338; 1995.

McCauley, E.D.; Murdoch, W.W. Predator–prey dynamics in environments rich and poor in nutrients. Nature 343:455–457; 1990.

Metz, J.A.J.; de Roos, A.M.; van den Bosch, F. Population models incorporating physiological structure: a quick survey of the basic concepts and an application to size-structured population dynamics in waterfleas. In: Ebenman, B.; Persson, L., eds. Size-structured populations: ecology and evolution. Heidelberg: Springer-Verlag; 1988:106–126.

Milinski, M. An evolutionarily stable strategy for sticklebacks. Z. Tierpsychol. 51:36–40; 1979.

Miller, T.J.; Crowder, L.B.; Rice, J.A.; Marschall, E.A. Larval size and recruitment mechanisms in fishes: toward a conceptual framework. Can. J. Fish. Aquat. Sci. 45:1657–1670; 1988.

Mittelbach, G.G. Foraging efficiency and body size: a study of optimal diet and habitat use by bluegills. Ecology 62:1370–1386; 1981.

Mittelbach, G.G. Predation-mediated habitat use: some consequences for species interactions. Biol. Fish. 16:159–169; 1986.

Mittelbach, G.G. Competition among refuging sunfishes and effects of fish density on littoral zone invertebrates. Ecology 69:614–623; 1988.

Mittelbach, G.G.; Chesson, P.L. Predation risk: indirect effects on fish populations. In: Kerfoot, W.C.; Sih, A., eds. Predation: direct and indirect impacts on aquatic communities. Hanover, NH: University Press of New England; 1987:537–555.

Mittelbach, G.G.; Osenberg, C.W. Stage-structured interactions in bluegill: consequences of adult resource variation. Ecology 74:2381–2394; 1993.

Mittelbach, G.G.; Osenberg, C.W.; Leibold, M.A. Trophic relations and ontogenetic niche shifts in aquatic systems. In: Ebenman, B.; Persson, L., eds. Size-structured populations: ecology and evolution. Heidelberg: Springer-Verlag; 1988:203–218.

Naud, M.; Magnan, P. Diel onshore-offshore migrations in northern redbelly dace, *Phoxinus eos* (Cope), in relation to prey distribution in a small oligotrophic lake. Can. J. Zool. 66:1249–1253; 1988.

Neill, W.E. Community responses to experimental nutrient perturbations in oligotrophic lakes: the importance of bottlenecks in size-structured populations. In: Ebenman, B.; Persson, L., eds. Size-structured populations: ecology and evolution. Heidelberg: Springer-Verlag; 1988:236–255.

Noakes, D.L.G.; Godin, J.-G.J. Ontogeny of behavior and concurrent developmental changes in sensory systems in teleost fishes. In: Hoar, W.S.; Randall, D.J., eds. Fish physiology. Vol. XIB. New York: Academic Press; 1988:345–395.

Oksanen, T.; Power, M.E.; Oksanen, L. Ideal free habitat selection and consumer-resource dynamics. Am. Nat. 146:565–585; 1995.

Olson, M.H.; Mittelbach, G.G.; Osenberg, C.W. Competition between predator and prey: resource-based mechanisms and implications for stage-structured dynamics. Ecology 76:1758–1771; 1995.

Osenberg, C.W.; Werner, E.E.; Mittelbach, G.G.; Hall, D.J. Growth patterns in bluegill (*Lepomis macrochirus*) and pumpkinseed (*L. gibbosus*) sunfish: environmental variation and the importance of ontogenetic niche shifts. Can. J. Fish. Aquat. Sci. 45:17–26; 1988.

Osenberg, C.W., Olson, M.H.; Mittelbach, G.G. 1994. Stage structure in fishes: resource productivity and competition gradients. In: Stouder, D.J.; Fresh, K.L.; Feller, R.J., eds. Theory and application in fish feeding ecology. Columbia: University of South Carolina Press; 1994:151–170.

Pepin, P.; Shears, T.H.; deLafontaine, Y. Significance of body size to the interaction between a larval fish (*Mallotus villosus*) and a vertebrate predator (*Gasterosteus aculeatus*). Mar. Ecol. Prog. Ser. 81:1–12; 1992.

Persson, L. Effects of reduced interspecific competition on resource utilization in perch (*Perca fluviatilis*). Ecology 67:355–364;1986.

Persson, L. Asymmetries in competitive and predatory interactions in fish populations evolution. In: Ebenman, B.; Persson, L., eds. Size-structured populations: ecology and evolution. Heidelberg: Springer-Verlag; 1988:203–218.

Persson, L. Behavioral response to predators reverses the outcome of competition between prey species. Behav. Ecol. Sociobiol. 28:101–105; 1991.

Persson, L. Predator-mediated competition in prey refuges: the importance of habitat dependent prey resources. Oikos 68:12–22; 1993.

Persson, L.; Eklöv, P. Prey refuges affecting interactions between piscivorous perch (*Perca fluviatilis*) and juvenile perch and roach (*Rutilus rutilus*). Ecology 76:70–81; 1995.

Persson, L.; Greenberg, L.A. Juvenile competitive bottlenecks: the perch (*Perca fluviatilis*)–roach (*Rutilus rutilus*) interaction. Ecology 71:44–56; 1990.

Persson, L.; Leonardsson, K.; Gyllenberg, M.; De Roos, A.; Christensen, B. Ontogenetic scaling and the dynamics of a size-structured consumer-resource model. Theor. Pop. Biol. (submitted).

Persson, L.; Diehl, S.; Johansson, L.; Andersson, G.; Hamrin, S.F. Shifts in fish communities along the productivity gradient of temperate lakes—patterns and the importance of size-structured interactions. J. Fish Biol. 38:281–293; 1991.

Persson, L.; Johansson, L.; Diehl, S.; Andersson, G.; Hamrin, S.F. Trophic interactions in temperate lake ecosystems—a test of food chain theory. Am. Nat. 140:59–84; 1992.

Peters, R.H. The ecological implications of body size. Cambridge: Cambridge University Press; 1983.

Prejs, A. Herbivory by temperate freshwater fishes and its consequences. Environ. Biol. Fish. 10:281–296; 1984.

Reinertsen, H.; Jensen, A.; Langeland, A.; Olson, Y. Algal competition for phosphorus: the influence of zooplankton and fish. Can. J. Fish. Aquat. Sci. 43:1135–1141; 1986.

Ryer, C.H. Studies on the foraging ecology of pipefish: prey selection and habitat selection. Dissertation, College of William and Mary, Williamsburg, VA; 1987.

Ryer, C.H. Pipefish foraging and the effects of altered habitat complexity. Mar. Ecol. Progr. Ser. 48:37–45; 1988.

Savino, J.F.; Stein, R.A. Predator–prey interaction between largemouth bass and bluegills as influenced by simulated, submersed vegetation. Trans. Am. Fish. Soc. 111:255–266; 1982.

Savino, J.F.; Stein, R.A. Behavioural interactions between fish predators and their prey: effects of plant density. Anim. Behav. 37:11–321; 1989a.

Savino, J.F.; Stein, R.A. Behavior of fish predators and their prey: habitat choice between open water and dense vegetation. Environ. Biol. Fish. 24:287–293; 1989b.

Schindler, D.E. Nutrient regeneration by sockeye salmon (*Oncorhynchus nerka*) fry and subsequent effects on zooplankton and phytoplankton. Can. J. Fish. Aquat. Sci. 49: 2498–2506; 1992.

Schindler, D.E.; Kitchell, J.F.; He, X.; Carpenter, S.R.; Hodgson, J.R.; Cottingham, K.L. Food web structure and phosphorus cycling in lakes. Trans. Am. Fish. Soc. 122:756–772; 1993.

Schindler, D.E.; Carpenter, S.R.; Cottingham, K.L.; He, X.; Hodgson, J.R.; Kitchell, J.F.; Soranno, P.A. Food web structure and littoral zone coupling to pelagic trophic cascades. In: Polis, G.A.; Winemiller, K.O., eds. Food webs: integration of patterns and dynamics. New York: Chapman & Hall; 1996:95–105.

Sih, A. The trophic level ideal free distributions: a game theory approach to understanding the predator–prey behavioral response race. In: Dugatkin, L.A.; Reeve, H.K., eds. Advances in game theory and the study of animal behavior. Oxford: Oxford University Press (in press).

Stein, R.A.; Threlkeld, S.T.; Sandgren, C.D.; Sprules, W.G.; Persson, L.; Neill, W.E.; Werner, E.E.; Dodson, S.I. Size-structured interactions in lake communities. In: Carpenter, S.R., ed. Complex interactions in lake communities. Heidelberg: Springer-Verlag; 1988:161–179.

Tonn, W.M.; Paszkowski, C.A.; Holopainen, I.J. Piscivory and recruitment: mechanisms structuring prey populations in small lakes. Ecology 73:951–958; 1992.

Tripet, F.; Perrin, N. Size-dependent predation by *Dugesia lugubris* (Turbellaria) on *Physa acuta* (Gastropoda): experiments and model. Funct. Ecol. 8:458–463; 1994.

Turner, A.M.; Mittelbach, G.G. Predator avoidance and community structure: interactions among piscivores, planktivores, and plankton. Ecology 71:2241–2254; 1990.

Vanni, M.J. Nutrient transport and recycling by consumers in lake food webs: implications for algal communities. In: Polis, G.A.; Winemiller, K.O., eds. Food webs: integration of patterns and dynamics. New York: Chapman & Hall; 1996:81–95.

Vanni, M.J.; Findlay, D.L. Trophic cascades and phytoplankton community structure. Ecology 71:921–937; 1990.

Vanni, M.J.; Layne, C.D.; Arnott, S.E. "Top-down" trophic interactions in lakes: effects of fish on nutrient dynamics. Ecology 78:21–40.

Vanni, M.J.; Layne, C.D. Nutrient recycling and herbivory as mechanisms in the "top down" effect of fish on phytoplankton in lakes. Ecology 78:1–20.

Webb, P.W. Hydromechanics: nonscombrid fish. In Hoar, W.S.; Randall, D.J., eds. Fish physiology. Vol. VII. Locomotion. New York: Academic Press; 1978:190–237.

Webb, P.W. Body and fin form and strike tactics of four teleost predators attacking fathead minnow *Pimephales promelas* prey. Can J. Fish. Aquat. Sci. 41:157–165; 1984.

Werner, E.E. Species packing and niche complementarity in three sunfishes. Am. Nat. 11:553–578; 1977.

Werner, E.E. Species interactions in freshwater fish communities. In: Diamond, J.; Case, T.J., eds. Community ecology. New York: Harper & Row; 1986:344–358.

Werner, E.E. Size, scaling and the evolution of life cycles. In: Ebenman, B.; Persson, L., eds. Size-structured populations: ecology and evolution. Heidelberg: Springer-Verlag; 1988:60–81.

Werner, E.E. Ontogenetic scaling of competitive relations: size-dependent effects and responses in two Anuran larvae. Ecology 75:197–231; 1994.

Werner, E.E.; Gilliam, J.F. The ontogenetic niche and species interactions in size-structured populations. Annu. Rev. Ecol. Syst. 15:393–426; 1984.

Werner, E.E.; Hall, D.J. Foraging efficiency and habitat switching in competing sunfishes. Ecology 60:256–264; 1979.

Werner, E.E.; Hall, D.J. Ontogenetic habitat shifts in bluegill: the foraging rate–predation risk trade-off. Ecology 69:1352–1366; 1988.

Werner, E.E.; Gilliam, J.F.; Hall, D.J.; Mittelbach, G.G. An experimental test of the effects of predation risk on habitat use in fish. Ecology 64:1540–1548; 1983.

Wetzel, R.G. The role of the littoral zone and detritus in lake metabolism. Arch. Hydrobiol. 13:145–161; 1979.

Wetzel, R.G. Land–water interfaces: metabolic and limnological indicators. Verh. Int. Verein. Theor. Angew. Limnol. 24:6–24; 1990.

Wilbur, H.M. Interactions between growing predators and growing prey. In: Ebenman, B.; Persson, L., eds. Size-structured populations: ecology and evolution. Heidelberg: Springer-Verlag; 1988:157–172.

Wilson, D.S. The adequacy of body size as a niche difference. Am. Nat. 109:769–784; 1975.

Winfield, I.J. The influence of simulated aquatic macrophytes on the zooplankton consumption rate of juvenile roach, *Rutilus rutilus,* rudd, *Scardinius erythrophthalmus,* and perch, *Perca fluviatilis.* J. Fish Biol. 29:37–48; 1986.

2. Influence of Submerged Macrophytes on Trophic Interactions Among Fish and Macroinvertebrates

Sebastian Diehl and Ryszard Kornijów

Introduction

Lentic macroinvertebrates have received comparatively little interest from lake managers and researchers for two reasons. First, with few exceptions (e.g., the control of mosquitos, the provision of forage for waterfowl), there is no public interest in managing ponds and lakes for macroinvertebrates per se. Macroinvertebrates are not a directly harvestable resource, such as fish, and are rarely perceived as a nuisance, such as excessively growing planktonic algae or macrophytes. Second, macroinvertebrates are impractical study organisms. They are taxonomically and functionally diverse, seasonally and geographically highly variable, strongly patchy in their local distributions, and tedious to sample and sort. This combination of high natural variability, high sampling effort, and low sampling precision makes it notoriously difficult to distinguish pattern from error and noise in estimates of macroinvertebrate responses to comparative or experimental gradients. Furthermore, many macroinvertebrates cannot be observed in their natural environments and are very difficult to manipulate in the field, which limits the approaches to study their behaviors and impacts on other organisms.

Despite these constraints to their study, macroinvertebrates perform important ecosystem and community functions in lakes, many of which are mediated through the interactions between macroinvertebrates and fish. For example, the consumption of sediment-living, detritivorous macroinvertebrates by fish con-

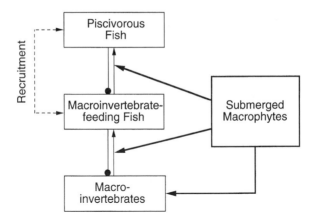

Figure 2.1. Effects of submerged macrophytes on fish–macroinvertebrate interactions considered in this chapter (thick solid arrows). Thin solid arrows paralleled by solid lines with vortices represent trophic interactions. Broken arrow represents recruitment processes.

stitutes a significant pathway of phosphorus recycling from sediments to the water column (Andersson et al., 1988). Macroinvertebrates are a major food source for many fish species during at least some ontogenetic stages, and the availability of macroinvertebrate resources has important consequences for the abundances and population size structures of various fish species (Persson, 1988; Osenberg et al., 1992; Olson et al., 1995). Conversely, fish predation acts as a major selective mortality agent on macroinvertebrates, with repercussions for the abundance, species composition, and size structure of macroinvertebrates (Erikson et al., 1980; Morin, 1984; Osenberg and Mittelbach, 1989; Diehl, 1995; Kornijów, 1997).

The abundance of various species of submerged macrophytes in the littoral zone of lakes influences fish–macroinvertebrate interactions in a variety of ways, affecting processes on a wide range of spatial and temporal scales (Fig. 2.1). For example, submerged macrophytes increase the diversity of habitats and resources for macroinvertebrates and, thus, the basis of secondary production; they reduce the susceptibility of macroinvertebrates to fish predators; they reduce the vulnerability of prey fish to piscivores (Diehl, 1988; Kornijów et al., 1990; Eklöv and Diehl, 1994). All of this affects the habitat choice, survival, and growth of macroinvertebrate-feeding fish, ultimately feeding back on the population dynamics of fish and the predation pressure by fish on macroinvertebrates and other components of the lake food web (Persson et al., 1992; Mittelbach and Osenberg, 1993; Diehl and Eklöv, 1995).

Experimental studies of fish–macroinvertebrate interactions have been largely restricted to small spatial and short temporal scales. The understanding of the large-scale, long-term effects of submerged vegetation on fish–macroinvertebrate interaction is still sketchy and will require increased efforts to identify and quantify feedback mechanisms among processes in littoral, benthic and pelagic habitats (Lodge et al., 1988; Osenberg et al., 1994; Persson et al., 1997). In this chapter, we review experimental and comparative studies of the influence of submerged macrophytes on trophic interactions among fish and macroinvertebrates (Fig. 2.1). We first review empirical relationships among submerged vegetation and the production, distribution, and abundance of macroinvertebrates in the littoral zone of lakes. Based on data from within-season experiments, we then discuss behavioral and trophic interactions among fish and macroinvertebrates and how these are affected by submerged macrophytes. Finally, we explore the effects of submerged vegetation on the long-term dynamics of fish and macroinvertebrates in the context of simple predator–prey models and discuss possible complications arising from ontogenetic, size-related habitat and diet shifts of fish. We conclude with a few suggestions for future research.

Relationships Between Submerged Macrophytes and Littoral Macroinvertebrates

Benthic and Epiphytic Macroinvertebrates

Littoral macroinvertebrates can be categorized according to the microhabitats in which they occur as being *benthic* (i.e., dwelling the bottom sediments), and *epiphytic* (i.e., associated with macrophytes). Benthic and epiphytic macroinvertebrates often belong to the same classes, orders, or families but rarely to the same genera (Pieczynski, 1977; Kornijów et al., 1990; Kornijów and Kairesalo, 1994a). Most littoral macroinvertebrates are typical of either benthic or epiphytic faunas, but there are some widespread opportunistic species such as *Asellus aquaticus* (Isopoda), *Helobdella stagnalis* (Hirudinea), and *Lymnaea peregra* (Gastropoda) that readily inhabit and move between both benthic and epiphytic habitats. Benthic and epiphytic macroinvertebrates differ with respect to their seasonal dynamics, food sources, and predators and should therefore be distinguished both conceptually and empirically (Pieczynski, 1977; Pardue and Webb, 1985; Kornijów et al., 1990; see Kajak et al., 1965; Mittelbach, 1981b; Downing, 1984; Kornijów and Kairesalo, 1994b, for recommendations on sampling techniques). To avoid confusion, in this chapter we generally use the inclusive term *macroinvertebrates* rather than *benthos* whenever it is either not necessary or not possible (based on the data reported in the primary literature) to distinguish between benthic and epiphytic macroinvertebrates.

Resource Use, Distribution, and Abundance of Macroinvertebrates in Relation to Submerged Macrophytes

Submerged macrophytes affect abiotic variables such as light, temperature, and oxygen concentration and provide additional habitat and resources for macroinvertebrates (Carpenter and Lodge, 1986; Kornijów et al., 1990; Lillie and Budd, 1992; Kornijów and Kairesalo, 1994a). For example, many herbivores benefit from the additional substrate area provided by submerged macrophytes to periphytic

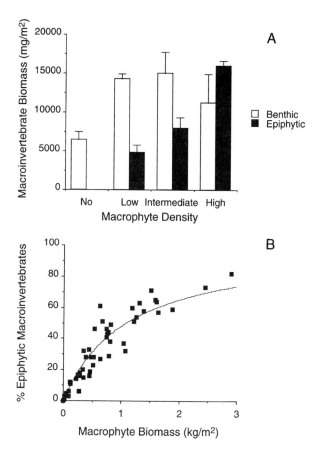

Figure 2.2. Relationships between macrophyte density and (A) the wet biomass of benthic and epiphytic macroinvertebrates in Lake Pääjärvi (with 95% confidence intervals), and (B) the percentage of total macroinvertebrate biomass composed of epiphytic macroinvertebrates in Lake Memphremagog. The fitted curve is described by $y = 100ax/(ax + b)$, with $a = 0.093$, $b = 0.102$, $r^2 = 0.87$. (Data from (A) Kornijów and Kairesalo, 1994a, and (B) Rasmussen, 1988.)

algae (Jones et al., this volume, Chapter 4; Brönmark and Vermaat, this volume, Chapter 3). Fresh tissue of macrophytes can also be important in the diets of some herbivores (Kornijów, 1996; Lodge et al., this volume, Chapter 8). After senescence and death, epiphytic algae and macrophytes become available as a food source to shredders and deposit feeders (Suren and Lake, 1989; Kornijów et al., 1995). Through these various pathways, submerged macrophytes have a strong, positive effect on the resource base of epiphytic and benthic macroinvertebrates, with the possible exception of some benthic collector-suspension feeders (e.g., chironomid larvae of the genus *Chironomus*) that may experience a reduced supply of fresh and dead planktonic algae under the cover of submerged macro-phytes (Kornijów and Moss, this volume, Chapter 12).

The taxonomic diversity and density of epiphytic macroinvertebrates are re-lated to the growth forms of the macrophytes on which they live, but generally there is a strong positive relationship between the abundances of submerged macrophytes and epiphytic macroinvertebrates (Figs. 2.2 and 2.6A; Cyr and Downing, 1988a,b; Jeffries, 1992; Lillie and Budd, 1992). Similarly, benthic macroinvertebrates are often more abundant in the sediments of vegetated areas compared with open areas (Fig. 2.2; Prejs, 1976; Pardue and Webb, 1985; Schramm and Jirka, 1989; Kornijów et al., 1990, but see the opposite pattern for benthos dominated by chironomids in Blindow et al., 1993). The relative contribution of epiphytic macroinvertebrates to total macroinvertebrates increases with increasing vegetation density, and in dense stands of submerged macrophytes, epiphytic macroinvertebrates are usually more abundant than benthic macroinvertebrates (Fig. 2.2; Pieczynski, 1977; Mittelbach, 1981b; Schramm and Jirka, 1989; Kor-nijów et al., 1990). The positive effects of submerged vegetation are usually smaller on the biomass than on the abundance of macroinvertebrates, because epiphytic macroinvertebrates are, on average, smaller than benthic macroinver-tebrates (Mittelbach, 1981b; Diehl, 1992; Rasmussen, 1993; Kornijów and Kaire-salo, 1994a; Diehl and Eklöv, 1995, but see Blindow et al., 1993, for a counter example). In absolute numbers or biomass, however, large, profitable (for fish) macroinvertebrates are more abundant in vegetated than unvegetated littoral habitats (Diehl, 1993a; Blindow et al., 1993).

Most of the data establishing the positive relationship between macroinver-tebrates and submerged macrophytes were collected in lakes and ponds containing fish. The relationship is therefore likely to be a result not only of the positive impact of submerged macrophytes on the productivity of macroinvertebrates, but also of the complex influence of vegetation on the interactions between macroin-vertebrates and their fish predators (see below).

Short-Term Interactions Between Fish and Macroinvertebrates

Behavioral Interactions

The interactions between fish and macroinvertebrates have been studied in the laboratory and in field experiments on relatively small spatial scales (labora-tory: < 1 m²; field: < 10 m²) and over relatively short temporal scales (laboratory: hours to days; field: weeks to months). These studies show four broad patterns, all of which suggest that fish–macroinvertebrate interactions may be strongly in-fluenced by the behavior of fish and macroinvertebrates.

First, many macroinvertebrates adjust their behavior in response to fish by seeking refuge and/or decreasing risky activities (Huang and Sih, 1990; McPeek, 1990b; Ball and Baker, 1995). Such risk-sensitive prey behavior can have pro-found consequences for the population dynamics of both prey and predators (Abrams, 1992; Werner, 1992).

Figure 2.3. Relationships between fish density treatments and (A) macroinvertebrate abundance, and (B) chironomid abundance at the end of 3-month field enclosure experiments. Submerged macrophytes were dense in A and absent in B. Fish were bluegill plus pumpkinseed sunfish (A) and perch (B). The fitted curves are described by $y = a + b \exp(-cx)$. For A: $a = 363$, $b = 867$, $c = 0.973$, $r^2 = 0.59$; for B: $a = 954$, $b = 6.05*10^8$, $c = 12.6$, $r^2 = 0.84$. (Data from [A] Mittelbach, 1988, and [B] Diehl, 1995.)

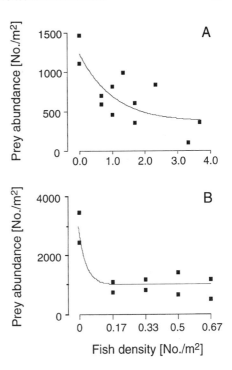

Second, predation by fish is selective. Generally, large macroinvertebrate taxa are more vulnerable to fish predation than small taxa (Crowder and Cooper, 1982; Morin, 1984; McPeek, 1990a; Diehl, 1992). This pattern appears to hold also on larger spatial and longer temporal scales as is indicated by comparative studies of lake systems with and without fish (Erikson et al., 1980; Bendell and McNicol, 1987; Evans, 1989; Bradford et al., in press). Exceptions are taxa with morphological or behavioral defenses such as snails with size refuges, odonates with cryptic behavior, or sediment-dwelling chironomids with refuges in deeper sediment layers (Osenberg and Mittelbach, 1989; McPeek, 1990b; Macchiusi and Baker, 1991; Kornijów, in press; Kornijów and Moss, this volume, Chapter 12).

Third, in studies spanning over more than two fish density treatments, negative effects of fish on the abundance or biomass of macroinvertebrates are usually strongest in the lowest range of fish densities, and the relationship between prey abundance and fish density can often be approximated by an exponentially declining curve (Fig. 2.3). In part, this may simply reflect the time scale of the studies. In a short-term study, if total prey consumption were directly proportional to predator and prey abundances, prey decline across predator densities would be expected to follow a simple exponential model (see Diehl, 1995; Osenberg and Mittelbach, 1996, for detailed discussions). In many studies, however, macroinvertebrate abundances appear to asymptote at nonzero values (Fig. 2.3), and behavioral processes are likely to be responsible for these asymptotes. For example, because the negative effects of fish are strongest on preferred taxa and

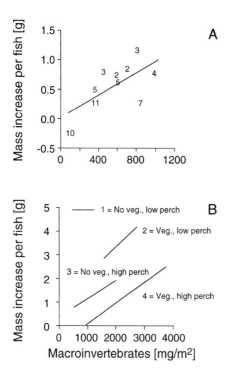

Figure 2.4. Relationships between macro-invertebrate prey density and the average individual growth rates of fish in field enclosures. (A) Linear regression statistics: $y = 0.023 + 0.00094x$, $r^2 = 0.43$. Data points in A are identified by their fish density treatments (number of bluegill plus pumpkin-seed sunfish/3 m^2). Note that, at a given prey density, fish growth is consistently lower at higher fish densities. Fish density explains 39% of the variation in the residuals of the shown relationship (residuals = 0.3 – 0.6*fish density, $r^2 = 0.39$, $P = .055$). (B) Lines 2 to 4 are linear regressions fitted to replicates of treatments varying in vegetation density (submerged macrophytes absent and present) and fish density (0.33 and 1 perch/m^2). In the treatment depicted by line 1, there was no relationship between prey density and fish growth, and the mean growth of fish is shown over the range of prey densities encountered. (Data from [A] Mittelbach, 1988, and [B] Diehl, 1993a.)

large size classes of macroinvertebrates, these prey items are frequently reduced already at low fish densities, and no further reduction may occur at higher fish densities. Similarly, because many macroinvertebrates adjust their behaviors to the risk of predation by fish, fish may be unable to reduce their prey below certain levels at which all prey make themselves unavailable to fish. Behavioral interactions among macroinvertebrates and fish may, in part, explain, why studies in which the lowest fish density treatments are nonzero often report only relatively subtle effects of fish density on macroinvertebrates (e.g., effects on selected taxa or on the size structure of the macroinvertebrate assemblage) (Persson and Green-berg, 1990; Bergman and Greenberg, 1994; Olson et al., 1995). Furthermore, nonzero asymptotes of macroinvertebrate density (Fig. 2.3) suggest that the degree to which macroinvertebrates can make themselves unavailable to fish may be limited by the availability of physical refuges.

Finally, the feeding and individual growth rates of many fish species are positively related to the abundance of macroinvertebrates (Fig. 2.4). Different fish species may, however, vary considerably in their efficiencies at exploiting macro-invertebrate prey (Diehl, 1988; Persson, 1988). Furthermore, at comparable macroinvertebrate densities, individual fish grow less at higher own densities (Fig. 2.4). This suggests that macroinvertebrate-feeding fish have negative effects on their own growth rates, which are stronger than would be expected if pure exploitative competition for macroinvertebrates were the only mechanism of density dependence. Again, the mechanisms that account for this additional density dependence are possibly behavioral (e.g., interference among fish and/or reduced activity and increased hiding of macroinvertebrates at higher fish densities).

Omnivory

Macroinvertebrate-feeding fish are omnivores (i.e., they consume prey from more than one trophic level) (Diehl, 1993b). Omnivory complicates the trophic relationships among fish and macroinvertebrates, because fish and invertebrate predators engage in both competitive and consumer–resource interactions. For example, fish could have an indirect positive effect on the abundance of nonpredatory macroinvertebrates by numerically releasing them from invertebrate predators (Diehl, 1992; Polis and Holt, 1992; Hill and Lodge, 1995; Prejs et al., 1977). Because most invertebrate predators are bigger than most nonpredatory invertebrates, preferential predation by fish on large macroinvertebrates provides another potential mechanism for an indirect positive effect of fish on nonpredatory macroinvertebrates. Decreased predation by fish on (smaller) nonpredatory macroinvertebrates in the presence of increased numbers of (larger) predatory macroinvertebrates has, in fact, been demonstrated (Fig. 2.5). Compared with fishless treatments, however, fish very rarely seem to have overall positive effects on nonpredatory macroinvertebrates and always have negative effects on predatory macroinvertebrates over the time scales of typical enclosure studies (Fig. 2.5, Diehl, 1993b). Unless additional mechanisms stabilize the intermediate predator (Diehl, 1993b), such a situation should not be sustainable; that is, predatory macroinvertebrates can only coexist with fish, if they, in the absence of fish, suppress nonpredatory macroinvertebrates below the levels at the three-trophic level equilibrium with fish present ($R^*_N < R^{**}$; Fig. 2.7C, Polis and Holt, 1992; Holt and Polis, 1997).

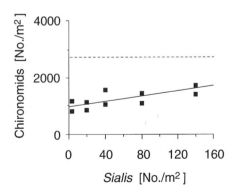

Figure 2.5. Relationship between density of chironomids at the end of a 3-month field enclosure experiment and initial density of *Sialis lutaria* (a predatory macroinvertebrate) in the presence (data points and solid line) and in the absence of omnivorous perch (dashed line; only the mean is shown, because there was no relationship between the densities of chironomids and *Sialis*). (Data from Diehl, 1995.)

Effects of Submerged Macrophytes on Short-Term Interactions Between Fish and Macroinvertebrates

Numerous laboratory experiments have demonstrated that the rates at which most fish species encounter and attack prey decline with increasing density of artificial or natural macrophytes (Diehl, 1988; Nelson and Bonsdorff, 1990, and references therein). This has two important consequences for the interactions between fish and macroinvertebrates. First, at a given prey density, individual growth rates of macroinvertebrate-feeding fish are negatively affected by vegetation density (see lines 1 vs. 2 and 3 vs. 4 in Fig. 2.4B). In other words, for individual fish to attain the same feeding and growth rates, prey densities have to be higher the higher the density of vegetation. Second, at a given fish density, survival and, subsequently, the density of macroinvertebrates are higher in densely compared with sparsely vegetated habitats (see below).

The negative effects of vegetation on the encounter rates of fish and the positive effects of vegetation on the standing stock and production of macroinvertebrates make it impossible to derive a general relationship between the growth rates of fish and the density of submerged vegetation. Positive, negative, and neutral effects of increased use of high-density vegetation patches on the individual growth of invertebrate-feeding fish have been reported (Crowder and Cooper, 1982; Turner and Mittelbach, 1990; Diehl, 1993a; Diehl and Eklöv, 1995; Persson and Eklöv, 1995).

It is likely that the relationship between the individual growth of fish and the density of submerged vegetation depends on the morphology, size, and sensory capacities of the fish as well as on the growth form of the macrophytes. For example, primarily visually searching fish such as Eurasian perch (*Perca fluviatilis*), bluegill sunfish (*Lepomis macrochirus*), and largemouth bass (*Micropterus salmoides*) are able to feed on epiphytic macroinvertebrates, which constitute the bulk of macro-invertebrate resources in vegetated habitats (Fig. 2.2B). In such species, maximum individual growth rates may be achieved at some intermediate level of vegetation density (Crowder and Cooper, 1982; Wiley et al., 1984; Savino et al., 1992). By contrast, primarily tactile bottom feeders such as bream (*Abramis brama*) are likely to be uniformly negatively affected by increasing vegetation density (Diehl, 1988).

Effects of Vegetation on the Long-Term Dynamics of Fish and Macroinvertebrates

Short-term and small-scale enclosure studies are insufficient to understand and predict the long-term effects of submerged vegetation on fish–macroinvertebrate interactions in whole-lake systems. The nonequilibrium nature of enclosure studies raises issues of feedback of fish predation on the long-term population dynamics of macroinvertebrates (including reproduction and dispersal) and of

Figure 2.6. Relationship among the average wet biomasses of macrophytes, macroinvertebrates, and fish in the littoral zones of ten southern Quebec lakes. Linear regression statistics are: (A) $y = 12.2 + 0.015x$, $r^2 = 0.42$, $P = .057$; (B): $y = 2.2 + 0.0051x$, $r^2 = 0.40$, $P = .066$. One lake (open triangles) was excluded from the analyses, because it had high biomasses (>12 g/m^2) of two fish species (*Moxostoma anisurum* and *M. valenciennesi*) that were absent from all other lakes. (Data from Pierce et al., 1994.)

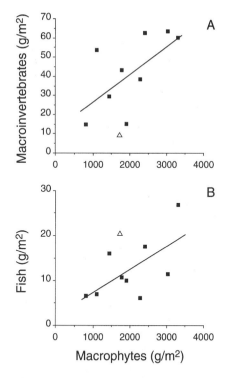

feedback of the abundances of macroinvertebrates and submerged macrophytes on the population dynamics of macroinvertebrate-feeding fish. An obvious and important question is: What are the effects of submerged macrophytes on the standing stock and production of macroinvertebrates and macroinvertebrate-feeding fish in natural systems? Data to address the question at the lake scale are so far scarce, but they suggest that submerged macrophyte density may have a positive effect on the abundances of both macroinvertebrates and fish (Fig. 2.6; Blindow et al., 1993). In the absence of an extensive body of empirical data at the whole-lake scale, our approach here is to derive plausible expectations on the long-term, large-scale dynamics of fish and macroinvertebrates based on knowledge gained at smaller spatial and shorter temporal scales. We first explore potential effects of submerged vegetation on the population dynamics of macroinvertebrates and fish in the context of simple predator–prey models, incorporating mechanisms observed in short-term experiments as model assumptions. The qualitative predictions derived from this exercise will then serve as reference points in the subsequent discussion of the more complex (and more realistic) situation in which fish populations are size-structured. It will become obvious that the vegetation–macroinvertebrate–fish interaction cannot be understood in isolation from other components of the lake food web and from processes that link the littoral zone to benthic and pelagic habitats (see also Lodge et al., 1988; Persson and Crowder, this volume, Chapter 1).

Expectations from Predator–Prey Theory

Considerations based on simple predator–prey theory suggest that increasing the density of submerged vegetation may often have positive effects on the long-term abundances of both macroinvertebrates and their fish predators and can stabilize the dynamic interaction between the two (Fig. 2.7A). Based on the evidence from short-term experiments, we can assume that increasing vegetation density increases macroinvertebrate productivity and reduces the search efficiency of macroinvertebrate-feeding fish. In a graphical analysis of a simple predator–prey model

Figure 2.7. Potential effects of vegetation on equilibrium densities of fish and macroinvertbrates. (A) Examples of macroinvertebrate (R) and fish (P) isoclines based on a differential equation predator–prey model assuming logistic prey growth with intrinsic growth rate r and carrying capacity K, a type II functional response of the predator $f(R)$, a constant conversion efficiency c of prey into predators, and a constant mortality rate m of the predator. The dynamical equations are

$$\frac{dR}{dt} = rR(1 - \frac{R}{K}) - f(R)P \quad \text{and} \quad \frac{dP}{dt} = cf(R)P - mP \quad \text{with } f(R) = \frac{aR}{1 + abR}$$

where a is the predator's search rate and b its handling time per prey. Isoclines connect all possible combinations of predator and prey densities at which the instantaneous rates of population change of either the prey (dR/dt) or the predator (dP/dt) are zero. Prey densities increase (decrease) anywhere below (above) the prey isocline. Predator densities increase (decrease) anywhere to the left (right) of the predator isocline. Intersections of the predator and prey isoclines denote local equilibria. Equilibria are locally stable when the prey isocline has a negative slope at the intersection with the predator isocline. Relative stability (measured as resilience [i.e., the reciprocal of the return time to equilibrium]) increases with the steepness of the negative slope of the prey isocline at equilibrium. Hatched lines and equilibrium 1 represent the baseline case. Equilibrium 2 illustrates a case in which an increase in vegetation density solely decreases the predator's search rate a. Equilibrium 3 illustrates a case in which an increase in vegetation density additionally increases the carrying capacity K of the prey. (B) Prey production at equilibrium in cases 1 and 2 of A. The heights of the vertical lines represent the harvestable prey production. (C) Example of isoclines of predatory (N) and nonpredatory (R) macroinvertebrates and fish (P) based on a differential equation food chain model assuming omnivory by the fish, logistic growth of nonpredatory macroinvertebrates, linear functional responses, and constant conversion efficiencies of resources into consumers. The intersection of the three planes denotes the equilibrium with all three trophic levels present. K denotes the carrying capacity of R in the absence of N and P, R^*_N and R^*_P the equilibria of R with only N or only P present, respectively, and R^{**} the equilibrium of R with all three trophic levels present. Note that the origin is in the lower front corner and that the isocline of R stretches from near the upper front corner down and back into the phase space. Increasing K will move the isocline of R up and to the right and push it back. The three-trophic level equilibrium (R^{**}, N^{**}, P^{**}) has to fall onto the intersection of the isoclines of N and P, which slants downward (toward lower N) with increasing P. Therefore, an increase in K will cause both R^{**} and P^{**} to increase but N^{**} to decrease.

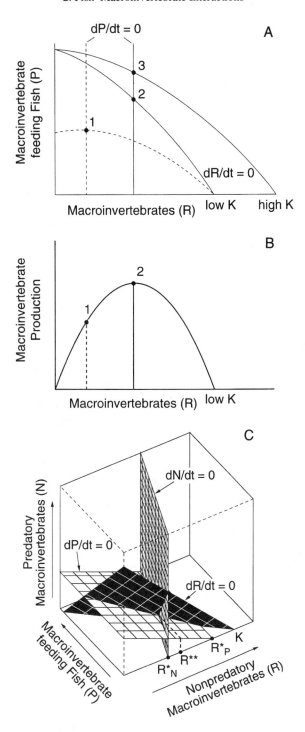

(Fig. 2.7A), reducing the predator's search efficiency moves the predator isocline to the right (because higher prey densities are required for the predators to attain the feeding rates at which predator production exactly offsets mortality) and the prey isocline up (because more predators are needed to consume a given number of prey). Increasing prey productivity moves the prey isocline further up and also to the right. Consequently, with an increase in vegetation density, prey will *always* equilibrate at a higher density, whereas the effects on the equilibrium density of predators and on the local stability of the equilibrium depend on parameter values. Under a broad range of circumstances, however, positive effects of vegetation density on both predator density and stability will be observed (compare equilibrium 1 with equilibria 2 and 3 in Fig. 2.7A; see figure caption for details of the model and the interpretation of isocline graphs).

Fig. 2.7A assumes constant prey behavior across varying densities of predators and vegetation. So far, only few studies have empirically addressed the long-term effects of risk-sensitive behavior on the population dynamics of macroinvertebrate prey, as mediated through altered individual growth and fecundity (Peckarsky et al., 1993; Ball and Baker, 1996). In theory, risk-sensitive prey behavior simultaneously bends the predator isocline to the right (prey hide more at higher predator densities) and reduces prey production (prey feed less at higher predator densities), both of which tend to stabilize the equilibrium (Ives and Dobson, 1987). Increasing vegetation density decreases the need for the prey to reduce its own activity in the presence of the predator, because vegetation decreases the abilities of fish predators to find prey. Because of its opposing effects on predator efficiency and prey activity, the net effect of increasing vegetation density on the equilibrium densities of predators and prey is not trivial to predict. Still, in models of risk-sensitive foraging by the prey, the equilibria of predators and prey may not depend on encounter rates (Abrams, 1984). Thus, vegetation may affect the densities of predators and prey mostly through its positive effect on the productivity of the prey's resources. For a family of models, increasing the productivity of the resources of a risk-sensitive prey increases the equilibrium densities of both prey and predator (e.g., Abrams, 1984, Appendix A, B; Abrams, 1992, Appendix B). We conclude that, if macroinvertebrates are risk-sensitive foragers, this will often increase the likelihood for both macroinvertebrates and fish to increase with vegetation density.

Macroinvertebrates are a heterogeneous collection of taxa and size classes that interact trophically with one another. Again, only few studies have, so far, empirically addressed the population dynamical consequences for macroinvertebrates of, for example, omnivory and cannibalism (Diehl, 1993b, 1995; Hopper et al., 1996). The accommodation of omnivory in the framework of Figure 2.7 requires, at minimum, the addition of a third dimension to the phase space; that is, fish (P), predatory (N), and nonpredatory (R) macroinvertebrates have to be represented separately (Fig. 2.7C). For various reasons, the prediction of potential effects of macrophytes on the dynamics of P, N, and R (as mediated through effects on resource productivity and encounter rates) becomes very complex, and no simple relationships can be derived for the (in nature extremely common) case of coexistence of all three trophic levels. First, the stability conditions of omnivory systems are complex, even in the simplest case of linear functional responses depicted in Figure 2.7C, and the stability of the three-trophic level equilibrium can change abruptly along smooth gradients of parameter changes (Holt and Polis, 1997). Second, for stable three-trophic level equilibria, vegetation-induced increases of the carrying capacity of R should increase both R and P but decrease N, and it cannot be generally known whether *total* macroinvertebrates ($R + N$) will increase or decrease with an increase in the carrying capacity of R (S. Diehl, unpublished data). Finally, the effects of vegetation on equilibria mediated through changes in encounter rates depend on how the benefit for N of reduced encounter rates with P balances with potential changes in encounter rates with R. Cannibalism among predatory macroinvertebrates can be qualitatively accommodated in the framework of Figure 2.7C by treating it as an extreme form of interference competition (Wollkind, 1976). Thus, the N isocline would bend over to the right in the R plane as cannibalism increases with the density of N. However, the prediction of possible effects of increasing vegetation density on equilibrium densities becomes very complex, because vegetation now also affects the encounter rates among conspecifics of N.

In conclusion, both scant empirical evidence and simple predator–prey models suggest that increasing the density of submerged vegetation enhances the densities of *both* macroinvertebrates and their fish predators. Assuming risk sensitivity in the prey behavior tends to reinforce this conclusion, but the inclusion of increased trophic complexity (omnivory, cannibalism) into models precludes the deduction of simple, theoretical relationships among the abundances of macrophytes, macroinvertebrates, and fish. In the following section, we explore the implications of an additional complexity (i.e., population size structure) for the interactions among submerged macrophytes, macroinvertebrates, and fish.

Vegetation, Macroinvertebrates, and Size Structure of Fish Populations

Most fish populations are size-structured, and many fish species feed on macroinvertebrates in littoral vegetation only during parts of their life histories (Werner, 1986; Persson, 1988). Unless all size classes of a fish species are affected similarly by vegetation, the accommodation of size structure in the fish population goes beyond the conceptual framework of Figure 2.7, because the density of macroinvertebrate-feeding fish would depend not only on the availability of macroinvertebrate resources but also on the availability of resources to all other life stages of the fish (Mittelbach and Chesson, 1987; Persson, 1988; Osenberg et al., 1992; Mittelbach and Osenberg, 1993).

Many macroinvertebrate-feeding fish are vulnerable to piscivores and use littoral vegetation as a predation refuge (Werner and Hall, 1988; Eklöv and Diehl, 1994; Persson and Eklöv, 1995). Consequently, the presence of submerged macrophytes in lakes creates the opportunity for different size classes of macroinvertebrate-feeding fish to segregate by habitat. In the presence of piscivores, small vulnerable-size classes tend to seek refuge in the vegetation,

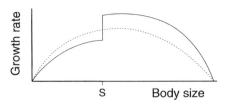

Figure 2.8. Schematic representation of the size specific individual growth rates of a fish species that shifts from a vegetated to an open habitat at size S (solid line). The open habitat is assumed to be more profitable than the vegetated habitat in terms of foraging return but more dangerous in terms of risk of piscivory. Furthermore, the risk of mortality from piscivores is assumed to decrease with increasing body size. At size S, the net benefit of using the vegetation habitat becomes smaller than the net benefit of using the open habitat. Also shown are the size-specific growth rates that would be observed if *all* size classes used the open habitat in the absence of piscivores (broken line). The difference between the two lines illustrates the instantaneous costs and benefits incurred by various size classes when small fish use the vegetation habitat. Abundance and population size structure are assumed to be identical in both cases (i.e., no feedbacks of differences in mortality and growth rates on the population dynamics are taken into account).

where they often experience reduced individual growth rates while simultaneously releasing larger invulnerable-size classes from intersize class competition in open habitats (Werner et al., 1983; Tonn et al., 1992; Mittelbach and Osenberg, 1993; Diehl and Eklöv, 1995). These interactive effects of vegetation and piscivores on the size-specific habitat use and growth rates of invertebrate-feeding fish suggest a sharp increase in individual growth rate at the size of habitat switch (Fig. 2.8). Such abrupt changes in growth rate have been documented for bluegill sunfish (Osenberg et al., 1988; Werner and Hall, 1988), and preliminary data indicate that Eurasian perch may show similar discontinuities in their growth rates at the sizes of ontogenetic niche shifts (L. Persson, personal communication).

The effects of submerged vegetation on patterns of habitat use and in-dividual growth of fish are likely to feed back on the overall population dynamics of the fish. For example, Blindow et al. (1991) observed positive effects of increased submerged vegetation cover on both the growth rates of macroinvertebrate-feeding perch and the abundance of piscivorous size classes of perch. This example indicates the possibility that habitat segregation among vulnerable and invulnerable size classes of fish in the presence of submerged vegetation may counteract the stunting of fish populations. This has been docu-mented for the bluegill sunfish, whose invulnerable stages attain high individual growth rates in the open water when vulnerable stages are restricted to the vegetation by piscivorous largemouth bass (Werner et al., 1983; Mittelbach and Osenberg, 1993). High individual growth rates of potentially piscivorous fish may, in turn, feed back on the abundance of piscivores and thus on the relative risks of vulnerable fish inside and outside the vegetation (Werner and Hall, 1988). How-ever, the dynamics of size-structured fish populations may be rather complex even in a single-habitat situation, and the population dynamical consequences of changes in the relative sizes and qualities of the habitats used by different size classes of fish are presently not well understood (Persson and Crowder, this volume, Chapter 1).

Both vegetation and piscivores have to be present to produce habitat segregation between vulnerable and invulnerable size classes of fish. However, the independent occurrence of either abundant submerged vegetation or abundant piscivores seems to be an uncommon and transient situation. Comparative data suggest, that the abundance of piscivores in temperate European lakes is correlated with the abundance of submerged macrophytes (Grimm and Backx, 1990; Persson et al., 1991; Blindow et al., 1993; Persson, 1994). This relationship is likely to be causal, and several positive feedback mechanisms between submerged vegetation and piscivores have been suggested (Blindow et al., 1993; Persson, 1994; Persson and Crowder, this volume, Chapter 1; Scheffer and Jeppesen, this volume, Chapter 31).

In conclusion, the question of which biomasses of macroinvertebrates and macroinvertebrate-feeding fish are to be expected in a lake of a given productivity and submerged macrophyte cover requires a complex answer. The relative and absolute abundances of macroinvertebrate-feeding fish in open and vegetated habitats depend on their habitat use as well as on the survival, growth, and reproductive rates of all size classes of fish, which, in turn, depend on the abundance of piscivores and on the relative sizes and productivities of open and vegetated habitats (Mittelbach, 1984; Osenberg et al., 1994; Persson et al., 1997). In a situation with both vegetation and piscivores, where predator-induced habitat segregation among competing size classes of fish promotes slow growth of vulnerable fish and fast growth of invulnerable fish, the densities of macroinvertebrate-feeding fish and the resulting predation pressure on macroinvertebrates may be higher in the vegetation and lower in open habitats than would be expected if each habitat was the only one available. This tendency may be reinforced through population dynamical feedback processes. For example, fast growth and high fecundity of invulnerable size classes of fish, which are supported by resources outside the vegetation, may result in far higher recruitment rates of small fish into the vegetation habitat than could be sustained by vegetation resources alone (Mittelbach and Chesson, 1987; Mittelbach and Osenberg, 1993).

Suggestions for Future Research

The relationships among submerged macrophytes, macroinvertebrates, and fish can be affected in complex ways by processes on a wide range of scales, from behavioral decisions of individual macroinvertebrates to the life histories of long-lived and wide-ranging piscivorous fish. Studying the interactions of processes at vastly differing scales is a challenging enterprise (Carpenter, 1988; Cooper et al., in press), and at present, the relative contributions of various processes to the distributional patterns of macroinvertebrates and fish are not well understood. To date, empirical data are available predominantly on patterns at intermediate scales of spatial, temporal, and organismal resolution (e.g., within-season population responses of macroinvertebrates to fish predation or diet and growth responses of macroinvertebrate-feeding fish to piscivores and vegetation). Our knowledge is especially deficient when it comes to the fine-scale mechanisms of fish–macroinvertebrate interactions and to the long-term population dynamics of size-structured fish populations and of macroinvertebrate species with terrestrial dispersal stages (e.g., insects).

In the future, high priority should be given to the study of behavioral inter-actions between macroinvertebrates and fish. This includes the systematic study of the effects of submerged vegetation on the search behavior, prey choice, and feeding and growth performances of different species and size classes of fish, as well as the study of behavioral responses of macroinvertebrates to their inver-tebrate and fish predators (see, e.g., Mittelbach, 1981a; Huang and Sih, 1990). The latter may provide crucial insights into the mechanisms of density dependence in the individual growth of macroinvertebrate-feeding fish (see above). The study of behavioral responses of macroinvertebrates to fish should be accompanied by the investigation of the costs and benefits of these behaviors and their population dynamical consequences to macroinvertebrate prey (Ball and Baker, 1996). In the same vein, the behavioral trade-offs and population dynamical consequences of predation and cannibalism among macroinvertebrates need to be studied (Peck-arsky et al., 1993; Hopper et al., 1996). The latter will require field experiments in which the densities of selected macroinvertebrates are manipulated (Diehl, 1995).

Although many of the above processes are most profitably studied at laboratory to field enclosure scales, long-term dynamics of fish and macroinvertebrates can only be studied at the whole-system scale. Unfortunately, for the reasons listed in the introduction, macroinvertebrates are most commonly not included in standard lake surveys or the monitoring of lake manipulations. What is strongly needed are both experimental and comparative studies of whole systems in which comprehen-sive data pertinent to the fish–macroinvertebrate interaction are collected. These data should include estimates of (1) the proportional lake area covered with macrophytes as well as macrophyte species composition and stand density; (2) the species composition, abundance, size structure, and production of macroinver-tebrates in habitats of varying vegetation cover and sediment structure; (3) the species composition, abundance, size structure, and habitat distribution of fish as well as their habitat- and size-specific diets and individual growth rates; (4) the abundance and size structure of other important resources of fish, especially zooplankton. Data should be gathered at several occasions during the year to allow the study of seasonal patterns (Butler, 1989; Collins and Hinch, 1993). Also, to increase statistical power and to facilitate comparisons among systems, it may often be useful to lump macroinvertebrates into guilds of similar resource use or vulnerability to fish predators (e.g., Diehl, 1992; Persson et al., 1996). Such sampling programs may seem daunting, but the payoff in terms of the detection of significant patterns can be high (see Blindow et al., 1993, for an excellent example).

Progress in the study of the long-term population dynamics of macroinver-tebrates and fish can be significantly enhanced if empirical studies are complemented by adequate modeling efforts. Because of the complexities of the life histories of most macroinvertebrates and fish, simple modeling approaches (e.g., Fig. 2.7) can merely function as heuristic tools. The dynamics of prey with often complex life cycles (e.g., insects) and predators that go through size-related ontogenetic niche shifts need to be described by structured population models (Crowley et al., 1987; Mittelbach and Chesson, 1987; Persson et al., 1997; Persson and Crowder, this volume, Chapter 1). Although structured population models are analytically far less tractable than simple strategic models, they hold promise to increase the understanding of any particular real system under study by allowing the incorporation of empirically derived (and testable) mechanistic detail (Murdoch et al., 1992). In this context, it is important to note that studies of fine-scale mechanisms should be performed along *gradients* of conditions to guide the choice of specific model functions (e.g., how does a specific parameter vary with body size of fish and macroinvertebrates or with density of macrophytes?).

In this chapter, we have treated submerged vegetation largely as an independent variable that sets the stage for the interactions between macroinvertebrates and fish (Fig. 2.1). There are, however, various pathways of feedback through which macroinvertebrates and fish affect submerged macrophytes (see Brönmark and Weisner, 1992; Scheffer et al., 1993; Lodge et al., 1994; Brönmark and Vermaat, this volume, Chapter 3; Jones et al., this volume, Chapter 4; Lodge et al., this volume, Chapter 8; Scheffer and Jeppesen, this volume, Chapter 31). We have also frequently simplified the macrophyte setting by contrasting two apparently discrete scenarios, absence and presence of submerged macrophytes, without specifying an exact level of the density or proportional cover of submerged macrophytes for the latter situation. Our intention, here, was to illustrate clearly distinct patterns and system behaviors. However, many of the processes discussed in this chapter are likely to vary nonlinearly with the proportion of lake area covered by submerged macrophytes (Scheffer et al., 1993; Meijer et al., 1994). An especially important area of future research is therefore the exploration of patterns and processes along gradients of vegetation density (e.g., Schriver et al., 1995). This bears the potential to identify critical levels of vegetation density at which feedback mechanisms such as the ones dicussed in this and other chapters of this book might become strong enough to stabilize lakes in states favored by managers.

Acknowledgments. We thank the workshop organizers for making the meeting very productive and enjoyable. We also thank Peter Abrams, Irmgard Blindow, Scott Cooper, Lennart Persson, Marten Scheffer, and an anonymous reviewer for comments on a previous draft of this chapter. Gary Mittelbach kindly provided the original data to produce Figures 2.3A and 2.4A.

Financial support for the research on which parts of this chapter are based was given by the Swedish Council for Forestry and Agricultural Research to S. Diehl and L. Persson and by the British Council, the Finnish Academy of Sciences, the Tor and Maj Nessling Foundation, and the International Agricultural Centre in Wageningen to R. Kornijów.

References

Abrams, P.A. Foraging time optimization and interactions in food webs. Am. Nat. 124:80–96; 1984.

Abrams, P.A. Predators that benefit prey and prey that harm predators: unusual effects of interacting foraging adaptations. Am. Nat. 140:573–600; 1992.

Andersson, G.; Granéli, W.; Stenson, J. The influence of animals on phosphorus cycling in lake ecosystems. Hydrobiologia 170:267–284; 1988.

Ball, S.L.; Baker, R.L. The non-lethal effects of predators and the influence of food availability on life history of adult *Chironomus tentans* (Diptera: Chironomidae). Freshwat. Biol. 34:1–12; 1995.

Ball, S.L.; Baker, R.L. Predator-induced life history changes: antipredator behavior costs or facultative life history shifts? Ecology 77:1116–1124; 1996.

Bendell, B.E.; McNicol, D.K. Fish predation, lake acidity and the composition of aquatic insect assemblages. Hydrobiologia 150:193–202; 1987.

Bergman, E.; Greenberg, L.A. Competition between a planktivore, a benthivore, and a species with ontogenetic niche shifts. Ecology 75:1233–1245; 1994.

Blindow, I.; Andersson, G.; Hargeby, A. Submerged macrophytes in shallow lakes and their importance for waterfowl. In: Finlayson, C.M.; Larsson, T., eds. Wetland management and restoration. Proceedings of a workshop, 1990, Sweden. Solna, Sweden: Swedish Environmental Protection Agency Report 3992; 1991:72–79.

Blindow, I.; Andersson, G.; Hargeby, A.; Johansson, S. Long-term pattern of alternative stable states in two shallow eutrophic lakes. Freshwat. Biol. 30:159–167; 1993.

Bradford, D.F.; Cooper, S.D; Jenkins, T.M., Jr.; Brown, A.D.; Kratz, K.; Sarnelle, O. Influences of natural acidity and introduced fish on amphibian, macroinvertebrate, and zooplankton assemblages in alpine lakes of the Sierra Nevada, California. Can. J. Fish. Aquat. Sci. (in press).

Brönmark, C.; Weisner, S.E.B. Indirect effects of fish community structure on submerged vegetation in shallow, eutrophic lakes: an alternative mechanism. Hydrobiologia 243/244: 293–301; 1992.

Butler, M.J. Community responses to variable predation: field studies with sunfish and freshwater macroinvertebrates. Ecol. Monog. 59:311–328; 1989.

Carpenter, S.R. Transmission of variance through lake food webs. In: Carpenter, S.R., ed. Complex interactions in lake communities. New York: Springer-Verlag; 1988:119–135.

Carpenter, S.R.; Lodge, D.M. Effects of submersed macrophytes on ecosystem processes. Aquat. Bot. 26:341–370; 1986.

Collins, N.C.; Hinch, S.G. Diel and seasonal variation in foraging activities of pumpkin-seeds in an Ontario pond. Trans. Am. Fish. Soc. 122:357–365; 1993.

Cooper, S.D.; Diehl, S.; Kratz, K.; Sarnelle, O. Implications of scale for patterns and processes in stream ecology. Aust. J. Ecol. (in press).

Crowder, L.B.; Cooper, W.E. Habitat structural complexity and the interaction between bluegills and their prey. Ecology 63:1802–1813; 1982.

Crowley, P.H.; Nisbet, R.M.; Guerney, W.S.C.; Lawton, J.H. Population regulation in animals with complex life-histories: formulation and analysis of a damselfly model. Adv. Ecol. Res. 17:1–59; 1987.

Cyr, H.; Downing, J.A. Empirical relationships of phytomacrofaunal abundance to plant biomass and macrophyte bed characteristics. Can. J. Fish. Aquat. Sci. 45:976–984; 1988a.

Cyr, H.; Downing, J.A. The abundance of phytophilous invertebrates on different species of submerged macrophytes. Freshwat. Biol. 20:365–374; 1988b.

Diehl, S. Foraging efficiency of three freshwater fish: effects of structural complexity and light. Oikos 53:207–214; 1988.

Diehl, S. Fish predation and benthic community structure: the role of omnivory and habitat complexity. Ecology 73:1646–1661; 1992.

Diehl, S. Effects of habitat structure on resource availability, diet and growth of benthivorous perch, *Perca fluviatilis*. Oikos 67:403–414; 1993a.

Diehl, S. Relative consumer sizes and the strengths of direct and indirect interactions in omnivorous feeding relationships. Oikos 68:151–157; 1993b.

Diehl, S. Direct and indirect effects of omnivory in a littoral lake community. Ecology 76:1727–1740; 1995.

Diehl, S.; Eklöv, P. Effects of piscivore-mediated habiat use on resources, diet, and growth of perch. Ecology 76:1712–1726; 1995.

Downing, J.A. Sampling the benthos dwelling on aquatic macrophytes. In: Downing, J.A.; Rigler, F.H., eds. A manual of methods for the assessment of secondary productivity in fresh waters. Oxford: Blackwell Scientific Publications; 1984:112–119.

Eklöv, P.; Diehl, S. Piscivore efficiency and refuging prey: the importance of predator search mode. Oecologia 98:344–353; 1994.

Erikson, M.O.G.; Henrikson, L.; Nilsson, B.-I.; Nyman, G.; Oscarson, H.G.; Stenson, J.A.E. Predator–prey relations important for the biotic changes in acidified lakes. Ambio 9:248–249; 1980.

Evans, R.A. Response of limnetic insect populations of two acidic, fishless lakes to liming and brook trout (*Salvelinus fontinalis*). Can. J. Fish. Aquat. Sci. 46:342–351; 1989.

Grimm, M.P.; Backx, J.J.G.M. The restauration of shallow eutrophic lakes, and the role of northern pike, aquatic vegetation and nutrient concentration. Hydrobiologia 200/201: 99–118; 1990.

Hill, A.M.; Lodge, D.M. Multi-trophic-level impact of sublethal interactions between bass and omnivorous crayfish. J. North Am. Benth. Soc. 14:306–314; 1995.

Holt, R.D.; Polis, G.A. A theoretical framework for intraguild predation. Am. Nat. 149:745–764; 1997.

Hopper, K.R.; Crowley, P.H.; Kielman, D. Density-dependence, hatching synchrony, and within-cohort cannibalism in young dragonfly larvae. Ecology 77:191–200; 1996.

Huang, C.; Sih, A. Experimental studies on behaviourally-mediated, indirect interactions through a shared predator. Ecology 71:1515–1522; 1990.

Ives, A.R.; Dobson, A.P. Antipredator behavior and the population dynamics of simple predator–prey systems. Am. Nat. 130:431–447; 1987.

Jeffries, M. Invertebrate colonization of artificial pondweeds of differing fractal dimension. Oikos 67:142–148; 1992.

Kajak, Z.; Kacprzak, K.; Polkowski, R. Tubular bottom sampler. Ekol. Pol. Ser. B 11:159–165; 1965.

Kornijów, R. Cumulative consumption of the lake macrophyte *Elodea* by abundant generalist invertebrate herbivores. Hydrobiologia 319:185–190; 1996.

Kornijów, R. The impact of predation by perch on the size-structure of *Chironomus* larvae—the role of vertical distribution of the prey in the bottom sediments, and habitat complexity. Hydrobiologia (in press).

Kornijów, R.; Kairesalo, T. *Elodea canadensis* sustains rich environment for macroinvertebrates. Verh. Int. Verein. Limnol. 25:4098–4111; 1994a.

Kornijów, R.; Kairesalo, T. A simple apparatus for sampling epiphytic communities associated with macrophytes. Hydrobiologia 294:141–143; 1994b.

Kornijów, R.; Gulati, R.D.; van Donk, E. Hydrophyte–macroinvertebrate interactions in Zwemhulst, a lake undergoing biomanipulation. Hydrobiologia 200/201:467–474; 1990.

Kornijów, R.; Gulati, R.D.; Ozimek, T. Food preference of some freshwater macroinvertebrates: comparing fresh and decomposed vascular plants and a filamentous alga. Freshwat. Biol. 33:205–212; 1995.

Lillie, R.; Budd, J. Habitat architecture of *Myriophyllum spicatum* L. as an index to habitat quality for fish and macroinvertebrates. J. Freshwat. Ecol. 7:113–125; 1992.

Lodge, D.M.; Barko, J.W.; Strayer, D.; Melack, J.M.; Mittelbach, G.G.; Howarth, R.W.; Menge, B.; Titus, J.E. Spatial heterogeneity and habitat interactions in lake communities. In: Carpenter, S.R., ed. Complex interactions in lake communities. New York: Springer-Verlag; 1988:181–208.

Lodge, D.M.; Kershner, M.W.; Aloi, J.E.; Covich, A.P. Effects of omnivorous crayfish (*Oronectes rusticus*) on a freshwater littoral food web. Ecology 75:1265–1281; 1994.

Macchiusi, F.; Baker, R.L. Prey behaviour and size-selective predation by fish. Freshwat. Biol. 25:533–538; 1991.

McPeek, M.A. Determination of species composition in the *Enallagma* damselfly assemblages of permanent lakes. Ecology 71:83–98; 1990a.

McPeek, M.A. Behavioural differences between *Enallagma* species (Odonata) influencing differential vulnerability to predators. Ecology 71:1714–1726; 1990b.

Meijer, M.-L.; Jeppesen, E.; van Donk, E.; Moss, B.; Scheffer, M.; Lammens, E.; van Nes, E.; van Berkum, J.A.; de Jong, G.J.; Faafeng, B.A.; Jensen, J.P. Long-term responses to fish stock reduction in small shallow lakes: interpretation of five-year results of four biomanipulation cases in The Netherlands and Denmark. Hydrobiologia 275/276:457–466; 1994.

Mittelbach, G.G. Foraging efficiency and body size: a study of optimal diet and habitat use by bluegills. Ecology 62:1370–1386; 1981a.

Mittelbach, G.G. Patterns of invertebrate size and abundance in aquatic habitats. Can. J. Fish. Aquat. Sci. 38:896–904; 1981b.

Mittelbach, G.G. Predation and resource partitioning in two sunfishes (Centrarchidae). Ecology 65:499–513; 1984.

Mittelbach, G.G. Competition among refuging sunfishes and effects of fish density on littoral zone invertebrates. Ecology 69:614–623; 1988.

Mittelbach, G.G.; Chesson, P.L. Predation risk: indirect effects on fish populations. In: Kerfoot, W.C.; Sih, A., eds. Predation: direct and indirect impacts on aquatic communities. Hanover, NH: University Press of New England; 1987:315–332.

Mittelbach, G.G.; Osenberg, C.W. Stage-structured interactions in bluegill: consequences of adult resource variation. Ecology 74:2381–2394; 1993.

Morin, P.J. The impact of fish exclusion on the abundance and species composition of larval odonates: results of short-term experiments in a North Carolina farm pond. Ecology 65:53–60; 1984.

Murdoch, W.W.; McCauley, E.; Nisbet, R.M.; Guerney, W.S.C.; de Roos, A.M. Individual-based models: combining testability and generality. In: DeAngelis, D.L.; Gross, L.J., eds. Individual-based models and approaches in ecology. London: Chapman & Hall; 1992:18–35.

Nelson, W.G.; Bonsdorff, E. Fish predation and habitat complexity: are complexity thresholds real? J. Exp. Mar. Biol. Ecol. 141:183–194; 1990.

Olson, M.H.; Mittelbach, G.G.; Osenberg, C.W. Competition between predator and prey: resource-based mechanisms and implications for stage-structured dynamics. Ecology 76:1758–1771; 1995.

Osenberg, C.W.; Mittelbach, G.G. Effects of body size on the predator–prey interaction between pumpkinseed sunfish and gastropods. Ecol. Monogr. 59:405–432; 1989.

Osenberg, C.W.; Mittelbach, G.G. The relative importance of resource limitation and predator limitation in food chains. In: Polis, G.A.; Winemiller, K.O., eds. Food webs: integration of patterns and dynamics. New York: Chapman & Hall; 1996:134–148.

Osenberg, C.W.; Werner, E.E.; Mittelbach, G.G.; Hall, D.J. Growth patterns in bluegill (*Lepomis macrochirus*) and pumpkinseed (*L. gibbosus*) sunfish: environmental variation and the importance of ontogenetic niche shifts. Can. J. Fish. Aquat. Sci. 45:17–26; 1988.

Osenberg, C.W.; Mittelbach, G.G.; Wainwright, P.C. Two-stage life histories in fish: the interaction between juvenile competition and adult performance. Ecology 73:255–267; 1992.

Osenberg, C.W.; Olson, M.H.; Mittelbach, G.G. Stage structure in fishes: resource productivity and competition gradients. In: Stouder, D.J.; Fresh, K.L.; Feller, R.J., eds. Theory and application of fish feeding ecology. Columbia, SC: University of South Carolina Press; 1994:151–170.

Pardue, W.J.; Webb, D.H. A comparison of aquatic macroinvertebrates occuring in association with eurasian watermilfoil (*Myriophyllum spicatum*) with those found in the open littoral zone. J. Freshwat. Ecol. 3:69–79; 1985.

Peckarsky, B.L.; Cowan, C.A.; Penton, M.A.; Anderson, C. Sublethal consequences of stream-dwelling predatory stoneflies on mayfly growth and fecundity. Ecology 74: 1836–1846; 1993.

Persson, L. Asymmetries in competitive and predatory interactions in fish populations. In: Ebenman, B.; Persson, L., eds. Size-structured populations—ecology and evolution. Heidelberg: Springer-Verlag; 1988:203–218.

Persson, L. Natural shifts in the structure of fish communities: mechanisms and constraints on perturbation sustenance. In: Cowx, I.G., ed. Rehabilitation of freshwater fisheries. Fishing News Books. Oxford: Blackwell Scientific Publishers; 1994:421–434.

Persson, L.; Eklöv, P. Prey refuges affecting interactions between piscivorous perch and juvenile perch and roach. Ecology 76:70–81; 1995.

Persson, L.; Greenberg, L.A. Juvenile competitive bottlenecks: the perch (*Perca fluviatilis*)–roach (*Rutilus rutilus*) interaction. Ecology 71:44–56; 1990.

Persson, L.; Andersson, J.; Wahlström, E.; Eklöv, P. Size-specific interactions in lake systems: predator gape limitation and prey growth rate and mortality. Ecology 77:900–911; 1996.

Persson, L.; Diehl, S.; Eklöv, P.; Christensen, B. Size specific trade-offs and individual behaviour—consequences at the population and community levels. In: Godin, J.-G.J., ed. Behavioural ecology of fishes. Oxford: Oxford University Press; 1997:316–343.

Persson, L.; Diehl, S.; Johansson, L.; Andersson, G.; Hamrin, S.F. Shifts in fish communities along the productivity gradient of temperate lakes—patterns and the importance of size-structured interactions. J. Fish Biol. 38:281–293; 1991.

Persson, L.; Diehl, S.; Johansson, L.; Andersson, G.; Hamrin, S.F. Trophic interactions in lake ecosystems: a test of food chain theory. Am. Nat. 140:59–84; 1992.

Pieczynski, E. Numbers and biomass of the littoral fauna in Mikolajskie Lake and in some other Masurian lakes. Ekol. Pol. 25:45–57; 1977.

Pierce, C.L.; Rasmussen, R.B.; Leggett, W.C. Littoral fish communities in southern Quebec lakes: relationships with limnological and prey resource variables. Can. J. Fish. Aquat. Sci. 51:1128–1138; 1994.

Polis, G.A.; Holt, R.D. Intraguild predation: the dynamics of complex trophic interactions. Trends Ecol. Evol. 7:151–154; 1992.

Prejs, A.; Koperski, P.; Prejs, K. Food web manipulation in small, eutrophic Lake Wirbel, Poland: the effect of replacement of key predators on epiphytic fauna. Hydrobiologia 342/343:377–381; 1997.

Prejs, K. Bottom fauna. In: Pieczynski, E., ed. Selected problems of lake littoral ecology. Warsaw: Warsaw University Press; 1976:23–144.

Rasmussen, J.B. Littoral zoobenthic biomass in lakes, and its relationships to physical, chemical, and trophic factors. Can. J. Fish. Aquat. Sci. 45:1438–1447; 1988.

Rasmussen, J.B. Patterns in the size structure of littoral zone macroinvertebrate communities. Can. J. Fish. Aquat. Sci. 50:2192–2207; 1993.

Savino, J.F.; Marshall, E.A.; Stein, R.A. Bluegill growth as modified by plant density: an exploration of underlying mechanisms. Oecologia 89:153–160; 1992.

Scheffer, M.; Hosper, S.H.; Meijer, M.-L.; Moss, B.; Jeppesen, E. Alternative equilibria in shallow lakes. Trends Ecol. Evol. 8:275–279; 1993.

Schramm, H.L.; Jirka, K.J. Effect of aquatic macrophytes on benthic macroinverytebrates in two Florida lakes. J. Freshwat. Ecol. 5:1–12; 1989.

Schriver, P.; Bøgestrand, J.; Jeppesen, E.; Søndergaard, M. Impact of submerged macrophytes on fish–zooplankton–phytoplankton interactions: large-scale enclosure experiments in a shallow eutrophic lake. Freshwat. Biol. 33:255–270; 1995.

Suren, A.M.; Lake, P.S. Edibility of fresh and decomposing macrophytes to three species of freshwater invertebrate herbivores. Hydrobiologia 178:165–178; 1989.

Tonn, W.M.; Paskowski, C.A.; Holopainen, I.J. Piscivory and recruitment: mechanisms structuring prey populations in small lakes. Ecology 73:951–958; 1992.

Turner, A.M.; Mittelbach, G.G. Predator avoidance and community structure: interactions among piscivores, planktivores, and plankton. Ecology 71:2241–2254; 1990.

Werner, E.E. Species interactions in freshwater fish communities. In: Diamond, J.; Case, T., eds. Community ecology. New York: Harper & Row; 1986:344–358.

Werner, E.E. Individual behavior and higher-order species interactions. Am. Nat. 140:S5-S32; 1992.

Werner, E.E.; Gilliam, J.F.; Hall, D.J.; Mittelbach, G.G. An experimental test of the effects of predation risk on habitat use in fish. Ecology 64:1540–1548; 1983.

Werner, E.E.; Hall, D.J. Ontogenetic habitat shifts in bluegill: the foraging rate-predation risk trade-off. Ecology 69:1352–1366; 1988.

Wiley, M.J.; Gorden, R.W.; Waite, S.W.; Powless, T. The relationship between aquatic macrophytes and sport fish production in Illinois ponds: a simple model. North Am. J. Fish. Manage. 3:111–119; 1984.

Wollkind, D.J. Exploitation in three trophic levels: an extension allowing intraspecific carnivore interaction. Am. Nat. 110:431–447; 1976.

3. Complex Fish–Snail–Epiphyton Interactions and Their Effects on Submerged Freshwater Macrophytes

Christer Brönmark and Jan E. Vermaat

Introduction

Eutrophication of shallow freshwater and marine ecosystems has often resulted in a drastic decline in the areal extension and biomass of submerged macrophytes and a concomitant increase in the biomass of phytoplankton (e.g., Phillips et al., 1978; Cambridge et al., 1986; Hough et al., 1989; Shepherd et al., 1989). Light availability is usually the most important factor determining the distribution pattern, biomass, and production of submerged macrophytes (e.g., Chambers and Kalff, 1985; Duarte et al., 1986), and it has been suggested that increasing phytoplankton biomass due to higher nutrient input results in a reduction of available light to a level at which net photosynthesis by submerged macrophytes is impossible (e.g., Jupp and Spence, 1977; Jones et al., 1983). However, Phillips et al. (1978) suggested that macrophytes may disappear even when the bottom is within the euphotic zone where light availability is adequate for photosynthesis. Instead of shading by phytoplankton, they argued that increasing nutrient levels stimulate epiphyton growth, which has a negative effect on the macrophyte host through shading and competition for nutrients. Recent modifications of the model of Phillips et al. (1978) also invoke epiphyton as a key factor in the decline of submerged macrophytes (e.g., Silberstein et al., 1986; Moss, 1989; Brönmark and Weisner, 1992; Van Vierssen et al., 1994).

An increase in nutrient input during eutrophication does not necessarily result in a gradual decline of submerged macrophytes (Moss and Leah, 1982). Instead, observations (e.g., Timms and Moss, 1984; Balls et al., 1989; Blindow et al., 1992) as well as theoretical models (e.g., Scheffer et al., 1993) suggest the possibility of alternate dominance of macrophytes or phytoplankton at comparable nutrient loadings. These alternative states would need self-inforcing feedbacks to have a reasonable degree of stability (Scheffer, 1989). Efficient periphyton removal by grazers may be one of these feedback mechanisms: numerical and/or functional responses in the grazer guild may prevent a massive increase of epiphytic growth. However, factors such as complex trophic interactions in food chains involving molluscivores, snails, and epiphyton, changes in nutrient turnover rates, and abiotic disturbances may affect the strength of the grazer–epiphyton interaction and thus eventually affect the distribution of submerged macrophytes.

In this chapter, we evaluate whether epiphyton removal by freshwater snails can be a sufficiently powerful feedback mechanism. Several recent reviews cover parts of the broader field of epiphyton interactions (e.g., Brönmark [1989] on snail–epiphyton–macrophyte interactions and Stevenson et al. [1996] on key factors in benthic algal ecology). Here, we briefly dwell on snail–epiphyton–macrophyte interactions as a necessary basis for our main focus: complex indirect interactions in realistic food chains including molluscivorous and piscivorous fish. We consider the effect of biotic processes and nutrient availability as potential feedback mechanisms affecting macrophyte growth and distribution.

Snail–Epiphyton Interactions

Epiphyton is grazed by an array of herbivorous invertebrates ranging in size from minute epiphytic copepods to rather large pulmonate snails and crayfish (e.g., Cattaneo and Kalff, 1986; Botts, 1993). Snails are an important component of the benthic community in many freshwater systems, and several experimental studies have shown that snail grazing has strong effects on epiphyton biomass, species composition, architecture, and productivity (see Brönmark, 1989, and Stevenson et al., 1996, for reviews). However, other taxa may also affect epiphyton through herbivory (Table 3.1; see also Jones et al., this volume, Chapter 4).

Different grazers may have different efficiencies in removing epiphyton, but few comparative studies exist. Larger snails were effectively replaced by smaller cladocerans, oligochaetes, and chironomids in exclosures with decreasing mesh widths (Cattaneo and Kalff, 1986). Using P^{32}, Kairesalo and Koskimies (1987) estimated grazing rates of snails, oligochaetes, and chironomids separately and found that oligochaetes were the most important herbivore in terms of overall grazing impact. In this case, the smaller oligochaetes were much more numerous than the snails (Table 3.1). However, in other lakes snail densities are often higher than the ones Kairesalo and Koskimies (1987) found and may equal those of oligochaetes (Soszka, 1975; Lalonde and Downing, 1992), and several papers have suggested that grazers other than snails have a negligible impact on epiphytic cover (Brönmark et al., 1992; Underwood et al., 1992; Daldorph and Thomas,

Table 3.1. Epiphyton Grazing Rates and Community Grazing Impact by Snails and Other Grazer Types[a]

Reference	Grazer taxon	Grazing rate (mg AFDW /ind/day)	Density (n/m^2)	Grazing impact (mg AFDW /m^2/day)
Kairesalo & Kos-kimies (1987)	*Lymnaea peregra*	0.85	25	213
	Oligochaetes	0.07	17,150	866
	Chironomids	0.03	4,300	93
Jacoby (1985)	*Theodoxus fluviatilis*	0.6	450	270
Vermaat (1994)	Several snail species	0.6 ± 0.1	400	240

[a]Epiphyton dry weight (DW) was converted to ash-free dry weight (AFDW) by using a conversion factor of 0.5. Grazing rate from Vermaat (1994) is a pooled mean ± 1 SE over different species and sizes (see Fig. 3.1); the snail density reported for Vermaat (1994) is the one estimated to be necessary to suppress a epiphyton spring bloom in Lake Veluwe.

1995). Hence, at this time a simple generalization on the quantitative significance of these different grazer guilds is not possible.

The seasonal timing of the presence of different grazers may differ substantially and consequently affect their impact on epiphyton. Chironomids often dominate in early spring (Mason and Bryant, 1975; Cattaneo, 1983; Cattaneo and Kalff, 1986), but their impact is nullified after pupation. Snails, however, are present year-round and may also remain active at low temperatures (Calow, 1975; Vermaat, 1994). The limited data available on snail community grazing rates suggest that the overall grazing impact of snails is in the order of 200–300 mg ash-free dry mass (AFDM)/m^2/day (i.e., per unit periphyton covered surface), which should be sufficient to suppress epiphytic spring blooms in eutrophic lakes (Van Vierssen et al., 1994). Ingestion rates in general are to a considerable extent governed by consumer body size (e.g., Cammen, 1980), and this should also be expected for freshwater snails. Cattaneo and Mousseau (1995) analyzed what factors affect periphyton removal rates by invertebrate grazers and showed that grazer body mass was most important. Similarly, Vermaat (1994) found that most of the variation in periphyton removal rates between four snail species (*Bithynia tentaculata, Lymnaea peregra, Physa fontinalis,* and *Valvata piscinalis*) was explained by snail size (Fig. 3.1; ANOVA: P (species) = .762, P (covariable size) = .037).

Intense grazing pressure may also reduce algal species diversity and affect the three-dimensional structure of the epiphytic community (e.g., Steinman, 1996). Typically, large overstory species such as stalked diatoms and filamentous algae decline in response to grazing, whereas the proportion of small, tightly adhering understory algae increases. Thus, grazing may halt or reverse the successional process, keeping the algal community at an early seral stage of small prostate algae.

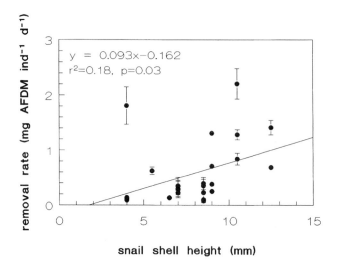

Figure 3.1. Periphyton removal rate as a function of shell height of four common freshwater snail species: *Bithynia tentaculata* (L.), *Lymnaea peregra* (Müll.), *Physa fontinalis* (L.) and *Valvata piscinalis* (Müll.). Shell lengths for these species were 8.3 ± 0.4, 9.3 ± 0.8, 6.3 ± 0.8 and 4.0 ± 0.1, replication was 11, 9, 2, and 5, respectively. AFDM, ash-free dry mass. (Data from Vermaat, 1994.)

Selective foraging may be a mechanism that could affect epiphyton species composition when subjected to grazing snails. Several studies have shown that snails show a preference for certain algal taxa (e.g., Calow, 1973, 1974), and it has been suggested that specific algal preferences coupled to species-specific differences in activity of digestive enzymes are a way for freshwater snails to partition a limiting resource between species (Calow and Calow, 1975). Kesler et al. (1986) instead suggested that differences in the dentition of the snail rasping tongue, the radula, are important in resource partitioning among sympatric snails. Barnese et al. (1990) examined the grazing trails of pulmonate and prosobranch snails and found no significant difference in grazing efficiency among pulmonate snails, whereas the prosobranchs were less effective grazers. Limited differences in radula morphology, as within pulmonate species, may thus not be important in affecting grazing efficiency, whereas the larger differences found between pulmonates and prosobranchs may affect the impact of snails on periphyton algal communities. However, the foraging apparatus of snails seems to operate at a different spatial scale compared with individual algal cells (i.e., snails are not able to discriminate and select specific components of the epiphyton community but, rather, are indiscriminate browsers). Given a patchiness in epiphyton composition, snails may choose to feed in patches or in microhabitats with a high proportion of a preferred food item (Lodge, 1985, 1986; Vermaat, 1994).

Snail grazing may also affect the productivity of epiphyton, and several studies have shown that a decrease in epiphyte biomass due to grazing often is followed by an increasing biomass-specific productivity (see review in Steinman, 1996). Herbivores may enhance primary productivity by recycling limiting nutrients, selecting for a shift in species composition toward faster-growing taxa, reduced competition for light and nutrients for understory algae, and removal of dead and senescent algal cells (e.g., Lamberti et al., 1989; Steinman, 1996).

Epiphyton–Macrophyte Interactions

The significance of epiphyton–macrophyte interactions has received considerable attention, especially the possibility of a symbiotic interaction between epiphytes and their host plant (e.g., Wetzel, 1983, 1996; Burkholder, 1996). Epiphytic algae have clear advantages of being associated with macrophytes. For example, epiphyton benefits from the macrophyte a priori by being provided with a large surface area for colonization and an elevated position in the water column with a higher availability of light. A more controversial issue has been the possibility of exchange of different dissolved substances between epiphytic algae and macrophytes. Macrophytes have been found to leak appreciable amounts of inorganic nutrients and dissolved organic compounds (reviewed by Burkholder, 1996), which should be readily used and beneficial to the algal and bacterial components of the epiphytic community, especially in oligotrophic waters. The rate of leakage changes with plant age, with an increasing release of high-quality substances as the leaves start to senesce and decompose. The taxonomic divergence of epiphytic communities on different species of macrophytes under low nutrient availability has been suggested to be due to host-specific composition of the released organic and inorganic compounds (Eminson and Moss, 1980). Thus, the epiphytic community clearly benefits by being associated with macrophytes, but it is less obvious what the benefits of this association are for the macrophytes. Epiphytes have been suggested to provide macrophytes with organic micronutrients (Wetzel, 1983), protection from pathogenic bacteria (Rogers and Breen, 1983), shading under intense Mediterranean irradiance (Van Vierssen, 1983), or diversion of grazers away from the macrophyte tissue (Hutchinson, 1975), but so far not much experimental evidence exists to support these hypotheses.

On the contrary, increased epiphytic growth is generally associated with negative impacts to the hosting macrophyte, often resulting in reduced growth. Macrophyte photosynthesis and growth may be negatively affected by an epiphytic cover through reduced light availability, an increased diffusive boundary layer hampering access to carbon and nutrient pools in the surrounding water mass, detrimental oxygen and pH levels at the leaf surface, or anoxia during darkness (e.g., Sand-Jensen, 1977; Bulthuis and Woelkerling, 1983; Sand-Jensen and Borum, 1984; Silberstein et al., 1986; Vermaat, 1994; Vermaat and Hootsmans, 1994).

Snail–Epiphyton–Macrophyte Interactions

As described above, experimental studies have shown that snail grazing often reduces the biomass of periphytic algae and that epiphytes may have an adverse effect on the growth of their macrophyte host. Thus, here exists a potential for complex indirect interactions between snails, epiphytes, and macrophytes. This led J.D. Thomas and co-workers to suggest a close mutualistic relationship between freshwater snails and submerged macrophytes (e.g., Thomas, 1982; Thomas et al., 1985; see also Carpenter and Lodge, 1986). It can therefore be hypothesized that macrophytes should attract grazing snails that remove the epiphytic cover and thereby benefit the macrophytes through decreasing competition for light and nutrients. Grazing snails would not only benefit from the macrophyte association by being provided with epiphytic food but may also be provided with a refuge from predators, a substrate for oviposition, and an accessway to atmospheric oxygen. Observational and experimental studies have verified several predictions emanating from this hypothesis. Complex habitats provide a refuge from predators by decreasing predation efficiency (Persson and Crowder, this volume, Chapter 1), and snail communities in macrophyte beds have been shown to have higher species richness as well as higher densities than surrounding nonvegetated littoral areas (Fig. 3.2; Brönmark, 1985a; Lodge, 1985; Lodge and Kelly, 1985; Lodge et al., 1987; Brown and Lodge, 1993).

Brönmark (1985b) showed in a laboratory experiment that the snail *L. peregra* enhanced the growth rate of the submerged macrophyte *Ceratophyllum demersum* by almost 30% compared with controls without snails. A small-scale field experi-

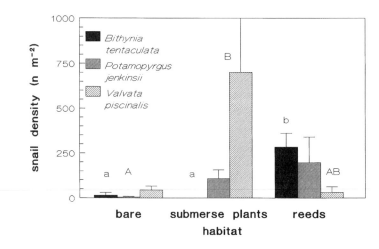

Figure 3.2. Habitat associations in late winter in freshwater snails. Mean adult densities ± SE are given. Significantly different habitat means are indicated with different lettering, lower case for *B. tentaculata,* upper case for the other two species. Data were collected in early 1994 from paired littoral transects with or without macrophytes in 13 lakes, ponds, and canals in the western part of The Netherlands.

ment by Underwood et al. (1992) provided similar results, whereas Underwood (1991) suggested that increased turnover rates of nutrients by snails may explain the increase in the macrophyte growth rate. Vermaat (1994) found no apparent effect of snail grazing on growth enhancement of *Potamogeton pectinatus,* but a more detailed analysis revealed that ungrazed plants with a thick epiphyton cover produced more leaves by re-allocating resources from tubers. A high leaf production rate may compensate for reduced photosynthesis by older leaves heavily burdened with epiphyton (cf. Sand-Jensen, 1977). Studies on marine seagrasses (Hootsmans and Vermaat, 1985; Howard and Short, 1986; Williams and Ruckelshaus, 1993), and stream macroalgae (Dudley, 1992) have shown that epiphyton grazing snails, insects, and crustaceans may enhance host growth in these environments as well.

Complex Food Chain Interactions

In recent years, several theoretical and empirical studies have emphasized the importance of complex interactions for structuring freshwater communities (e.g., Kerfoot and Sih, 1987; Carpenter, 1988; Gulati et al., 1990). Most research on indirect interactions in freshwater habitats has focused on interactions in pelagic food chains. Experimental increase of piscivore density has often resulted in a decrease in phytoplankton mediated through a decrease in the density of planktivorous fish and an increase in large efficient grazers (i.e., cladoceran zooplankton). Fewer studies on complex interactions have involved benthic food chains, but lately several studies have demonstrated the existence of trophic cascades and other complex interactions in these food chains as well (reviewed by Brönmark et al., 1997). Below, we evaluate the importance of complex interactions for submerged macrophytes and start with the effects of predation on freshwater snails, a key process in this respect.

Direct and Indirect Effects of Molluscivores on Snails

In a conceptual model, Lodge et al. (1987) emphasized the importance of predation for structuring freshwater snail assemblages in permanent ponds and lakes where abiotic constraints such as calcium concentration and/or disturbance events (e.g., drying out) are not limiting. A range of predators prey on freshwater snails, including invertebrates such as belostomatid bugs, dytiscid larvae, leeches, flatworms, and crayfish and vertebrate predators such as fish and birds. Experimental manipulations of predator density have shown that predators may reduce the density of freshwater snails and also induce a shift in species composition toward smaller, more hard-shelled species (Brown and DeVries, 1985; Weber and Lodge, 1990; Merrick et al., 1991; Brönmark et al., 1992; Martin et al., 1992; Osenberg et al., 1992; Brönmark 1994). These studies have documented a direct lethal effect of predation, but freshwater snails may also be able to evaluate local predation risk and respond behaviorally. In laboratory studies, predatory crayfish have been

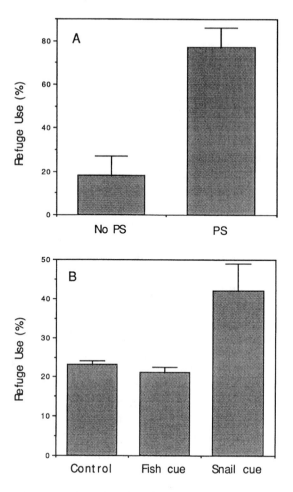

Figure 3.3. (A) The proportion of snails (all species) occupying the covered part of artificial substrates placed in lakes with (PS) or without pumpkinseed sunfish (no PS). Vertical bars denote 1 SE. (Modified from Turner, 1996.) (B) Proportion of *Physella* snails using refuges when exposed to water with chemical cues from pumpkinseed sunfish (fish cues), to water in which snails had been crushed (snail cue), and to water without cues (control). Vertical bars denote 1 SE. (Modified from Turner, 1996.)

shown to induce pulmonate snails to crawl out of the water to decrease encounter rates with the predator (Alexander and Covich, 1991), and further, snails respond to some leeches by crawl-outs, shell-shaking, closing their operculum, or decreasing their activity (Townsend and McCarthy, 1980; Brönmark and Malmqvist, 1986). Changes in activity patterns may be state-dependent, however, as in *B. tentaculata,* which showed a higher reduction of activity in response to leeches when food was absent compared with when periphyton was present in the experimental tanks (Brönmark and Malmqvist, 1986). Turner (1996) compared snail habitat use in lakes with and without pumpkinseed sunfish (*Lepomis gibbosus*), a specialized snail predator, and found that snails spent more time under cover in lakes with pumpkinseeds (Fig. 3.3A). In a pool experiment, he was able to confirm that presence of cover increased snail survival, and further, a laboratory experiment showed that the snail *Physella* sp. changed its behavior toward increased refuge use in response to chemical cues from crushed conspecifics (Fig. 3.3B). Fish cue did not in itself elicit a change in behavior. McCollum et al. (unpublished data; see also Crowder et al., this volume, Chapter 14) also found that chemical cues from a snail-eating fish resulted in reduced activity levels of snails. Besides changes in behavior, presence of predators may also affect the life history strategies of prey. Crowl and Covich (1990) found that waterborne chemical cues from crayfish feeding on snails induced a change in snail size at first reproduction and also affected their longevity. Snails exposed to cues from predatory crayfish started to reproduce later and at a larger size than snails that were not exposed to predatory crayfish. Presence of molluscivores may thus cause phenotypic changes in life history strategies in snails, affecting their population density and structure.

Food Chain Effects of Molluscivores

Several recent studies have tested whether changes in the density of molluscivorous predators will cascade down and eventually affect the biomass of primary producers. Brönmark et al. (1992) manipulated the density of a highly specialized molluscivorous fish, the pumpkinseed sunfish, in enclosures in two North American lakes. The density and biomass of snails and periphyton were monitored over two seasons and compared with fish exclosures and cageless controls. Pumpkinseed sunfish dramatically reduced the density and biomass of snails, which in turn resulted in an increased biomass of epiphyton due to the reduced grazing pressure. Martin et al. (1992) reported similar results from a study designed to test the effects of large and small sunfish (especially *Lepomis microlophus,* redear sunfish) on littoral macroinvertebrates. They found strong negative effects of fish on the biomass of snails and a positive effect on periphytic algae. The effect was independent of sunfish size, which was counter to the original predictions. The authors had predicted that small sunfish would only affect smaller macroinvertebrates such as soft-bodied insect larvae and that only larger sunfish were expected to be able to decimate snail populations (cf. Mittelbach, 1984; Stein et al., 1984). The strong effect from even small sunfish was suggested to be due to predation on newly hatched snails early in the season, preventing the recruitment of snails to larger sizes, and thus, a predator may have long-term effects on lower trophic levels through a short-term predation event by influencing the recruitment of a dominant herbivore (snails). What is even more interesting in this context is that Martin et al. (1992) found an effect of sunfish on submerged macrophytes as well. In the second year, submerged macrophytes (*Najas* and *Potamogeton*) sprouted in cages without fish, resulting in a dramatic difference in comparison with cages with fish and with the lake bottom outside the cages where hardly any macrophytes were found. The change in macrophyte abundance was attributed to reduced negative effects of epiphyton when snails became abundant in the cages without molluscivorous fish. Brönmark (1994) found analogous results in a enclosure/exclosure experiment involving a cyprinid fish, the tench (*Tinca tinca*). Tench has the morphological capacity to crush snail shells with its molari-form pharyngeal teeth, but previous studies had suggested that tench diets only include a minor proportion of snails (e.g., Kennedy and Fitzmaurice, 1970), indicating that tench effects on lower trophic levels would be weak. The field experiment clearly showed, however, that tench may have strong effects on inter-actions in benthic food chains. Snails were dramatically reduced in tench en-closures compared with either cages with perch (*Perca fluviatilis*) or cages with no fish, and the resulting decrease in grazing pressure allowed for an increase in the biomass of epiphytic algae (Fig. 3.4). Further, growth of the submerged macro-phyte *Elodea canadensis* was significantly reduced in tench cages, possibly due to increased competition for light and nutrients with the epiphytic algae.

Crayfish is another important and common snail predator that may affect interactions in benthic food chains involving snails and epiphyton. However, crayfish are more omnivorous, feeding on macroinvertebrates, macrophytes, peri-phyton, and detritus. Omnivory should decrease interaction strength in interaction chains, as direct effects may be counteracted by indirect effects (e.g., Strong, 1992). Given the omnivorous diet, it is not intuitively obvious how manipulations in crayfish density would affect lower trophic levels. A reduction of snails in the presence of crayfish may be predicted to have a positive effect on the biomass of epiphytic algae, but crayfish may, however, have a direct negative effect on epiphyton through grazing. Further, the effect of crayfish on submerged macro-phytes may be through indirect effects on epiphyton or by direct consumption. Recently, Lodge et al. (1994) tested the strength of direct and indirect interactions imposed by crayfish in littoral benthic food chains. Crayfish were found to have a strong negative effect on snails and submerged macrophytes, whereas there was a positive effect on epiphyton biomass per unit macrophyte surface area, indicating that crayfish grazing effects were not as strong as the indirect snail–epiphyton effect. Although an increased epiphyton biomass would be expected to have a negative effect on the growth of submerged macrophytes, a high occurrence of floating fragments of submerged macrophytes suggested that direct consumption explained most of the decline of macrophytes in crayfish enclosures.

The above studies have shown strong cascading effects in benthic food chains based on a direct lethal effect of predators on herbivores. However, as seen above, predators may also affect the behavior of prey, and a change in activity or habitat use in snails in response to presence of molluscivores may be expected to affect snail–epiphyton interaction strength. Turner (1977) manipulated the perceived risk of predation by adding different amounts of crushed snails to experimental pools. The perceived increase in risk of predation resulted in an increased use of cover by the snail *Physella,* and this had positive effects on the biomass of periphytic algae in habitats outside the cover, whereas periphyton biomass in refuge was kept at low levels (Fig. 3.5). Increased risk of predation also resulted in a reduced growth of *Physella.*

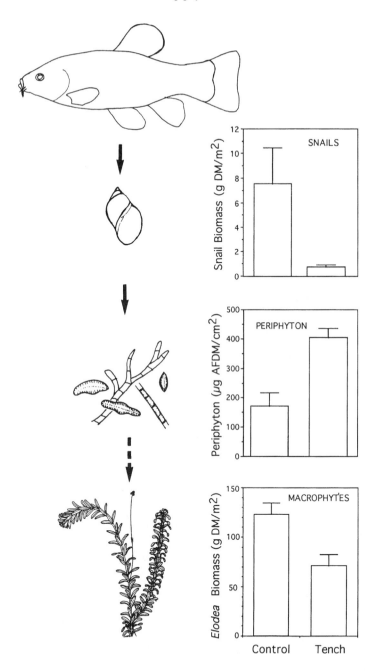

Figure 3.4. Direct and indirect effects of molluscivorous tench (*Tinca tinca*) on snails, periphyton, and the submerged macrophyte *Elodea canadensis*. (Modified from Brönmark, 1994.)

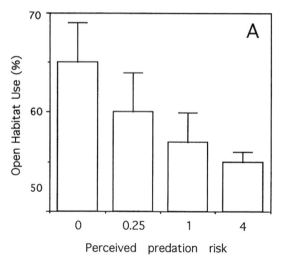

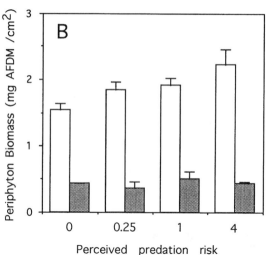

Figure 3.5. (A) Snail habitat use and (B) periphyton biomass in open and covered habitats in relation to perceived mortality risk. Predation risk was manipulated by adding different amounts of crushed snails to experimental pools (zero to four snails per day).

Piscivore Effects Down Benthic Food Chains

Brönmark and Weisner (1992) suggested that the change between alternative stable states in shallow eutrophic lakes was due to stochastic abiotic disturbance events acting on fish community structure. Inherent in this model was the importance of strong direct links between piscivorous and molluscivorous fish, which then cascaded down the food chain and affected macrophytes through changes in abundance of herbivorous snails and epiphyton. Very few studies have evaluated the effect of piscivores down benthic food chains with four trophic levels. The only experimental study of piscivore effects in such food chains was performed by Power (1990), who found strong cascade effects in a stream system with a food chain involving piscivorous fish, benthivorous fish, grazing invertebrates, and periphytic algae. Piscivores reduced the abundance of predatory fish fry and invertebrates, which resulted in an increase of a herbivorous chironomid and a decrease of algae. Brönmark and Weisner (1996) took another approach. Instead of experimental manipulations, they surveyed a large number of natural ponds for patterns in the distribution of piscivorous and molluscivorous fish, snails, and periphytic algae. The ponds differed only with respect to fish assemblage structure and were divided into three categories: ponds without fish, ponds with mollusci-vorous fish, and ponds with both molluscivorous and piscivorous fish. Ponds with two trophic levels (no fish) had a high density and biomass of snails and a low biomass of periphyton, whereas ponds with three trophic levels (with mollusci-vorous fish) had reduced densities and biomasses of snails and a high biomass of periphyton. This is in accordance with predictions from food chain theory and the earlier findings from experimental manipulations of molluscivore density (see above). However, ponds with four trophic levels (with piscivores) deviated from the predicted pattern. Although densities of molluscivorous fish were low and densities of snails were high, the biomass of periphyton was high and did not differ from ponds with only molluscivorous fish. A closer examination of the size structure of molluscivorous fish in ponds with and without piscivores suggested a possible explanation. Ponds with piscivores were dominated by sparse populations of large-bodied molluscivorous tench and crucian carp, whereas in ponds without piscivores, the tench and crucian carp populations were very dense and dominated by small individuals (Brönmark et al., 1995). Piscivores are gape-limited pred-ators, and thus there exists an upper prey size above which prey no longer are vulnerable to predators (i.e., prey have reached an absolute size refuge). Selective predation by large molluscivores that had reached an absolute size refuge from piscivory reduced the density of large snail species (lymneids) that are effective grazers, resulting in a snail community dominated by small detritivorous snails. Thus, decoupling at the piscivore-molluscivore level resulted in no cascading effects of piscivores on periphytic algae in this system (see Hambright et al., 1991; Hambright, 1994, for examples of decoupling due to size refuges in pelagic food chains), and consequently, changes in the abundance of piscivores may have no, or negligible, effects on the biomass of submerged macrophytes through indirect interactions in the benthic food chain.

Nutrient Effects

Nutrient availability is an important determinant of epiphyton biomass, species composition, and productivity. In some systems, nutrient availability is the pri-mary determinant of algal biomass (bottom-up control), whereas in others grazing is the major factor controlling algae (top-down control). Other factors such as light availability and disturbance events may be of great importance in yet other systems (see Stevenson et al., 1996, for a thorough review of factors affecting freshwater benthic algae). Here, we first consider how biotic processes may interact with nutrient availability by modifying turnover rates and then look at a larger scale and see how changes in the productivity of a system interact with food chain interactions and what the consequences might be.

Nutrient Turnover

Recent studies have shown that consumers, besides having direct lethal effects, may affect primary producers through a change in the availability and quality of nutrients (reviewed by Vanni, 1996). Consumers make nutrients that were immobilized in prey biomass available to primary producers by excretion or defecation. However, consumers may not only affect the quantity of available nutrients but may also through their excretions change the relative availability or the ratio of nutrients, and this may affect algal species differently (Vanni, 1996). Several studies have experimentally shown that grazing by snails and other benthic macro-invertebrates may affect periphyton through changes in nutrient turnover rates. Mulholland et al. (1983) showed that grazing on epiphyton increased nutrient turnover rates, and Underwood (1991) found that presence of snails enhanced the growth of the submerged macrophyte *Ceratophyllum* even if they were not in contact with the plant. The release of phosphate, nitrate, nitrite ammonia, and urea by snails was suggested to increase phosphorus and nitrogen levels, resulting in an increase in plant growth. Underwood (1991) estimated the release rates to be 3.7×10^{-4} µg phosphate, 23.9×10^{-4} µg nitrate, 4.4×10^{-4} µg ammonia, and 0.8×10^{-4} µg urea per hour and milligrams total wet weight of snail. For comparison, at a density of 100 individuals/m^2, adult *Planorbis planorbis* (Underwood, 1991; Underwood et al., 1992) would then release up to 24 µmol phosphate/m^2/day, which is in the same order of magnitude as phosphate release from oxic sediments (10 µmol/m^2/day) (Nürnberg, 1984). Cuker (1983), however, found no increase in nutrient availability by grazing snails and suggested that nutrients may be recycled in the grazer gut system by algal cells resistant to gut passage. However, the significance of nutrient release from grazer feces must depend on grazer density, grazer nutrient conservation possibilities, and the nutrient content of the food. Therefore, in oligotrophic arctic lakes (Cuker, 1983; Merrick et al., 1991) with low snail densities, fecal nutrient release must be low and difficult to quantify as compared with more eutrophic systems (e.g., Underwood, 1991; Underwood et al., 1992).

A laboratory experiment has also suggested that nutrient recycling by fish may affect periphyton biomass and species composition (McCollum and Crowder, unpublished data), and fish-mediated changes in phytoplankton have been ascribed to changes in nutrient recycling and availability (e.g., Threlkeld, 1987). Observations by Brönmark et al. (1992) and Brönmark (1994) may, however, suggest that this mechanism can be refuted. Comparisons of periphyton biomass between cages with pumpkinseed sunfish and yellow perch (Brönmark et al., 1992) or tench and perch (Brönmark, 1994) showed a strong cascade effect (fish-snails algae) in pumpkinseed and tench cages, whereas the biomass of periphyton in perch cages was identical to cages without fish. Thus, increased nutrient turnover rates due to fish excretion did not affect periphyton biomass.

Environmental Productivity

The overall productivity of an environment is one of the most important factors determining the abiotic template that biotic processes can operate within. Theoretical models have suggested that potential length of food chains in an environment is set by environmental productivity (e.g., Oksanen et al., 1981; Fretwell, 1987). The Oksanen model also predicts that effects of increases in environmental productivity on equilibrium biomass will depend on the total number of trophic levels in the environment and if the population under consideration is a primary producer, herbivore, or primary or secondary carnivore (i.e., the trophic level).

Several studies in the laboratory and in the field have shown that enrichment with phosphorus and/or nitrogen results in increased biomass of benthic algae (reviewed by Borchardt, 1996). For example, phosphorus enrichment of small areas in the littoral zone of an oligotrophic North American lake resulted in a dramatic increase in epiphytic biomass (Osenberg, 1989; see also Wetzel, 1996). Growth of snails increased in response to the nutrient enhancement, and it was concluded that both snails and epiphytes were resource limited (Osenberg, 1989). Other studies have shown a smaller effect of nutrient enhancement on algal biomass. Instead, the increase in nutrients seems to have been channeled into increased grazer production, resulting in grazers maintaining constant algal biomasses despite nutrient enhancement (e.g., Hill et al., 1992). The observed differences in response to nutrient enhancement may be explained by at least two mechanisms. First, there is a problem with the temporal scale of experimental manipulations. Algae and their grazers operate under very different temporal scales, with algae responding numerically to changes in resource levels in the course of days to weeks, whereas the reproductive cycle of snails and other herbivores is more in the order of months up to years and thus sets limits to the numerical responses that could be observed in short-term nutrient enrichment experiments. The importance of temporal scale was clearly shown in an experiment in which a pristine tundra river was fertilized for four consecutive summers (Peterson et al., 1993). A dramatic increase of benthic algal biomass and productivity was found during the first two summers, but this response was modified in the last years due to an increase of herbivorous insects that prevented a buildup of periphytic algae. Further, the number of trophic levels in the food chain may affect the response of the primary producers to nutrient enhancement. In food chains with an even number of trophic levels, primary producers are controlled by herbivores, and addition of nutrients should be transferred to an increased herbivore biomass, whereas plant biomass should remain constant (e.g., Oksanen et al., 1981; Fretwell, 1987). In food chains with only one level, grazers obviously cannot control plant biomass, and in three-level systems predators keep herbivores at such a low level that they do not control plants. Thus, in odd-numbered food chains nutrient enhancement is predicted to result in increasing plant biomass. No such manipulation has been performed with benthic systems in lakes, but Rosemond et al. (1993) found that nutrient enhancement in a stream system resulted in increased algal biomass in treatments without grazers (one trophic level) and no effect on algal biomass in treatments with grazers (two trophic levels). Similarly, Wootton and Power (1993) manipulated primary productivity in a stream system, not by nutrient enhancement but through changes in light availability, and found that algae responded positively to increased productivity in food chains with three trophic levels, but when a fourth trophic level (piscivores) was added grazers increased and algae decreased. Thus, temporal scale and food chain configuration may both affect the response to nutrient enhancement in freshwater systems.

Conclusions

Many conceptual models have been proposed to explain the changes in submerged macrophyte distribution in shallow eutrophic lakes and in several complex inter-

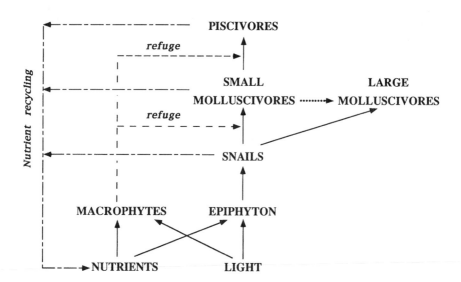

Figure 3.6. Complex interactions in a benthic freshwater food chain. Filled lines denote direct consumer–resource interactions. Hatched lines emanating in macrophytes represent an interaction modification (i.e., macrophytes add increased habitat complexity, which modifies the strength of the consumer–resource interaction by decreasing predator efficiency). Hatched lines from consumers denote nutrient recycling (e.g., excretions, defecation), whereas punctuated line show recruitment of small molluscivorous fish into the larger size classes where they have reached an absolute size refuge from predation by gape-limited piscivores.

actions between grazers, epiphyton, and plants are central components. As reviewed above, several empirical studies have shown that many of the direct interactions between these actors in benthic freshwater food chains are strong and may even give rise to strong indirect interactions as predicted by food chain theory. However, these predictions seem to hold only for systems with up to three trophic levels. This is problematic because most lakes hold populations of piscivorous fish. Experiments on pelagic food chains have suggested that changes in piscivore densities may cascade all the way down to primary producers, but it seems that absolute size refuge in the primary carnivore and changes in the species composition of the dominant herbivores may decouple strong cascading interactions from piscivores in benthic food chains (Brönmark and Weisner, 1996). Further, feedback loops such as nutrient recycling by herbivores and predators and the provision of prey refuge by structurally complex macrophyte beds (Fig. 3.6) may further decrease our ability to understand or predict the outcome of changes in benthic food chains. Thus, despite considerable progress in understanding food chain interactions in benthic freshwater systems, our possibilities to predict food chain dynamics in complex littorals are still rather limited. Future experimental studies including simultaneous manipulations of resource and consumer levels, as well as experiments performed at larger temporal and spatial scales, are of key importance.

Acknowledgments. We thank Francien Boessenkool for access to unpublished data (Fig. 3.2); Larry Crowder, Beth McCollum, and Andy Turner for access to unpublished manuscripts; and Dave Lodge and J.D. Thomas for comments on an earlier manuscript. And, of course, the organizers of the workshop! Financial support was received from the Swedish Council for Forestry and Agricultural Research (CB).

References

Alexander, J.E., Jr; Covich, A.P. Predator avoidance by the freshwater snail *Physella virgata* in response to the crayfish *Procambarus simulans.* Oecologia 87:435–442; 1991.

Balls, H.; Moss, B.; Irvine, K. The loss of submerged plants with eutrophication. I. Experimental design, water chemistry, aquatic plant and phytoplankton biomass in experiments carried out in ponds in the Norfolk Broadlands. Freshwat. Biol. 22:71–87; 1989.

Barnese, L.E.; Lowe, R.L.; Hunter, R.D. Comparative grazing efficiency of pulmonate and prosobranch snails. J. North Am. Benth. Soc. 9:35–44; 1990.

Blindow, I.; Andersson, G.; Hargeby, A.; Johansson, S. Long-term patterns of alternative stable states in two shallow eutrophic lakes. Freshwat. Biol. 30:159–167; 1992.

Borchardt, M.A. Nutrients. In: Stevenson, R.J.; Bothwell, M.L.; Lowe, R.L., eds. Algal ecology. Freshwater benthic systems. San Diego, CA: Academic Press; 1996; 183–227.

Botts, P. The impact of small chironomid grazers on epiphytic algal abundance and dispersion. Freshwat. Biol. 30:25–33; 1993.

Brönmark, C. Freshwater snail diversity: effects of pond area, habitat heterogeneity and isolation. Oecologia 67:127–131; 1985a.

Brönmark, C. Interactions between epiphytes, macrophytes and herbivores: an experimental approach. Oikos 45:26–30; 1985b.

Brönmark, C. Interactions between epiphytes, macrophytes and freshwater snails: a review. J. Moll. Stud. 55:299–311; 1989.

Brönmark, C. Effects of tench and perch on interactions in a freshwater, benthic food chain. Ecology 75:1818–1824; 1994.

Brönmark, C.; Malmqvist, B. Interactions between the leech *Glossiphonia complanata* and its gastropod prey. Oecologia 69:268–276; 1986.

Brönmark, C.; Weisner, S.E.B. Indirect effects of fish community structure on submerged vegetation in shallow, eutrophic lakes: an alternative mechanism. Hydrobiologia 243/244:293–301; 1992.

Brönmark, C.; Weisner, S.E.B. Decoupling of cascading trophic interactions in a freshwater, benthic food chain. Oecologia 108:534–541; 1996.

Brönmark, C.; Paszkowski, C.A.; Tonn, W.M.; Hargeby, A. Predation as a determinant of size structure in populations of crucian carp (*Carassius carassius*) and tench (*Tinca tinca*). Ecol. Freshwat. Fish. 4:85–92; 1995.

Brönmark, C.; Klosiewski, S.P., Stein, R.A. Indirect effects of predation in a freshwater, benthic food chain. Ecology 73:1662–1674; 1992.

Brönmark, C.; Dahl, J.; Greenberg, L.A. Complex trophic interactions in freshwater benthic food chains. In: Streit, B.; Städler, T.; Lively, C., eds. Ecology and evolution of freshwater animals. Basel: Birkhauser Publishers; 1997:55–88.

Brown, K.M.; DeVries, D.R. Predation and the distribution and abundance of a pulmonate pond snail. Oecologia 66:93–99; 1985.

Brown, K.M.; Lodge, D.L. Gastropod abundance in vegetated habitats: the importance of specifying null models. Limnol. Oceanogr. 38:217–225; 1993.

Bulthuis, D.A.; Woelkerling, W.J. Biomass accumulation and shading effects of epiphytes on leaves of the seagrass, *Heterozostera tasmanica,* in Victoria, Australia. Aquat. Bot. 16:137–148; 1983.

Burkholder, J.M. Interactions of benthic algae with their substrata. In: Stevenson, R.J.; Bothwell, M.L.; Lowe, R.L., eds. Algal ecology. Freshwater benthic systems. San Diego, CA: Academic Press; 1996:253–297.

Calow, P. The food of *Ancylus fluviatilis* (Müll), a littoral, stone-dwelling herbivore. Oecologia 13:113–133; 1973.

Calow, P. Evidence for bacterial feeding in *Planorbis contortus* Linn. (Gastropoda: Pulmonata). Proc. Malac. Soc. Lond. 41:145–156; 1974.

Calow, P. The feeding strategies of two freshwater gastropods, *Ancylus fluviatilis* (Müll.) and *Planorbis contortus* Linn. (Pulmonata), in terms of ingestion rates and absorption efficiencies. Oecologia 20:33–49; 1975.

Calow, P.; Calow, L.J. Cellulase activity and niche separation in freshwater gastropods. Nature 255:478–480; 1975.

Cambridge, M.C.; Chiffings, A.W.; Brittan, C.; Moore, L.; McCombe, A.J. The loss of seagrass in Cockburn Sound, Western Australia. II. Possible causes of seagrass decline. Aquat. Bot. 24:269–285; 1986.

Cammen, L.M. Ingestion rate: an empirical model for aquatic deposit feeders and detritivores. Oecologia 44:303–310; 1980.

Carpenter, S.R., ed. Complex interactions in lake communities. New York: Springer-Verlag; 1988.

Carpenter, S.R.; Lodge, D.M. Effects of submersed macrophytes on ecosystem processes. Aquat. Bot. 26:341–370; 1986.

Cattaneo, A. Grazing on epiphytes. Limnol. Oceanogr. 28:124–132; 1983.

Cattaneo, A.; Kalff, J. The effect of grazer size manipulation on periphyton communities. Oecologia 69:612–617; 1986.

Cattaneo, A.; Mousseau, B. Empirical analysis of the removal rate of periphyton by grazers. Oecologia 103:249–254; 1995.

Chambers, P.A.; Kalff, J. Depth distribution and biomass of submersed macrophyte communities in relation to Secchi depth. Can. J. Fish. Aquat. Sci. 42:701–709; 1985.

Crowl, T.A.; Covich, A.P. Predator-induced life-history shifts in a freshwater snail. Science 247:949–951; 1990.

Cuker, B.E. Grazing and nutrient interactions in controlling the activity and composition of the epilithic algal community of an arctic lake. Limnol. Oceanogr. 28:133–141; 1983.

Daldorph, P.W.G.; Thomas, J.D. Factors influencing the stability of nutrient-enriched freshwater macrophyte communities: the role of sticklebacks *Pungitius pungitius* and freshwater snails. Freshwat. Biol. 33:271–289; 1995.

Duarte, C.M.; Kalff, J.; Peters, R.H. Patterns in biomass and cover of aquatic macrophytes. Can. J. Fish. Aquat. Sci. 43:1900–1908; 1986.

Dudley, T.L. Beneficial effects of herbivores on stream macroalgae via epiphyte removal. Oikos 65:121–127; 1992.

Eminson, D.; Moss, B. The composition and ecology of periphyton communities in freshwaters. 1. The influence of host type and external environment on community composition. Br. Phycol. J. 15:429–446; 1980.

Fretwell, S.D. Food chain dynamics: the central theory of ecology? Oikos 50:291–301; 1987.

Gulati, R.D.; Lammens, E.H.R.R.; Meijer, M.-L.; van Donk, E., eds. Biomanipulation. Tool for water management. Dordrecht: Kluwer Academic Publishers; 1990.

Hambright, K.D. Morphological constraints in the piscivore–planktivore interaction: Implications for the trophic cascade hypothesis. Limnol. Oceanogr. 39:897–912; 1994.

Hambright, K.D.; Drenner, R.W.; McComas, S.R.; Hairston, N.G., Jr. Gape-limited piscivores, planktivore size refuges, and the trophic cascade hypothesis. Hydrobiologia 121:389–404; 1991.

Hill, W.R.; Boston, H.L.; Steinman, A.D. Grazers and nutrients simultaneously limit lotic primary productivity. Can. J. Fish. Aquat. Sci. 49:504–512; 1992.

Hootsmans, M.J.M., Vermaat, J.E. The effects of periphyton-grazing by three epifaunal species on the growth of *Zostera marina* L. under experimental conditions Aquat. Bot. 22:83–88; 1985.

Hough, R.A.; Fornwall, M.D.; Negele, B.J.; Thompson, R.L.; Putt, D.A. Plant community dynamics in a chain of lakes: principal factors in the decline of rooted macrophytes with eutrophication. Hydrobiologia 173:199–217; 1989.

Howard, R.K.; Short, F.T. Seagrass growth and survivorship under the influence of epiphyte grazers. Aquat. Bot. 24:287–302; 1986.

Hutchinson, G.E. A treatise on limnology. III. Limnological botany. New York: Wiley; 1975.

Jacoby, J.M. Grazing effects on periphyton by *Theodoxus fluviatilis* (Gastropoda) in a lowland stream. J. Freshwat. Ecol. 3:265–274; 1985.

Jones, R.C.; Walti, K.; Adams, M.S. Phytoplankton as a factor in the decline of the submersed macrophyte *Myriophyllum spicatum* L. in Lake Wingra, Wisconsin, USA. Hydrobiologia 107:213–219;1983.

Jupp, B.; Spence, D.H.W. Limitations on macrophytes in a eutrophic lake, Loch Leven. I. Effects of phytoplankton. J. Ecol. 65:175–186; 1977.

Kairesalo, T.; Koskimies, I. Grazing by oligochaetes and snails on epiphytes. Freshwat. Biol. 17:317–324; 1987.

Kennedy, M.; Fitzmaurice, P. The biology of Tench, *Tinca tinca* (L.) in Irish waters. Proc. R. Irish Acad. 69:31–82; 1970.

Kerfoot, W.C.; Sih, A., eds. Predation. Direct and indirect impacts on aquatic communities. Hanover, NH: University Press of New England; 1987.

Kesler, D.H.; Jokinen, E.H.; Munns, W.R., Jr. Trophic preferences and feeding morphology of two pulmonate snails species from a small New England pond, USA. Can. J. Zool. 64:2570–2575; 1986.

Lalonde, S.; Downing, J.A. Phytofauna of eleven macrophyte beds of differing trophic status, depth and composition. Can. J. Fish. Aquat. Sci. 49:992–1000; 1992.

Lamberti, G.A.; Ashkenas, L.R.; Gregory, S.V.; Steinman, A.D.; McIntire, C.D. Productive capacity of periphyton as a determinant of plant–herbivore interactions in streams. Ecology 70:1840–1856; 1989.

Lodge, D.M. Macrophyte gastropod associations: observations and experiments on macrophyte choice by gastropods. Freshwat. Biol. 15:695–708; 1985.

Lodge, D.M. Selective grazing on periphyton: a determinant of freshwater gastropod microdistribution. Freshwat. Biol. 16:831–841; 1986.

Lodge, D.M.; Kelly, P. Habitat disturbance and the stability of freshwater gastropod populations. Oecologia 68:111–117; 1985.

Lodge, D.M.; Brown, K.M.; Klosiewski, S.P.; Stein, R.A.; Covich, A.P.; Leathers, B.K.; Brönmark, C. Distribution of freshwater snails: spatial scale and the relative importance of physicochemical and biotic factors. Am. Malacol. Bull. 5:73–84; 1987.

Lodge, D.M.; Kershner, M.W.; Aloi, J. Effects of an omnivorous crayfish (*Orconectes rusticus*) on a freshwater littoral food web. Ecology 75:1265–1281; 1994.

Martin, T.H.; Crowder, L.B.; Dumas, C.F.; Burkholder, J.M. Indirect effects of fish on macrophytes in Bays Mountain Lake: evidence for a littoral trophic cascade. Oecologia 89:476–481; 1992.

Mason, C.F.; Bryant, R.J. Changes in the ecology of the Norfolk Broads. Freshwat. Biol. 5:257–270;1975.

Merrick, G.W.; Hershey, A.E.; McDonald, M.E. Lake trout (*Salvelinus namaycush*) control of snail density and size distribution in an arctic lake. Can. J. Fish. Aquat. Sci. 48:498–502; 1991.

Mittelbach, G.G. Predation and resource partitioning in two sunfishes (Centrarchidae). Ecology 65:499–513; 1984.

Moss, B. Water pollution and the management of ecosystems: a case study of science and scientists. In: Grubb, P.J.; Whittaker, J.B., eds. Toward a more exact ecology. Oxford: Blackwell; 1989:401–422.

Moss, B.; Leah, R.T. Changes in the ecosystem of a guanotrophic and brackish shallow lake in Eastern England: potential problems in its restoration. Int. Rev. Ges. Hydrobiol. 67:625–659; 1982.

Mulholland, P.J.; Newbold, J.D.; Elwood, J.W.; Hom, C.L. The effect of grazing intensity on phosphorus spiralling in autotrophic streams. Oecologia 58:358–366; 1983.

Nürnberg, G. The prediction of internal phosphorus load in lakes with anoxic hypolimnion. Limnol. Oceanogr. 29:111–124; 1984.

Oksanen, L.; Fretwell, S.D.; Arruda, J.; Niemela, P. Exploitation ecosystems in gradients of primary productivity. Am. Nat. 118:240–261; 1981.

Osenberg, C.W. Resource limitation, competition and the influence of life history in a freshwater snail community. Oecologia 79:512–519; 1989.

Osenberg, C.W.; Mittelbach, G.G.; Wainwright, P.C. Two-stage life histories in fish: the interaction between juvenile competition and adult performance. Ecology 73:255–267; 1992.

Peterson, B.J.; Deegan, L.; Helfrich, J.; Hobbie, J.E.; Hullar, M.; Moller, B.; Ford, T.E.; Hershey, A.; Hiltner, A.; Kipphut, G.; Lock, M.; Fiebig, D.M.; McKinley, V.; Miller, M.C.; Vestal, J.R.; Ventullo, R.; Volk, G. Biological responses of a tundra river to fertilization. Ecology 74:653–672; 1993.

Phillips, G.L.; Eminson, D.F.; Moss, B. A mechanism to account for macrophyte decline in progressively eutrophicated waters. Aquat. Bot. 4:103–125; 1978.

Power, M.E. Effects of fish in river food webs. Science 250:811–814; 1990.

Rogers, K.H.; Breen, C.M. An investigation of macrophyte, epiphyte and grazer interactions. In: Wetzel, R.G., ed. Periphyton of freshwater ecosystems. The Hague: Junk Publishers; 1983:217–226.

Rosemond, A.D., Mulholland, P.J.; Elwood, J.W. Top-down and bottom-up control of stream periphyton: effects of nutrients and herbivores. Ecology 74:1264–1280; 1993.

Sand Jensen, K. Effect of epiphytes on eelgrass photosynthesis. Aquat. Bot. 3:55–63; 1977.

Sand Jensen, K.; Borum, J. Epiphyte shading and its effect on photosynthesis and diel metabolism of *Lobelia dortmanna* L. during the spring bloom in a Danish lake. Aquat. Bot. 20:109–119; 1984.

Scheffer, M. Alternative stable states in eutrophic, shallow freshwater systems: a minimal model. Hydrobiol. Bull. 23:73–83; 1989.

Scheffer, M.; Hosper, S.H.; Meijer, M.-L.; Moss, B.; Jeppesen, E. Alternative equilibria in shallow lakes. Trends Ecol. Evol. 8:275–279; 1993.

Shepherd, S.A.; McComb, A.J.; Bulthuis, D.A.; Neverauskas, V.; Steffensen, D.A.; West, R. Decline of seagrasses. In: Larkum, A.W.D.; McComb, A.J.; Shepherd, S.A., eds. Biology of seagrasses, a treatise on the biology of seagrasses with special reference to the Australian region. Aquatic Plant Studies. Vol. 2. Amsterdam: Elsevier; 1989:346–393.

Silberstein, K.; Chiffings, A.W.; McComb, A.J. The loss of seagrass in Cockburn Sound, Western Australia. III. The effect of epiphytes on the productivity of *Posidonia australis* Hook. F. Aquat. Bot. 24:355–371; 1986.

Soszka, G.J. The invertebrates on submerged macrcophytes in three Masurian lakes. Ekol. Pol. 23:371–391; 1975.

Stein, R.A.; Goodman, C.G.; Marshall, E.A. Using time and energetic measures of cost in estimating prey value for fish predators. Ecology 65:702–715; 1984.

Steinman, A.D. Effects of grazers on freshwater benthic algae. In: Stevenson, R.J.; Bothwell, M.L.; Lowe, R.L., eds. Algal ecology. Freshwater benthic systems. San Diego, CA: Academic Press; 1996:341–373.

Stevenson, R.J.; Bothwell, M.L.; Lowe, R.L., eds. Algal ecology. Freshwater benthic systems. San Diego, CA: Academic Press; 1996.

Strong, D.R. Are trophic cascades all wet? Differentiation and donor-control in speciose ecosystems. Ecology 73:747–754; 1992.

Thomas, J.D. Chemical ecology of the snail hosts of schistosomiasis: snail–snail and snail–plant interactions. Malacologia 22:81–91; 1982.

Thomas, J.D.; Nwanko, D.I.; Sterry, P.R. The feeding strategies of juvenile and adult *Biomphalaria glabrata* (Say) under simulated natural conditions and their relevance to ecological theory and snail control. Proc. R. Soc. Lond. B 226:177–209; 1985.

Threlkeld, S.T. Experimental evaluation of trophic-cascade and nutrient mediated effects of planktivorous fish on plankton community structure. In: Kerfoot, C.W.; Sih, A., eds. Predation: direct and indirect impacts on aquatic communities. Hanover, NH: University Press of New England; 1987:161–173.

Timms, R.M.; Moss, B. Prevention of growth of potentially dense phytoplankton populations by zooplankton grazing, in the presence of zooplanktivorous fish, in a shallow wetland ecosystem. Limnol. Oceanogr. 29:472–486; 1984.

Townsend, C.R.; McCarthy, T.K. On the defence strategy of *Physa fontinalis* (L.), a freshwater pulmonate snail. Oecologia 46:75–79; 1980.

Turner, A.M. Freshwater snails alter habitat use in response to predation. Anim. Behav. 51:747–756; 1996.

Turner, A.M. Contrasting short-term and long-term effects of predation risk on consumer habitat use and resource dynamics. Behav. Ecol. 8:120–125;1997.

Underwood, G.J.C. Growth enhancement of the macrophyte *Ceratophyllum demersum* in the presence of the snail *Planorbis planorbis:* the effects of grazing and chemical conditioning. Freshwat. Biol. 26:325–334; 1991.

Underwood, G.J.C.; Thomas, J.D.; Baker, J.H. An experimental investigation of interactions in snail-macrophyte-epiphyte systems. Oecologia 91:587–595; 1992.

Vanni, M.J. Nutrient transport and recycling by consumers in lake food webs: implications for algal communities. In: Polis, G.A.; Winemiller, K.O., eds. Food webs. Integration of patterns and dynamics. New York: Chapman & Hall; 1996:81–95.

Van Vierssen, W. The influence of human activities on the functioning of macrophyte-dominated aquatic ecosystems in the coastal area of Western Europe. Proc. EWRS/AAB Int. Symp. Aquat. Macrophytes Nijmegen; 1983:273–281.

Van Vierssen, W.; Hootsmans, M.J.M.; Vermaat, J.E., eds. Lake Veluwe, a macrophyte-dominated system under eutrophication stress. Dordrecht: Kluwer Academic Publishers; 1994.

Vermaat, J.E. Periphyton removal by freshwater micrograzers. In: van Vierssen, W.; Hootsmans, M.J.M.; Vermaat, J.E., eds. Lake Veluwe, a macrophyte-dominated system under eutrophication stress. Dordrecht: Kluwer Academic Publishers; 1994:213–249.

Vermaat, J.E.; Hootsmans, M.J.M. Periphyton dynamics in a temperature-light gradient. In: Van Vierssen, W.; Hootsmans, M.J.M.; Vermaat, J.E., eds. Lake Veluwe, a macrophyte-dominated system under eutrophication stress. Dordrecht: Kluwer Academic Publishers; 1994:193–212.

Weber, L.M.; Lodge, D.M. Periphytic food and predatory crayfish: relative roles in determining snail distribution. Oecologia 82:33–39; 1990.

Wetzel, R.G. Attached algal-substrata interactions: fact or myth, and when and how? In: Wetzel, R.G., ed. Periphyton of freshwater ecosystems. The Hague: Dr. W. Junk Publishers; 1983:208–215.

Wetzel, R.G. Benthic algae and nutrient cycling in lentic freshwater systems. In: Stevenson, R.J.; Bothwell, M.L.; Lowe, R.L., eds. Algal ecology. Freshwater benthic systems. San Diego, CA: Academic Press; 1996:641–667.

Williams, S.L.; Ruckelshaus, M.H. Effects of nitrogen availability on eelgrass (*Zostera marina*) and epiphytes. Ecology 74:904–918; 1993.

Wootton, J.T.; Power, M.E. Productivity, consumers, and the structure of a river food chain. Proc. Natl. Acad. Sci. U.S.A. 90:1384–1387; 1993.

4. Interactions between Periphyton, Nonmolluscan Invertebrates, and Fish in Standing Freshwaters

John I. Jones, Brian Moss, and Johnstone O. Young

Introduction

The Importance of Periphyton

Densities of several hundreds of micrograms of algal chlorophyll *a* per square meter are not unusual on submerged plant surfaces. Frequently, the supporting surface cannot be seen through the covering mass, which may compete with the plant for light, inorganic carbon, and nutrients. Given the problems posed by the overlying water column as well as the periphyton (syn: epiphyton) for such plants, it is remarkable that submerged plants develop at all in other than the clearest, most nutrient-deficient waters.

That they do so abundantly, even in nutrient-rich waters, may therefore depend on balancing mechanisms that prevent the development of both periphyton and phytoplankton to the potentials set by the nutrient supply. Phytoplankton growth can be lessened by grazing, allelopathy, washout, and mixing to depths where respiration dominates photosynthesis. Given a substratum on which to develop, only the former two mechanisms are likely to be important controls on the periphyton, and to date there has been no unequivocal evidence in support of allelopathy (Forsberg et al., 1990). It is with the first, grazing, that this chapter is concerned.

Macrophytes greatly increase the surface area available for colonization by animals when compared with the lake bed beneath (see Diehl and Kornijów, this volume, Chapter 2). Furthermore, periphyton provides more nutritious substrates than sediments or vascular plant tissues. Consequently, periphyton is an important energy source for both detritus and grazing food chains (Gressens, 1995). However, grazer links are likely to be complex for they involve a system of several metabolically active components, including the plants themselves, the periphyton, invertebrate and vertebrate grazers of a wide range of sizes and types, and the invertebrate and vertebrate predators of these grazers. This review is concerned with the control potentially exerted by such a system.

Organization and Scope of This Review

The review is organized in sections that take components of the periphyton grazer system in turn: (1) the nature of the periphyton and (2) of the grazers; (3) grazer effects on the periphyton; (4) periphyton effects on the grazers; (5) effects of grazers on grazers; and (6) effects of predators on grazers. Finally, some general observations are made.

Most studies on grazer–periphyton relationships have been in streams (Feminella and Hawkins, 1995) with less information from still waterbodies, where emphasis has been on snail grazing. The impact of smaller invertebrates such as microcrustacea, oligochaetes, and chironomid larvae on periphyton is not so well known, although their high numbers and production in the littoral zone (Prejs, 1976; Sarvala et al., 1981) suggest a high potential grazing pressure (Cattaneo, 1983; Kairesalo, 1984; Kairesalo and Koskimies, 1987). This review excludes work on snails, which are considered by Brönmark and Vermaat (this volume, Chapter 3). References to work on streams are indicated by asterisks. Aloi (1990) has recently reviewed the methodology used in investigations of periphyton.

Nature of the Periphyton

Communities colonizing new substrates often consist of low posture, prostrate, or apically attached cells. With time, communities become progressively more three-dimensional as protrusive, stalked, and/or filamentous forms grow (Patterson and Wright, 1986*; Steinmann and McIntire, 1986*, 1987*). The resultant periphytic communities are usually several layered, with an adherent understory of basal cells of eventually filamentous algae, encrusting green algae, and tightly attached diatoms such as *Cocconeis*. Overtopping it are protrusive long diatoms, stalked diatoms, and short filaments. There may be further longer, loosely associated filaments and colonies, and large unicells, sometimes motile, entangled in the matrix. Different growth forms will have different susceptibilities to grazing. The overstory is usually most vulnerable to grazers (Kessler, 1981; Jacoby, 1985*, 1987*; Steinmann et al., 1987*).

Clouds of filamentous green algae (e.g., of *Cladophora*, *Ulothrix*, or Zygnematales) can form by proliferation of filaments formerly attached to substrates.

Although more precisely termed *metaphyton,* they often develop from and are closely linked with epiphytic communities and are considered here also.

Nature of the Grazers

General Survey

Invertebrate grazers other than molluscs, in still waterbodies, include oligochaetes (e.g., Young, 1945; Kairesalo and Koskimies, 1987; Hann, 1991), nematodes (Kairesalo, 1984), microcrustacea (Cladocera, Copepoda, Ostracoda [e.g., Young, 1945; Fairchild et al., 1989; Gressens, 1995]), amphipod crustaceans (e.g., Mazumder et al., 1989; Dodds, 1991), isopod crustaceans (e.g., Sozska, 1975), crayfish (Flint and Goldman, 1975), mysids (Irvine et al., 1993), chironomid larvae (e.g., Young, 1945; Mason and Bryant, 1975; Hann, 1991), caddisfly larvae (e.g., Sozska, 1975), and mayfly larvae (e.g., Moss, 1976; Sozska, 1975).

The impact of other littoral invertebrates grazers such as protozoans, haliplid coleopterans, and corixid hemipterans on periphyton communities has not been studied. McCormick (1991*) and Sleigh et al. (1992*) have studied the first of these groups in streams and found annual production to be of the same order as macroinvertebrates and fish. Only a few papers on grazing on periphyton by fish in still waterbodies are available (Cattaneo, 1983; Cattaneo and Kalff, 1986). Amphibians at the larval stage may be major grazers (Osborne and McLachlan, 1985; Seale, 1980; Brönmark et al., 1991).

Most grazers are mobile, focusing their grazing on distinct patches of periphyton (e.g., Kohler, 1984*; Vaughn, 1986*; Scimgeoar et al., 1991*). A few are sedentary, including some chironomid, caddisfly, and lepidopteran species, which live in constructed retreats or shallow mines in macrophytes, and feed in a localized area (Lamberti and Moore, 1984*; Hart, 1985*; Bergey, 1995*). Studies of these species have been limited to lotic habitats.

Mouthpart Adaptations and the Nature of the Food Eaten

The relative success or failure of herbivores in cropping periphyton is linked with the suitability of their mouthparts to the physiognomy of the algal assemblage (Fig. 4.1). Most grazers are scrapers (e.g., snails, which have a radula, and chironomids, caddisflies, and mayflies, which have blade-like mandibles and other mouthparts with limited setation). Some grazers are collector-gatherers (some species of mayfly larvae), with mouthparts covered with dense brushes of hairs or setae, most associated with the maxillae and labium, which allow them to browse on periphyton. Other taxa use additional limbs to remove periphyton from the substrate before feeding on it. In stream studies, differences in mouthpart morphology have been used to explain different effects on periphyton (e.g., Lamberti et al., 1987a*,b*; Karouna and Fuller, 1992*).

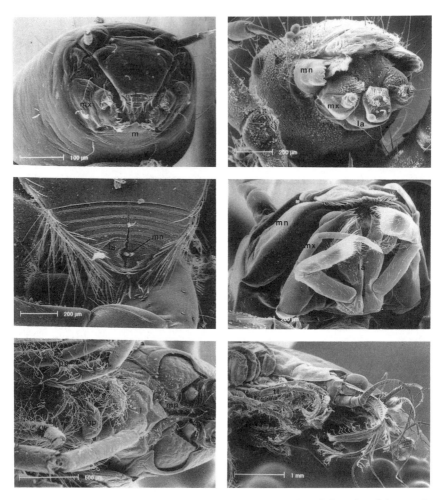

Figure 4.1. Scanning electron micrographs showing the scale and diversity of the mouthparts and associated feeding structures of six potential grazers. (Top left) Blade-like mouthparts of chironomid larvae (*Cricotopus sylvestris*) and (top right) caddis larvae (Limnephilidae); (middle left) reduced beak-like mouthparts and front limbs of corixid hemipteran (*Sigara dorsalis*); (middle right) setation of mouthparts of mayfly larvae (*Cloëon dipterum*); (bottom left) additional thoracic limbs, which assist in feeding of amphipod (*Gammarus pulex*), and (bottom right) mysid (*Neomysis integer*). gp, gnathopod; l, labrum; la, labium; m, mentum; mn, mandible; mp, mandibular palp; mx, maxilla, pp, pereiopod; tl, thorax limbs. Scale bars are as shown.

Grazer Effects on the Periphyton

Biomass

Grazer effects may include direct consumption, alteration of community composition, nutrient mobilization, physical disruption of the mat, and an increase in growth rate and turnover of periphyton. Some stream studies have indicated that certain algae typical of the early stages of colonization can benefit from grazers under certain conditions (e.g., Lamberti and Resh, 1983*; Jacoby, 1987*; Petersen, 1987*) or that grazing may benefit the algal community as a whole and not just individual species (e.g., in increased primary productivity [Lamberti and Resh, 1983*] and species richness [Eichenberger and Schlatter, 1978*; Sumner and McIntire, 1982*]). However, much evidence shows that grazers can effectively remove periphyton, and it has been suggested that a summer minimum in periphyton biomass is brought about by grazers (Cattaneo and Kalff, 1986; Cattaneo, 1990).

One of the problems that exists in the interpretation of grazing studies is that they often infer grazer effects from the difference in biomass between grazed and ungrazed treatments. Due to the exponential nature of biomass accumulation, this may not be true and especially so, if other density-dependent factors are limiting the ungrazed biomass (Mitchell and Wass, 1996; Mitchell and Perrow, this volume, Chapter 9). This may explain findings such as those of Knudson (1957), who concluded that grazing had little effect on density of an epiphytic diatom, *Tabellaria flocculosa,* growing on emergent plants in three English Lake District lakes, as long as growth conditions were favorable; periods of constant density, however, may represent a balance when the rate of increase equals the erosion rate (a combination of grazing and wave action). Gressens (1995) also found that grazers (chironomids, snails, chydorid cladocerans) did not significantly decrease biomass of periphyton attached to artificial substrates. Hence, Gressens and Lowe (1994) advise caution in the interpretation of correlations of chironomid density with low periphyton biomass as they are not necessarily caused by grazing.

An accurate estimate of grazing rate is difficult to determine. Nevertheless, Kairesalo and Koskimies (1987) found that the daily consumption by oligochaetes and snails corresponded to 22–45% of the average phosphorus uptake by epiphytes on *Equisetum fluviatile* and concluded that grazing is an important mechanism affecting epiphytic abundance. Botts and Cowell (1992) found that a small chironomid, *Psectrocladius,* ate 13.9% of the available *Cosmarium* standing stock daily. Cattaneo (1983) used estimates of grazer standing stock (mainly, oligochaetes and chironomids) and changes in epiphyte biomass to calculate grazing rates sufficient to be a significant component of epiphyte decline. Data of Mason and Bryant (1975), reworked by Kessler (1981), suggest grazing rates of chironomids on *Typha* stems to have been 44% of the periphyton net accumulation rate. All these studies indicate that invertebrate grazers can play an extremely important role in controlling periphyton populations in lakes.

However, some reported cases do not fit with this general pattern. Fairchild et al. (1989a) found a positive relationship between invertebrate densities (especially

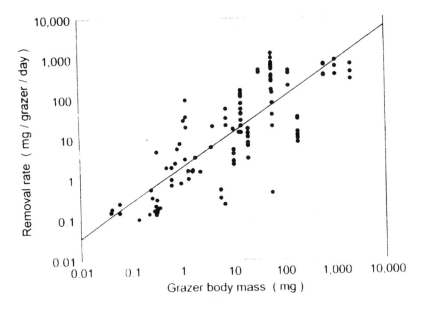

Figure 4.2. Relationship between estimated removal rate (see text for details) of peri-phyton and the grazer body mass. (From Cattaneo and Mousseau, 1995.)

chironomid larvae and chydorid cladocerans) and periphyton standing crop. They speculate that the moderate algal consumption by grazers in a softwater lake with low dissolved inorganic carbon (DIC) is accompanied by increased growth stimulated by regenerated carbon. Under similar conditions, Fairchild et al. (1989b) found that removal of grazers did not increase periphyton density on porous pots from which nutrients were diffusing.

While ignoring the problems associated with estimates of grazing rate, Cattaneo and Mousseau (1995) in a review of the literature found that uncorrected values of periphyton removal rate increased linearly with grazer body mass and food availability and decreased with grazer abundance (Fig. 4.2):

$$\log R = 0.99 \log(\text{bodymass}) - 0.71 \log(\text{grazer biomass}) \qquad (1)$$
$$+ 0.46 \log(\text{periphyton biomass}) \; r^2 = 0.78, P < .0001.$$

In order of importance, the correlates were grazer body mass (65% of variation), crowding (7%), and food availability (6%). High grazer densities were associated with competition for a limited food resource and increased aggressive behavior. Temperature effects were not significant within the range of study (9 – 26°C). Removal rate, corrected for body size, was similar among all grazer taxa except amphibians, in which it tended to be lower (Fig. 4.3). Rates calculated from field experiments were similar among stream and standing water sites but lower than those from experimental systems. This is most likely to be a result of density-

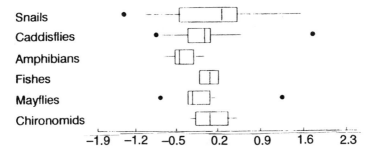

Figure 4.3. Box-and-whisker plots of the residuals from Equation 1 (see text) separated by taxa. The median value is marked by the central line; 25 and 75 percentile values form the ends of the box; whiskers delimit the range of the observations except for extreme values, represented by points. (From Cattaneo and Mousseau, 1995.)

dependent factors limiting the ungrazed biomass in the field, but other factors such as less active grazing in the presence of predators or alternative food sources may have played a role.

Community Composition

Herbivore diet depends on attributes of both herbivore (size, motility, morphological specializations for feeding, digestive capabilities) and periphyton (size, ease of harvest, density, palatability, nutritional value, and perhaps also reproductive traits and chemical defenses) (Gregory, 1983*). These factors result in a general selectivity by different animals for particular types of periphyton.

Early studies indicated preference for particular types of periphyton by grazers (Fryer, 1968; Whiteside et al., 1978; Moore, 1979; Titmus and Badcock, 1981). The effects depend on the initial periphyton community (Hart, 1981*; Kohler, 1984*), the nature and size of the grazers (scrapers, for example, are more devastating than browsers [Mason and Bryant, 1975; Bowker et al., 1983; Karouna and Fuller, 1992*]), their abundance (Dodds, 1991*), and the environmental conditions (Dodds, 1991*; Poff and Ward, 1992*).

Selection may be made on the basis of growth form (e.g., adnate diatoms vs. arborescent species [Peer, 1986; Fryer, 1957, 1968, 1974; Fulton, 1988]), on particular species (Mills and Wyatt, 1974; Botts and Cowell, 1992), or on size of particle, irrespective of its taxonomy (Patrick, 1970*; Kessler 1981; Pinder 1992*). The consequences include the development of different, sometimes less easily grazeable forms, such as coarse filaments (Eichenberger and Schlatter, 1978*; Cattaneo, 1983; Hart, 1985*; Cattaneo and Kalff, 1986; Garland and Buikema, 1986; Botts, 1993) which may subsequently be grazed by larger animals such as crayfish or vertebrates (Dickman, 1968; Osborne and McLachlan, 1985), or restriction of the periphyton community to closely adpressed diatoms (Hann, 1991). Development of less edible cyanophyte growths may occur but is not a prominent effect.

Communities that persist under intense grazing pressure have taxa that have high growth or recruitment rates, are buffered from overexploitation by their growth form (small, flat, or adnate), pass through the guts of the grazer unharmed, or are rejected because of unpalatability or handling dificulties (Hann, 1991). Removal of particular algal taxa may alter the assemblage by directly reducing the dominant algae or by increasing other algal species with poor competitive ability. The consequences of grazing may differ in subtropical and tropical lakes. Botts and Cowell (1993) found that abundances of epiphytic algae and invertebrate grazers were only weakly correlated in a subtropical lake, whereas in temperate lakes, they usually show strong temporal correlation.

Grazing by medium-sized fauna on filaments is mechanically possible with long chain diatoms anchored to a substratum but impossible for long filamentous blue-green and green algae where there is cohesion between strands (Nicotri, 1977). However, large grazers are the most effective at eating filaments. Anuran larvae eat periphyton (Dickman, 1968; Osborne and McLachlan, 1985) and can remove massive floating clumps of *Mougeotia.* Many American fish are algivores (Scott and Crossman, 1973) or use filamentous algae as a substantial food source, particularly in late summer (France et al., 1991). Power and Matthews (1983*) and Power et al. (1985*) showed strong complementarity between the distribution of a herbivorous minnow, *Campostoma anomalum,* and filamentous green algae in streams. Cattaneo (1983) and Cattaneo and Kalff (1986) excluded minnows, crayfish, and tadpoles from cages and observed increases in *Mougeotia* and *Oedogonium* over diatoms. Crayfish are opportunistic feeders on animals, plants, and algal filaments (Flint and Goldman, 1975), and it has been shown they can dramatically reduce macroalgae (Schindler et al., 1985; Aloi, 1988; Creed, 1988).

Grazer control on metaphyton flocs or clouds appears to be effective only at the start of growth. Once the cloud develops, it becomes effectively "immune" and functions like a macrophyte, with grazers removing the algae attached on its surfaces (Dodds, 1991*). Dudley (1986*) found mayflies able to control *Clado-phora* when it was sparse, but they had no effect once it was established. The often-abundant presence of invertebrates in clouds may be because they act as a predator refuge rather than a food source, although entrapment of detrital particles may attract some invertebrates (Dudley et al., 1986*; Dodds, 1991*). Even ver-tebrates may have little effect once the algal mass is well established (France et al., 1991) and grazing is opportunistic and not preferential (Tallman et al., 1984).

Productivity

Studies on the effects of grazing by animals other than mollusks on the produc-tivity of periphyton are rare and their results somewhat contradictory. Grazing by snails and by stream animals has been shown to reduce periphyton productivity (Rosemond et al., 1993). Other studies have indicated a decrease on an areal basis but an increase on a biomass-specific basis (Lamberti and Resh, 1983*; Gelwick and Matthews, 1992*), or an increase of areal primary production (Lamberti et al., 1989*; Power, 1990a*). It would appear that enhancement of production is due to nutrient renewal by algal cell disruption and excretion by grazers (Underwood 1991), reduced competition for nutrients (Lamberti and Resh, 1983*), increased light levels to lower strata of periphyton, and removal of senescent cells by consumption or dislodgment (Lamberti et al., 1987a*), but these gains must outweigh the losses to grazing for effects to be seen. This will depend on both the intensity of grazing and what factors are limiting algal growth. Hence, Feminella and Hawkins (1995*) suggest that grazers in streams have a less consistent effect on productivity than on abundance and biomass.

Disturbance

Periphytic algae may be dislodged or suffer mechanical damage by grazers during feeding or, in the case of caddisflies and chironomids, case building (Cattaneo, 1983; Pringle, 1985*; Botts, 1993). In streams, this organic matter may be important for browsers (metaphyton), filter feeders, or once settled, deposit-collecting detrivores (Lamberti et al., 1987a*, 1989*, 1995*). The proportion of periphyton removed in this way may be large (Cattaneo and Mousseau, 1995).

Differences in the efficiency by which algal taxa are digested can also alter algal dominance patterns. Certain algae, particularly small diatoms, pass intact through the grazer gut (Brown, 1960; Peterson, 1987), to be deposited in nutrient-rich feces (Vaughn, 1986*).

Nutrient Enhancement due to the Presence of Grazers

Mollusks and nonmolluscan stream grazers can positively affect the supply of nutrients to the grazer-resistant algal understory (e.g., McCormick and Stevenson, 1991*; McCormick, 1994*), either by excreting nitrogen directly or by removing the periphyton overstory and facilitating diffusion from the water column to the understory. The only nonmolluscan lake study (Flint and Goldman, 1975), however, suggested that increased productivity of periphyton due to the presence of crayfish was not due to crayfish excretion. On a local scale, motile diatoms may migrate to chironomid tubes and use excreted nutrients (Botts, 1993).

Periphyton Effects on the Grazers

The quantity, quality, and composition of periphyton can have an important effect on grazer distribution, abundance, growth rate, life history, and probably, production and community composition. Information on the last two is lacking. Numerical response of invertebrates to changes in periphyton abundance may be attributable to migration, emergence, or reproduction.

Distribution

Mason and Bryant (1975) and Kornijów (1992) found some species of chironomid larvae moved from mud in late spring as temperatures rose onto plants and grazed periphyton; they returned to the mud in autumn as temperatures fell. Gressens and Lowe (1994) observed dispersal of chironomid larvae among algal patches differing in abundances of several diatom species and the green alga *Stigeoclonium*. Patch preference of the larvae was negatively correlated with abundance of *Stigeoclonium*, chlorophyll a concentration, and algal biovolume and positively correlated with algal diversity. Periphyton quality was more important than quantity. In streams, caddisfly and mayfly species respond to variations in periphyton abundance by selectively colonizing and spending more time on patches with more periphyton (Hart, 1981*; Feminella and Resh, 1991*). Selection for algal abundance was of overriding importance, even though algal type was shown to have marked effects on the development of the grazers (Vaughn, 1986*). Presumably, selection of food source will depend on food availability and competition from other grazers.

Abundance

In their review of stream work, Feminella and Hawkins (1995*) found higher grazer densities at high periphyton abundance (caddisflies, mayflies, and chironomids were the most frequently observed taxa). Those studies that have been conducted in lakes seem to show a similar situation. Numbers of chydorid cladocerans and chironomids were found to be highly correlated with diatom numbers (Fairchild, 1981), and in another study, chironomid density increased with an

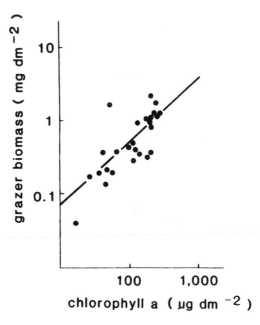

Figure 4.4. Correlation between periphyton biomass collected from sites in Lake Memphremagog, Quebec, and the biomass of oligochaetes and chironomids from the same sites 4–5 days later. (From Cattaneo, 1983.)

increase in algal biomass (Fairchild and Lowe, 1984). However, there may be a lag in the response of invertebrates (Fig. 4.4); chironomid larvae, oligochaetes, and nematodes were most numerous in epiphytic communities living on *Equisetum* soon after the peak of algal abundance (Kairesalo, 1984). This may explain reports of higher grazer density in low-periphyton situations. Herbivore populations appear to be controlled from the bottom up by algal availability, which may occur in conjunction with strong top-down control of periphyton biomass by grazing.

Growth

No studies on the effects of periphyton on the growth of grazers in lakes have been made. The review by Feminella and Hawkins (1995*) showed that most studies in streams found highest growth at high periphyton biomass. Only two studies reported highest growth in low periphyton biomass.

The nature of the periphyton food source may also influence growth (Lamberti et al., 1989*). Assimilation efficiency of 20 aquatic insects, using 45 published values was 30–60% for periphyton (70–95% for animal food; 5–30% for detritus). For a snail, *Juga silicula,* it was 70–80% when the snail was first added to a stream but declined with time (to 40%). This coincided with a shift in composition from diatoms and unicellular greens to filamentous green and blue-green algae. Variation in protein and lipid content and cell wall thickness among periphyton is likely to influence palatability. High C:N (>17:1) is thought to be deleterious. Periphyton has C:N ratios of 4:1–8:1; cf. 13:1–69:1 in macrophytes (Gregory, 1983*).

In the slow-flowing rivers of southeast England young-of-the-year fish (roach, *Rutilus rutilus,* and chubb, *Leuciscus cephalus*) switch to feeding on periphyton when preferred zooplankton are unavailable, either due to low numbers (Garner, 1996a*) or at night when sight feeding is precluded (Garner, 1996b*). This switch coincides with a decline in the growth rate of these fish (Garner et al., 1995*).

Although they are not prominent in many periphyton communities, cyanobacteria are generally poor food. They contain much protein, but their mucosaccharide sheath is indigestible and toxins may be produced. This was reflected in a slower development of caddisflies when fed cyanobacteria than other algae (Vaughn, 1986*). Evidence from grazing experiments is mixed. *Gammarus* would not consume *Phormidium* (Moore, 1975), but orthoclad midges have suppressed *Phormidium* and *Oscillatoria* in outdoor channels (Eichenberger and Schlatter, 1978*). Most studies have been on filaments. The situation may be different for colonies or unicells.

Life History

As well as the effects on growth, invertebrate development is influenced by the food source. The rate at which invertebrates pass through their life cycle is influenced by both the quality (Vaughn, 1986*) and quantity (our own results, unpublished) of the available periphyton. There is also a considerable effect on fecundity (our own results, unpublished), and subsequently population size. It is not surprising, therefore, that invertebrate life cycles are often timed to coincide with food availability (Tokeshi, 1986a*).

Effects of Grazers on Other Grazers

Different grazer species may interfere with each other. Intense competition for food can occur (Tokeshi, 1986b*), which will lead to less selection as food availability is reduced. The processes of hatching, migration from mud to plants and vice versa, pupation, cocoon formation, and emergence will affect the species pool of invertebrates. Often these processes are closely coupled with food availability, leading to populations synchronized with the spring peak of periphyton (Tokeshi, 1986a*).

Both inter- and intraspecific competition have been shown to occur between grazers in lakes. Kajak (1963) found that increased density of *Chironomus* reduced the densities of snails and other midges, whereas Cattaneo (1983) showed the reverse, with increases in midges, ostracods, and chydorids in the absence of snails. Negative effects of snails on chironomids, snails on chydorids, snails on snails, and chironomids on chironomids have been detailed, inter alia, by Gressens (1995) (Fig. 4.5) and Cuker (1983). Both positive and negative effects have been recorded between snails and tadpoles (Brönmark et al., 1991). Snails feed on less nutritious algae in the presence of tadpoles but encourage algal communites beneficial to tadpole growth, whereas tadpoles compete with one another for food

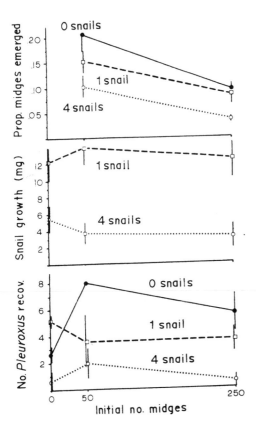

Figure 4.5. Response of grazers to variations in midge and snail densities in enclosures in Gas Station Pond, Illinois. The response of midges was measured as the proportion of midge larvae that emerged. Change in weight of snails indicates the snail response. The number of *Pleuroxus* recovered was used to estimate chydorid populations. (From Gressens, 1995.)

(Brönmark et al., 1991). Interference with feeding activity and dwelling construction appears to be the predominant form of competition between grazers.

Field exclosure experiments, using different mesh sizes, have shown that the largest grazers normally dominate (Cattaneo and Kalff, 1986). Small grazers such as oligochaetes, chironomids, and Cladocera replaced snails in finer-mesh enclosures, producing increased numbers of individuals but a similar total biomass. This reduction in grazer size but not biomass did not result in reduced algal biomass (with one exception) but in a dominance by small algae, characterized by a high turnover rate. A similar situation has been found in laboratory microcosms (Brock et al., 1995), where chemical removal of arthropod grazers produced only a short-term increase in periphyton, before snail and oligochaete populations responded to competitive release. Subsequent periphyton biomass was no different from untreated controls.

Effects of Predators on Grazers

Studies on the importance of cascading trophic interactions (top-down) effects in freshwater benthic food chains to the periphyton trophic level have focused on fish and not on invertebrate predators such as triclads, leeches, coleopterans, hemipterans, and odonatans. Moss (1976) found that the addition of bluegill sunfish to fertilized ponds caused an increase in biomass of certain macrophytes and epiphytes, which was probably due to fish predation on the grazing invertebrates (amphipods and mayflies). Mazumder et al. (1989) showed that yellow perch was associated with lower abundances of amphipods and chironomids and higher concentrations of periphyton particulate phosphorus. In the absence of fish, the invertebrates were more abundant, and periphyton productivity values were lower (Fig. 4.6). In experiments in a Swedish eutrophic pond, Brönmark (1994) concluded that nonmolluscan benthic macroinvertebrates were not greatly affected by the presence of tench or perch, whereas tench produced a positive effect on the biomass of periphyton by eating snails and reducing grazing pressure.

Batzer and Resh (1991) studied interactions among a predaceous hydrophilid beetle larva, a grazing chironomid larva, and periphyton in 11 experimental ponds. Abundant beetle larvae reduced autumn densities of chironomid larvae, which allowed periphyton to grow independently of midge density. In winter, densities of beetle larvae declined and midge populations increased. At a lower density of beetles, a close relationship was found between the densities of the chironomid and periphyton biomass, which was not seen when beetles were abundant (Fig. 4.7). Grazing by chironomids reduced periphyton biomass, which sometimes resulted in insufficient food reserves to support high chironomid production. Although many chironomids were eaten, their seasonal production was slightly greater when beetles were abundant than when they were few.

Interactions between piscivorous and herbivorous fish can also have consequences for periphyton. The addition of piscivores to streams containing the stone

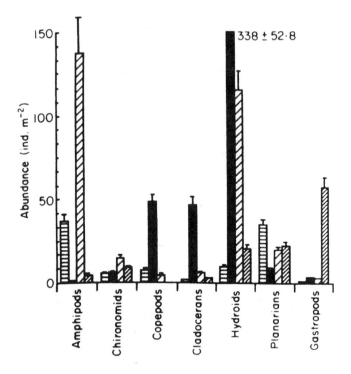

Figure 4.6. Abundance of invertebrates from enclosures in Lake St. George, Ontario, subjected to four treatments: control (horizontal lines), fish added (solid), nutrients added (wide hatched), and nutrients and fish added (narrow hatched). (From Mazumder et al., 1989.)

roller (*Campostoma*) and locariid catfish, which are effective removers of algae, led to ultimate increases in filamentous periphyton (Power and Matthews, 1983*).

In streams, the effects of fish can be pronounced. In the absence of gravel refuge, selective feeding on invertebrate predators such as damselfly larvae, can lead to increased populations of small algivores and hence fewer algae (Power, 1990b*; Power, 1992*).

The most important aspect of the periphyton–grazer–fish interaction in lakes is that these herbivorous invertebrates provide an alternative food source for fish and hence reduce the pressure on zooplankton. Without such invertebrates, the ontogenetic shifts seen in the feeding of adult fish would not occur (see Persson and Crowder, this volume, Chapter 1; Diehl and Kornijów, this volume, Chapter 2), and all fish would compete for the same food resource, a situation seen in plantless eutrophic lakes. Such feeding on invertebrates by fish will result to some extent in an increase in periphyton, but the overall cost to the plants may be outweighed by gains resulting from increased zooplankton abundance.

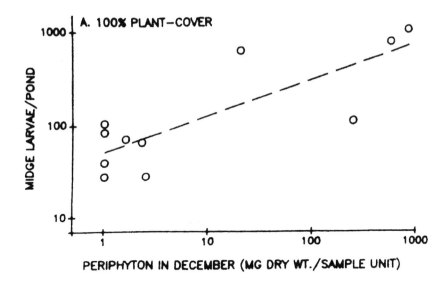

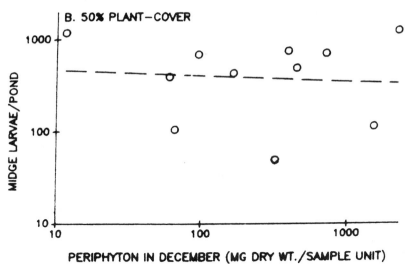

Figure 4.7. Comparison of periphyton biomass and densities of midge larvae (*Cricotopus sylvestris*) from ponds in a manipulated Californian wetland collected over the subsequent 3 months. Hydrophilid larvae were scarce in (A) the 100% plant-covered areas but abundant in (B) the 50% plant-covered areas. (From Batzer and Resh, 1991.)

Overview

There seems no doubt that invertebrate grazers can severely crop periphyton growths and that, in temperate habitats at least, they are frequently allowed to do so by predation rates that allow considerable buildup of their populations. This has been acknowledged for snail populations for some time; the current review suggests that it is no less true for nonmolluscan grazers, particularly the larvae and nymphs of insects. It is thus possible that epiphytic grazers are as important as planktonic ones in maintaining conditions appropriate to macrophyte dominance of the primary production in shallow lakes.

Much of the literature centers on individual species or restricted groups, however, and there is a shortage of studies of whole communities, containing more than two trophic levels. Occasional references to instances when the periphyton is not controlled by invertebrate grazers suggests that, sometimes, predation on these grazers may be considerable, but an overall picture of the importance of such relationships, as is available for the temperate plankton, is not yet possible. The fact that grazing rates measured in experimental enclosures from which predators are deliberately or inadvertently excluded are greater than in the open habitat also suggests an important role for top-down effects.

In continental areas with rich vertebrate faunas, there may be an important role for vertebrate grazers on periphyton, including both fish and amphibians. There do not appear to be specific periphyton grazers among birds or mammals. Although such vertebrates frequently eat macrophytes and hence the associated periphyton, their essentially aerial adaptations do not allow the delicate positioning that is needed to graze the periphyton without damaging the host plant. Experimental work on vertebrate periphyton grazers is scarce; they appear to be particularly important in grazing the coarse filamentous algae that may form large blanketing flocs or clouds in the water. Small invertebrate grazers appear to be able to tackle these only when the flocs are sparse but can prevent their development if grazing sets in early enough. Such filamentous flocs may be more damaging to macrophyte growth than thin adnate layers of periphyton, and the gap in the literature concerning them and their vertebrate grazers is an important one.

Currently, the literature is dominated by studies on periphyton films on rocks and stones in streams and, in standing waters, by work on mollusks. Grazing rates in streams and standing waters appear broadly similar, as do those of snails and other invertebrates. It is not yet clear whether it matters much that the periphyton substrate is living or inert. Comprehensive work (our unpublished data) suggests that living substrates only minimally influence the potential size of the periphyton community; the extent to which its composition may be determined by the activities of the host is still not fully resolved, but the complexities may be far less great than hitherto thought. The architecture and growth rate of the macrophyte, rather than any chemical influence, may be most important, but determination of the problem is confounded by random effects of colonization into the periphyton community, coupled with selective effects of grazing after colonization.

Such selective effects undoubtedly occur, but they are not as simple as they appear to be in the plankton, where small species are more vulnerable than larger ones. Some periphyton grazers take large cells, others small cells. The effect on a previously ungrazed periphyton community will vary, dependent on the nature of the grazers present. In turn, this will depend on a large range of factors, including the architecture of the host plant, the degree of competition, the time of year and the phasing of different life histories, the action of predators, the vagaries of colonization, and probably other factors. Again, this is a very different situation to that in the plankton, in which the activities of a limited range of grazing animals, of similar size range and grazing methods (filter and suspension feeding), effectively hunted by one major sort of predator, are concerned. The possibilities of predicting and modeling the outcome of events in the plankton are much greater than are yet possible in periphyton grazer systems.

This review has highlighted the urgent need for research, both in the field and laboratory, due to the paucity of information concerning lacustrine nonmolluscan groups in the following areas: ingestion and digestion rates; assimilation efficiencies; effects of grazer type (including mouthpart morphology), selectivity, size, density, multiple species action, disturbance, and nutrient enhancement on individual species and whole communities of periphyton including standing crop, production, and composition; the effect of periphyton on grazer abundance, distribution, growth rates, production, and community composition; interrelationships between grazing species; and the effect of predation on grazers, and hence periphyton.

Overall, the macrophyte–periphyton–grazer–predator system offers considerable parallels to that well studied on a larger scale by range managers in prairies and savannahs. Rather similar elements are present. Frequently, we talk of the periphyton in terms merely of its chlorophyll, its grazers as numbers or biomass. No one would talk of an African savannah as a uniform growth of grass with a single sort of grazer, say, an antelope. The plant communities vary from long grasses, after the rains, on which several species of large antelope may feed to short grasses supporting small antelope, where there has been previous heavy grazing. A variety of rodents feeds on seeds, whereas grasshoppers and locusts also graze the fresh grasses, and a huge biomass of termites takes the dead culms and fragments of wood from thorn bushes and shrubs. Some of the grazers, browsing as individuals, also depend on these (e.g., the black rhinoceros). Others gather in herds (e.g., elephants), temporarily to devastate the vegetation before moving on and allowing a new succession to begin with a new set of grazers. The plant community will vary in species composition, dependent on slope and soil, previous history, and microclimate; a parallel array of specialist insects will graze the different communities. Streams crossing the landscape will bear riverine gallery forest with a completely different array of grazers, including smaller antelopes such as the puku and perhaps hippopotami in prominent pools.

If this seems like an unreasonable comparison with the periphyton system, then consider the scale. A periphyton diatom might be $5 \times 5 \times 30$ μm in size. If it be considered equivalent to a clump of grass of base 5 cm × 5 cm, a square centimeter of aquatic plant surface is equivalent to a hectare of land surface and a whole aquatic plant to several kilometers squared. It is by adjusting our thinking to the appropriate scale that we may best understand the functioning and importance of the periphyton system.

References

Aloi, J.E. Changes in periphyton biomass, primary productivity and community composition resulting from field manipulations of crayfish. Bull. North Am. Benth. Soc. 5:103; 1988.

Aloi, J.E. A critical review of recent freshwater periphyton field methods. Can. J. Fish. Aquat. Sci. 47:656–670; 1990.

Batzer, D.P.; Resh, V.H. Trophic interactions among a beetle predator, a chironomid grazer, and periphyton in a seasonal wetland. Oikos 60:251–257; 1991.

Bergey, E.A. Local effects of a sedentary grazer on stream algae. Freshwat. Biol. 33:401–409; 1995.

Botts, P.S. The impact of small chironomid grazers on epiphytic algal abundance and dispersion. Freshwat. Biol. 30:25–33; 1993.

Botts, P.S.; Cowell, B.C. Feeding electivity of two epiphytic chironomids in a subtropical lake. Oecologia 89:331–337; 1992.

Botts, P.S.; Cowell, B.C. Temporal patterns of abundance of epiphytic invertebrates on Typha shoots in a subtropical lake. J. North Am. Benth. Soc. 12:27–39; 1993.

Bowker, D.W.; Warehorn, M.T.; Learner, M.A. The selection and ingestion of epilithic algae by *Nais elinguis* (Oligochaeta: Naididae). Hydrobiologia 98:171–178; 1983.

Brock, T.C.M.; Roijackers, R.M.M.; Rollon, R.; Bransen, F.; Van der Heyden, L. Effects of nutrient loading and insecticide application on the ecology of *Elodea*-dominated freshwater microcosms. II. Responses of macrophytes, periphyton and macroinvertebrate grazers. Arch. Hydrobiol. 134:53–74; 1995.

Brönmark, C. Effects of tench and perch on interactions in a freshwater, benthic food chain. Ecology 75:1818–1828; 1994.

Brönmark, C.; Rundle, S.D.; Earlandsson, A. Interactions between freshwater snails and tadpoles: competition and facilitation. Oecologia 87:8–18; 1991.

Brown, D.S. The ingestion and digestion of *Cloëon dipterum* L. Hydrobiologia 16:81–96; 1960.

Cattaneo, A. Grazing on epiphytes. Limnol. Oceanogr. 28:124–132; 1983.

Cattaneo, A., The effect of fetch on periphyton variation. Hydrobiologia 206:1–10; 1990.

Cattaneo, A.; Kalff, J. The effect of grazer site manipulation on periphyton communities. Oecologia 69:612–617; 1986.

Cattaneo, A.; Mousseau, B. Empirical analysis of the removal rate of periphyton by grazers. Oecologia 103:249–254; 1995.

Creed, R.P. The influence of crayfish grazing on benthic community structure in a Michigan stream. Bull. North Am. Benth. Soc. 5:67; 1988.

Cuker, B.E. Competition and coexistence among the grazing snail *Lymnaea,* Chironomidae, and Microcrustacea in an arctic epilithic lacustrine community. Ecology 64:10–15; 1983.

Dickman, M. The effect of grazing by tadpoles on the structure of a periphyton community. Ecology 49:1188–1190; 1968.

Dodds, W.K. Community interactions between the filamentous alga *Cladophora glomerata* (L.) Kuetzing, its epiphytes, and epiphyte grazers. Oecologia 85:572–580; 1991.

Dudley, T. Beneficial effects of grazers on algal growth. Bull. North Am. Benth. Soc. 3:87; 1986.

Dudley, T.L.; Cooper, S.D.; Hemphill, N. Effects of macroalgae in a stream invertebrate community. J. North Am. Benth. Soc. 3:93–106; 1986.

Eichenberger, E.; Schlatter, F. The effect of herbivorous insects on the production of benthic algal vegetation in outdoor channels. Verh. Int. Verein. Limnol. 20:1806–1810; 1978.

Fairchild, G.W. Movement and distribution of *Sida crystallina* and other littoral microcrustacea. Ecology 62:1341–1352; 1981.

Fairchild, G.W.; Lowe, R.L. Artificial substrates that release nutrients: effects on periphyton and invertebrate succession. Hydrobiologia 114:29–37; 1984.

Fairchild, G.W.; Campbell, J.M.; Lowe, R.L. Numerical response of Chydorids (Cladocera) and Chironomids (Diptera) to nutrient-enhanced periphyton growth. Arch. Hydrobiol. 114:369–382; 1989a.

Fairchild, G.W.; Sherman, J.W.; Acker, F.W. Effects of nutrient (N,P,C) enrichment, grazing and depth upon littoral periphyton of a softwater lake. Hydrobiologia 173:69–83; 1989b.

Feminella, J.W.; Hawkins, C.P. Interactions between stream herbivores and periphyton: a quantitative analysis of past experiments. J. North Am. Benth. Soc. 14:465–509; 1995.

Feminella, J.W.; Resh, V.H. Herbivorous caddisflies, macroalgae, and epilithic microalgae: dynamic interactions in a stream grazing system. Oecologia 87:247–256; 1991.

Flint, R.W.; Goldman, C.R. The effects of a benthic grazer on the primary productivity of the littoral zone of Lake Tahoe. Limnol. Oceanogr. 20:935–944; 1975.

Forsberg, C.; Kleiven, S.; Willen, T. Absence of allelopathic effects of *Chara* on phytoplankton in situ. Aquat. Bot. 38:289–294; 1990.

France, R.L.; Howell, E.T.; Paterson, M.J.; Welbourn, P.M. Relationship between littoral grazers and metaphytic algae in five softwater lakes. Hydrobiologia 220:9–27; 1991.

Fryer, G. The feeding mechanism of some freshwater cyclopoid copepods. Proc. Zool. Soc. Lond. 129:1–25; 1957.

Fryer, G. Evolution and adaptive radiation in the Chydoridae (Crustacea: Cladocera): a study in comparative functional morphology and ecology. Philos. Trans. R. Soc. Lond. B 254:221–385; 1968.

Fryer, G. Evolution and adaptive radiation in the Macrothricidae (Crustacea: Cladocera): a study in comparative functional morphology and ecology. Philos. Trans. R. Soc. Lond. B 269:237–274; 1974.

Fulton, R.S. III. Grazing on filamentous algae by herbivorous zooplankton. Freshwat. Biol. 20:263–271; 1988.

Garland, J.L.; Buikema, A.L. Effects of meiofaunal grazing on detrital and epiphytic assemblages. Bull. North Am. Benth. Soc. 3:86; 1986.

Garner, P.; Bass, J.A.B.; Collett, G.D. The effect of weed cutting upon the biota of a large regulated river. Aquatic conservation: marine and freshwater ecosystems. 6:21–29; 1995.

Garner, P. Microhabitat use and diet of 0+ cyprinid fishes in a lentic regulated reach of the River Great Ouse. J. Fish Biol. 48:367–382; 1996a.

Garner, P. Diel patterns in the feeding and habitat use of 0 -group fishes in a regulated river the River Great Ouse, England. Ecol. Freshwat. Fish. 5:175–182; 1996b.

Gelwick, F.P.; Matthews, W.J. Effects of an algivorous minnow on temperate stream ecosystem properties. Ecology 72:1630–1645, 1992.

Gregory, S.V. Plant–herbivore interactions in stream systems. In: Barnes, J.R.; Minshall, G.W., eds. Stream ecology. New York: Plenum Press; 1983:155–187.

Gressens, S.E. Grazer density, competition and the response of the periphyton community. Oikos 73:336–346; 1995.

Gressens, S.E.; Lowe, R.L. Periphyton patch preference in grazing chironomid larvae. J. North Am. Benth. Soc. 13:89–99; 1994.

Hann, B.J. Invertebrate grazer–periphyton interactions in a eutrophic marsh pond. Freshwat. Biol. 26:87–96; 1991.

Hart, D.D. Foraging and resource patchiness: field experiments with a grazing stream insect. Oikos 37:46–52; 1981.

Hart, D.D. Grazing insects mediate algal interactions in a stream benthic community. Oikos 44:40–46; 1985.

Irvine, K.; Moss, B.; Bales, M.; Snook, D. The changing ecosystem of a shallow, brackish lake, Hickling Broad, Norfolk, U.K. I. Trophic relationships with special reference to the role of *Neomysis integer*. Freshwat. Biol. 29:119–139; 1993.

Jacoby, J.M. Grazing effects on periphyton by *Theodoxus fluviatilis* (Gastropoda) in a lowland stream. J. Freshwat. Ecol. 3:265–274; 1985.

Jacoby, J.M. Alterations in periphyton characteristics due to grazing in a Cascade foothill stream. Freshwat. Biol. 18:495–508; 1987.

Kairesalo, T. The seasonal succession of epiphytic communities within an *Equisetum fluviatile* L. stand in Lake Paajarvi, Southern Finland. Internationale Rev. Ges. Hydrobiol. 69:475–505; 1984.

Kairesalo, T.; Koskimies, I. Grazing by oligochaetes and snails on epiphytes. Freshwat. Biol. 17:317–324; 1987.

Kajak, Z. The effect of experimentaly induced variations in the abundance of *Tendipes plumosus* L. larvae on intraspecific and interspecific relations. Ekol. Pol. 11:355–367; 1963.

Karouna, N.K.; Fuller, R.L. Influence of four grazers on periphyton communities associated with clay tiles and leaves. Hydrobiologia 245:53–64; 1992.

Kessler, D.H. Grazing rate determination of *Corynoneura scutellata* Winnertz (Chironomidae: Diptera). Hydrobiologia 80:63–66; 1981.

Kohler, S.L. Search mechanism for a stream grazer in patchy environments: the role of food abundance. Oecologia (Berlin) 62:209–218; 1984.

Knudson, B.M. Ecology of the epiphytic diatom *Tabellaria flocculosa* (Roth) Kutz. var. *flocculosa* in three English lakes. J. Ecol. 45:93–112; 1957.

Kornijów, R. Seasonal migration by larvae of an epiphytic chironomid. Freshwat. Biol. 27:85–89; 1992.

Lamberti, G.A.; Resh, V.H. Stream periphyton and insect herbivores: an experimental study of grazing by a caddisfly population. Ecology 64:75–81; 1983.

Lamberti, G.A.; Moore, J.W. Aquatic insects as primary consumers. In: Resh, V.H.; Rosenberg, D.M., eds. The ecology of aquatic insects. New York: Praeger; 1984: 164–195.

Lamberti, G.A.; Feminella, J.W.; Resh, V.H. Herbivory and intraspecific competition in a stream caddisfly population. Oecologia 73:75–81; 1987a.

Lamberti, G.A.; Ashkenas, L.R.; Gregory, S.V.; Steinman, A.D. Effects of three herbivores on periphyton communities in laboratory streams. J. North Am. Benth. Soc. 6:92–104; 1987b.

Lamberti, G.A.; Gregory, S.V.; Ashkenas, L.R.; Steinman, A.D.; McIntire, C.D. Productive capacity of periphyton as a determinant of plant–herbivore interactions in streams. Ecology 70:1840–1856; 1989.

Lamberti, G.G.; Gregory, S.V.; Ashkenas, L.R.; Li, J.L.; Steinman, A.D.; McIntire, C.D. Influence of grazer type and abundance on plant–herbivore interactions in streams. Hydrobiologia 306:237–247; 1995.

Mason, C.F.; Bryant, R.J. Periphyton production and grazing by chironomids in Alderfen broad, Norfolk. Freshwat. Biol. 5:271–277; 1975.

Mazumder, A.; Taylor, W.D.; McQueen, D.J.; Lean, D.R.S. Effects of nutrients and grazers on periphyton phosphorus in lake enclosures. Freshwat. Biol. 22:405–415; 1989.

McCormick, P.V. Lotic protistan herbivore selectivity and its potential impact on benthic algal assemblages. J. North Am. Benth. Soc. 10:238–250; 1991.

McCormick, P.V. Evaluating the multiple mechanisms underlying herbivore–algal interactions in streams. Hydrobiologia 291:47–59; 1994.

McCormick, P.V.; Stevenson, R.J. Grazer control of nutrient availability in the periphyton. Oecologia 86:287–291; 1991.

Mills, D.H.; Wyatt, J.T. Ostracod reactions to non-toxic and toxic algae. Oecologia 17:171–177; 1974.

Mitchell, S.F.; Wass, R.T. Quantifying herbivory: grazing consumption and interaction strength. Oikos 76:573–576; 1996.

Moore, J.W. The role of algae in the diet of *Asellus aquaticus* L. and *Gammarus pulex* L. J. Anim. Ecol. 44:719–730; 1975.

Moore, J.W. Factors influencing algal consumption and feeding rate in *Heterotrissocladius changi* Saether and *Polypedilum nubeculosum.* Oecologia 40:219–227; 1979.

Moss, B. The effects of fertilization and fish on community structure and biomass of aquatic macrophytes and epiphytic algal populations: an ecosystem experiment. J. Ecol. 64: 313–342; 1976.

Nicotri, M.E. Grazing effects of four marine intertidal herbivores on the microflora. Ecology 58:1020–1032; 1977.

Osborne, P.L.; McLachlan, A.J. The effect of tadpoles on algal growth in temporary, rain-filled rock pools. Freshwat. Biol. 15:77–87; 1985.

Patrick, R. Benthic stream communities. Am. Scientist 58:546–549; 1970.

Patterson, D.M.; Wright, S.J.Z. The epiphylous algal colonization of *Elodea canadensis* Michx.: community structure and development. N. Phytol. 103:809–819; 1986.

Peer, R.L. The effects of microcrustaceans on succession and diversity of an algal microcosm community. Oecologia 68:308–314; 1986.

Peterson, G.P. Gut passage and insect grazer selectivity of lotic diatoms. Freshwat. Biol. 18:461–468; 1987.

Pinder, L.C.V. Biology of epiphytic Chironomidae (Diptera: Nematocera) in chalk streams. Hydrobiologia 248:39–51; 1992.

Poff, N.L.; Ward, J.V. Heterogeneous currents and algal resources mediate in situ foraging activity of a mobile stream grazer. Oikos 65:465–478; 1992.

Power, M.E. Resource enhancement by indirect effects of grazers: armored catfish, algae, and sediment. Ecology 71:897–904; 1990a.

Power, M.E. Effects of fish in river food webs. Science 250:811–814; 1990b.

Power, M.E. Habitat heterogeneity and the functional significance of fish in river food webs. Ecology 73:1675–1688; 1992.

Power, M.E.; Matthews, W.J. Algae-grazing minnows (*Campostoma anomulum*), piscivorous bass (*Micropterus* spp.), and the distribution of attached algae in a prairie-margin stream. Oecologia 60:328–332; 1983.

Power, M.E.; Matthews, W.J.; Stewart, A.J. Grazing minnows, piscivorous bass, and stream algae: dynamics of a strong interaction. Ecology 66:1448–1456; 1985.

Prejs, K. Bottom fauna. In: Pieczynska, E., ed. Selected problems of lake littoral ecology. Warsaw: University of Warsaw; 1976:123–144.

Pringle, C.M. Effects of chironomid (Insecta: Diptera) tube-building activities on stream diatom communities. J. Phycol. 21:185–194; 1985.

Rosemond, A.D.; Mulholland, P.J.; Elwood, J.W. Top-down and bottom-up control on stream periphyton: effects of nutrients and herbivores. Ecology 74:1264–1280; 1993.

Sarvala, J.; Ilmavirta, V.; Paasivirta, L.; Salonen, K. The ecosystem of the oligotrophic lake Paajarvi. 3. Secondary production and the ecological energy budget of the lake. Verh. Int. Verein. Limnol. 21:454–459; 1981.

Schindler, D.W.; Mills, K.H.; Malley, D.F.; Findlay, D.L.; Shearer, J.A.; Davies, I.J.; Turner, M.A.; Linsey, G.A.; Cruickshank, D.R. Long term ecosystem stress: the effects of years of exprimental acidification on a small lake. Science 228:1395–1401; 1985.

Scimgeour, G.J.; Culp, J.M.; Bothwell, M.L.; Wrona, F.J.; McKee, M.H. Mechanisms of algal patch depletion: importance of consumptive and nonconsumptive losses in mayfly diatom systems. Oecologia 85:343–348; 1991.

Scott, W.B.; Crossman, E.J. Freshwater fishes of Canada. Bull. Fish. Res. Bd. Can. 184:1–966; 1973.

Seale, D.B. Influence of amphibian larvae on primary production, nutrient flux, and competition in a pond ecosystem. Ecology 61:1531–1550; 1980.

Sleigh, M.A.; Baldock, B.M.; Baker, J.H. Protozoan communities in chalk streams. Hydrobiologia 248:53–64; 1992.

Sozska, G.J. Ecological relations between invertebrates and submerged macrophytes in the lake littoral. Ekol. Pol. 23:393–415; 1975.

Steinman, A.D.; McIntire, C.D. The effects of current velocity and light energy on the structure of periphyton assemblages in laboratory streams. J. Phycol. 22:352–361; 1986.

Steinman, A.D.; McIntire, C.D. Effects of irradiance on the community structure and biomass of algal assemblages in laboratory streams. Can. J. Fish. Aquat. Sci. 44:1640–1648; 1987.

Steinman, A.D.; McIntire, C.D.; Gregory, S.V.; Lamberti, G.A.; Ashkenas, L.R. Effects of herbivore type and density on taxonomic structure and physiognomy of algal assemblages in laboratory streams. J. North Am. Benth. Soc. 6:175–188; 1987.

Sumner, W.T.; McIntire, C.D. Grazer–periphyton interactions in laboratory streams. Arch. Hydrobiol. 93:135–157; 1982.

Tallman, R.F.; Mills, K.H.; Rotter, R.G. The comparative ecology of pearl dace (*Semotilus margarita*) and fathead minnow (*Pimephales promelas*) in Lake 114, the Experimental Lakes Area, northwestern Ontario, with an appended key to the cyprinids of the Experimental Lakes Area. Can. Manage. Rep. Fish. Aquat. Sci. 1756:1–27; 1984.

Titmus, G.; Badcock, R.M. Distribution and feeding of larval chironomidae in a gravel-pit lake. Freshwat. Biol. 11:263–271; 1981.

Tokeshi, M. Population dynamics, life histories and species richness in an epiphytic chironomid community. Freshwat. Biol. 16:431–441; 1986a.

Tokeshi, M. Resource utilization, overlap and temporal community dynamics: a null model analysis of an epiphytic chironomid community. J. Anim. Ecol. 55:491–506; 1986b.

Underwood, G.J.C. Growth enhancement of the macrophyte *Ceratophyllum demersum* in the presence of the snail *Planorbis planorbis:* the effect of grazing and chemical conditioning. Freshwat. Biol. 26:325–334; 1991.

Vaughn, C.C. The role of periphyton abundance and quality in the microdistribution of a stream grazer, *Helicopsyche borealis* (Trichoptera: Helicopsychidae). Freshwat. Biol. 16:485–493; 1986.

Whiteside, M.C.; Williams, J.B.; White, C.P. Seasonal abundance and pattern of chydorid Cladocera in mud and vegetative habitats. Ecology 59:1177–1188; 1978.

Young, O.W. A limnological investigation of periphyton in Douglas Lake, Michigan. Trans. Am. Microsc. Soc. 64:1–20; 1945.

5. Impact of Submerged Macrophytes on Fish–Zooplankton Interactions in Lakes

Erik Jeppesen, Torben L. Lauridsen, Timo Kairesalo, and
Martin R. Perrow

Introduction

Fish have a major structuring impact on the zooplankton communities in lakes (Hrbaček et al., 1961; Brooks and Dodson, 1965) that may cascade to the lower trophic levels and chemical environment (Carpenter et al., 1985; Carpenter and Kitchell, 1993). Ample evidence is available from enclosure experiments (e.g., Christoffersen et al., 1993), whole-lake experiments (e.g., Shapiro et al., 1975; Benndorf, 1987; Gulati et al., 1990; Carpenter and Kitchell, 1993), and empirical analyses (Jeppesen et al., 1990, 1997). More recently, it has become evident that 0^+ fish may play a key role in zooplankton population dynamics (Cryer et al., 1986; Mills et al., 1987), and some studies suggest that fish larvae are responsible for the midsummer decline in zooplankton (Luecke et al., 1990; Jeppesen et al., 1997), a phenomenon that is often attributed to increased density of inedible phytoplankton such as cyanobacteria (e.g., De Bernardi and Guisanni, 1990). Whole-lake (Søndergaard et al., 1997) and enclosure (He and Wright, 1992) experiments support the structuring role of 0^+ fish. How the importance of top-down control of zooplankton by fish varies along a trophic gradient is debated extensively. McQueen et al. (1986) suggested that the cascading effect of zooplanktivorous fish is stronger in oligotrophic lakes than in eutrophic lakes, but a growing body of literature argues that the cascading effect of fish is greater in eutrophic and hypertrophic lakes with respect to the food web in the classic sense

(Jeppesen et al., 1990, 1997; Leibold, 1990; Sarnelle 1992) and the microbial web (Jeppesen et al., 1992; Jürgens 1994; Jürgens and Jeppesen, this volume, Chapter 16).

In addition to nutrient-dependent fish effects, the role of fish seems also to change with lake depth. A cross-analysis of data and existing empirical relations suggests that plankti-benthivorous fish have a higher impact on the zooplankton in macrophyte-free shallow lakes than in corresponding deep lakes (Jeppesen et al., in press). This is because fish biomass per unit of area at any given nutrient level does not change with mean depth (Hanson and Leggett, 1982; Downing et al., 1990), implying that the biomass per unit of volume, and thus probably also the predation on zooplankton, increases with decreasing mean depth. In addition, vertical migration as an antipredator defense mechanism (Lampert, 1993) is less effective in shallow lakes. Furthermore, the availability of alternative food sources for the fish, such as sediment and bottom fauna, is higher in shallow lakes than in deep lakes because of detritus and thus of higher quality for the benthos of shallow lakes. In lakes with low oxygen in the hypolimnion, parts of the sediment may furthermore be inaccessible to the fish. Consequently, the shallow lake zoo-planktivorous fish biomass may be sustained at high levels by additional alternative food sources, and they can therefore maintain a higher predation pressure on zooplankton than in deep lakes (Jeppesen et al., 1997). In shallow lakes, however, submerged macrophytes may cover large areas, and if abundant, this may alter the interaction strength between fish and pelagic zooplankton as zooplankton may use macrophytes as refuge against predation from fish. The interactions are, however, complex, as the presence of macrophytes also influences the mutual interaction between piscivorous fish and prey fish. For instance, small planktivores use macro-phytes as an antipredator defense mechanism. Macrophytes may also influence the competition between various predatory species and, at the juvenile stages, competition between prey fish and predators (see Persson and Crowder, this volume, Chapter 1). Yet, we far from fully understand how the interactions between fish and zooplankton are influenced by macrophytes. However, certain patterns are emerging.

We first briefly describe the zooplankton communities in macrophyte beds. Thereafter, we discuss how macrophytes may influence the interactions between fish and zooplankton and show how these interactions may have cascading effects on phytoplankton, protozoans, and bacterioplankton. Then, we present a tentative model describing how fish–zooplankton interactions may alter along a nutrient gradient, and finally, we suggest future research needs. We intend to highlight key issues emerging particularly from our own results rather than present a full literature review. In this chapter, the term *littoral zone* does not only comprise the nearshore areas but all areas with plants and nearby open water. Thus, in very shallow lakes the littoral zone may extend to the lake center.

Zooplankton Community in Macrophyte Beds

Crustacean and rotifer communities in plant beds consist of epiphytic, benthic, and pelagic forms. Most plant-associated species in the littoral zone can be categorized as scrapers on solid substrate (e.g., *Eurycercus*), suckers (some rotifers), or graspers (cyclopoid copepods) (Gliwicz and Rybak, 1976). Some forms such as *Sida* collect seston by filtration while fixed to the plants, but they do show a certain plasticity, as they periodically appear free-swimming, especially at night (Vuille, 1991). Others are probably facultative filter feeders (e.g., *Chydorus, Eurycercus,* and *Acroperus*), and others are predators (*Polyphemus*). The ecological role of these plant-associated cladocerans is, however, virtually unknown.

Most, if not all, pelagic zooplankton species may periodically occur in the plant beds. It is characteristic that if a littoral zone is present, cladocerans such as *Ceriodaphnia* spp., *Chydorus sphaericus, Diaphanosoma brachyurum,* and cyclopoid copepods are often more abundant here than in the pelagic zone, whereas rotifer and calanoid copepod densities show the opposite pattern (Gliwicz and Rybak, 1976; Vuille, 1991; Lauridsen et al., 1996). In particular, *Ceriodaphnia* spp. and *D. brachyurum* seem well adapted to plant beds, probably because they are efficient microfiltrators (of bacterioplankton, etc.) (Gliwicz and Rybak, 1976; DeMott, 1986). In addition, *Ceriodaphnia* tolerates low-oxygen conditions (Gliwicz and Rybak, 1976).

The study by Jeppesen et al. (unpublished results) illustrates how habitat choice depends on plant density. In fish-free enclosures in which zooplankton could select between open water and three different plant densities (artificial plastic plants), major differences in relative zooplankton composition and total abundance were found. *Bosnuna longirostris* preferred open water. *Daphnia* spp. were widespread at all plant densities and in open water. *D. brachyurum* abundance was highest at intermediate plant volume infested (PVI), and *Ceriodaphnia* spp. and cyclopoid copepods abundance was highest in the most dense vegetation (Fig. 5.1). In accordance with these results, a cross-analysis of data from 13 lakes conducted by Cyr and Downing (1988) revealed a negative relationship between the abundance of *B. longirostris* and plant density but a positive one with the density or biomass of cyclopoid copepods and other cladocerans. Other studies, however, showed high abundance of *B. longirostris* in the vegetation (Pennak, 1966; Gliwicz and Rybak, 1976; Jakobsen and Johnsen, 1987; Lauridsen et al., 1996). These differences may reflect variations in fish predation pressure (see below). The relative contribution of pelagic zooplankton may also depend on plant bed size, being lower in large beds (Lauridsen et al., 1996) as pelagic zooplankton often aggregate in the transitional zone between plant beds and open water (Lauridsen and Buenk, 1996). In addition. light intensity and water currents may influence the zooplankton distribution (e.g., Kairesalo and Penttilä, 1990). In accordance with Figure 5.1, the abundance (Cyr and Downing, 1988; Paterson, 1993; Phillips et al., 1996) and biomass (Cyr and Downing, 1988) of microcrustaceans often increase with plant biomass but to a varying degree depending on plant type (Paterson, 1993). In the study by Paterson (1993), the number of microcrustaceans per unit of area was an order of magnitude higher in the plant beds (0.5×10^6 m^2) than in open water and plant-free sediment in the littoral zone. Likewise, the average size of the different species is often larger inside the vegetation than outside (Vuille and Maurer, 1991; Lauridsen and Buenk, 1996).

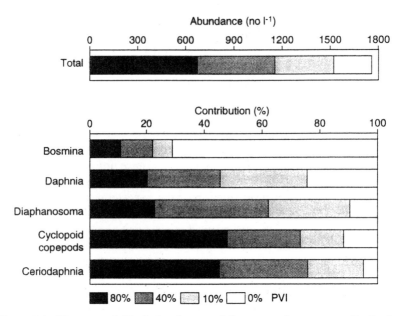

Figure 5.1. (Upper panel) Total abundance and (lower panel) percentage distribution in terms of numbers of various zooplankton in fish-free enclosures with contrasting densities of artificial plants (ivy imitations) (0, 10, 40, and 80% plant volume infested [PVI]) in Lake Stigsholm July 1995. The data represent a day–night average. (Modified from Jeppesen et al., in preparation.)

Increasing invertebrate density with increasing plant density and high densities per square meter in the plant beds have also been observed in the case of macroinvertebrates (Cyr and Downing, 1988; Diehl and Kornijów, this volume, Chapter 2).

Diel Horizontal and Littoral Vertical Migration of Zooplankton

The presence of planktivorous fish in both the pelagic and the littoral zone may alter the horizontal distribution of zooplankton. Some recent studies show daytime aggregation of pelagic zooplankton in the littoral zone (Lauridsen and Buenk, 1996; Stansfield et al., 1997), whereas other studies support the "shore avoidance hypothesis" proposed by Hutchinson (1967). These differences seem to be highly dependent on season, reflecting differences in density and horizontal distribution of fish. Cryer and Townsend (1988) studied the horizontal distribution of pelagic zooplankton during 2 years with contrasting densities of 0⁺ fish and found that in July densities of *Daphnia hyalina* and *B. longirostris* were an order of magnitude higher in the pelagic than the littoral zone in the low-fish year but up to 100-fold more abundant in the littoral than in the pelagic zone when fish density was high (*D. hyalina*). Likewise, Jakobsen and Johnson (1987) recorded an even distribution in the pelagic and littoral zones of *Daphnia longispina* and *Bosmina longispina* in Lake Kvernavatn, Norway, early in the season, but a segregation took place when sticklebacks aggregated in the littoral zone, *D. longispina* moving to the pelagic zone, while *B. longispina,* being smaller and less predation vulnerable, moved to the littoral zone.

The horizontal distribution of zooplankton also varies from day to night. Based on zooplankton studies in shallow Hoveton Great Broad, Timms and Moss (1984) proposed that pelagic zooplankton move into macrophyte beds during daytime, using it as spatial refuge against fish predation, but migrate into open water at night to feed on phytoplankton. Since this study, the existence of diel horizontal migration (DHM) has been confirmed by several studies (Davies, 1985; De Meester et al., 1993; De Stasio, 1993; Lauridsen and Buenk, 1996; Lauridsen et al., 1996; Jeppesen et al., 1992; Lauridsen et al., this volume, Chapter 13). Diel and seasonal studies undertaken in the littoral zone by Kairesalo (1980), Lehtovaara and Sarvala (1984), Walls et al. (1990), and Stansfield et al. (1992) lend further support to DHM. Studies undertaken in several lakes with varying fish densities suggest that the extent of DHM is positively related to the density of planktivorous fish in the pelagic zone (Lauridsen et al., this volume, Chapter 13). Likewise, fish density seems to decide the migrating species and size classes. In Lake Ring, Denmark, where the abundance of planktivorous fish was low, Lauridsen and Buenk (1996) thus found that *Daphnia magna* showed higher levels of DHM than the smaller *Daphnia galeata*. In the more fish-rich Lake Stigsholm, even small species such as *B. longirostris* and *Ceriodaphnia* spp. migrated (Lauridsen et al., 1996; Jeppesen et al., submitted) (Fig. 5.2). In addition, other studies have shown that it is the largest individuals of the various species in particular that undergo DHM (Lauridsen and Buenk, 1996; Pedersen, unpublished results). All these observations are in accordance with the size-efficiency hypothesis (Brooks and Dodson, 1965). The empirical evidence of fish-mediated horizontal migration is confirmed by controlled laboratory and field experiments. Laboratory experiments by Lauridsen and Lodge (1996) showed that the presence of fish (*Lepomis cyanellus*) or kairiomone cue in open water initiated a migration of *D. magna* toward the plant bed. A preliminary experiment in 80 m^2 enclosures in Lake Stigsholm showed that in the presence of planktivorous fish in high densities (13–16 m^2) *Daphnia* spp. sought refuge in the vegetation (Jacobsen et al., 1997). Later, more detailed enclosure studies in this lake have shown that addition of 0^+ perch (*Perca fluviatilis*) to fishless enclosures resulted in a differential degree of habitat shift of zooplankton, dependent on species and development stage (copepods), being particularly high for *Daphnia* spp. (Jeppesen et al., in preparation). DHM can also be an antipredator defense mechanism against invertebrate predators such as *Chaoborus flavicans* (Kvam and Kleiven, 1995), involving chemical cues (Kleiven et al., 1996). Reverse migration (highest densities in the plant beds at night) may be observed if plant-associated invertebrate predators such as odonates are important, as discussed by Lauridsen et al. (this volume, Chapter 13). DHM induced by invertebrate predators will probably be most important in lakes with low fish densities.

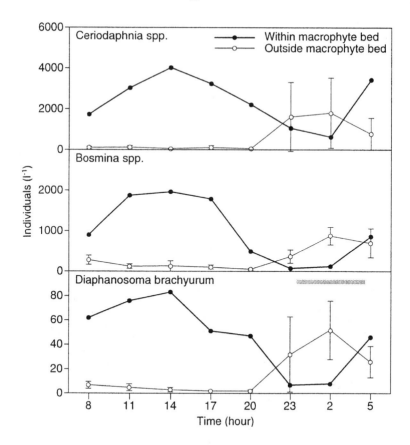

Figure 5.2. Diel variations in the abundance of various cladocerans in a 2-m exclosure (open to small fish and zooplankton) with dense coverage of submerged macrophytes and at a reference station outside the macrophyte bed in Lake Stigsholm in August. Hatched area shows the dark period. (Modified from Lauridsen et al., 1996, by permission of Oxford University Press.)

Macrophytes may also have a repellent effect on various pelagic zooplankters (Hasler and Jones, 1949; Pennak, 1966; Pennak, 1973; Dorgelo and Heykoop, 1985; Lauridsen and Lodge, 1996). This seems appropriate as the plant beds host several plant-associated facultative filtrators that most likely are superior competitors to pelagic zooplankton in a plant community in which the production of periphytic algae is substantially higher than that of phytoplankton (Wetzel, 1975; Wetzel and Søndergaard, this volume, Chapter 7). Habitat choice of pelagic zooplankton therefore seems to be a trade-off between the risk of predation and optimum food conditions that at least partly seem to be regulated by chemical cues from either fish or plants.

Diel vertical migration in the pelagic zone (DVM) is well documented and is most frequently interpreted as being an antipredator defense against predation

(Ringelberg, 1991; Lampert, 1993) that may involve chemical cues (von Elert and Loose, 1996). DVM of both pelagic and benthic crustaceans (e.g., Szlaur, 1963; Whiteside, 1974) may also occur in the plant-rich littoral zone, and various explanations have been offered. In some studies, it has been interpreted as a night-time escape from low oxygen concentrations within the macrophyte beds (Meyers, 1980; Timson and Laybourn-Parry, 1985). Others have argued that DVM in the littoral zone reflects diel variation in predation risk, the predation pressure being released at night due to reduced visibility and offshore migration by fish (Jeppesen et al., in preparation). Further evidence for the importance of fish for DVM is given by De Stasio (1993), who showed that fish removal in a lake almost eliminated DVM in the pelagic as well as in the littoral zone. Likewise, Jeppesen et al. (in preparation) have shown that in the littoral zone especially large-bodied predation-vulnerable zooplankton such as *Daphnia* spp. and adult cyclopoid copepods exhibited DVM, whereas smaller forms such as nauplii and rotifers did not. As expected, DVM was higher in the macrophyte-free area than in the plant bed. In addition, the degree of DVM increased with increasing density of planktivorous fish and decreasing plant density. The studies by De Stasio (1993) and Jeppesen et al. (in preparation) showed substantial DVM despite the samples integrating the entire water column except the lower few centimeters. This suggests that the zooplankton in the littoral zone sought refuge at the sediment surface and in plant beds perhaps also close to plant surfaces.

DVM in the littoral zone has, however, also been observed in the absence of fish by Paterson (1993), who argued that DVM of plant-associated forms was lower in fishless lakes than in fish-rich lakes, which may indicate a lower predation pressure on zooplankton by invertebrate predators than fish. Accordingly, Paterson (1994) found only marginal differences in the density of cladocerans and cyclopoid copepods in experiments run at different densities of some important littoral invertebrate predators (*Odonata, Acari, Tanypodinae*). Also, Johnson et al. (1987) and Blois-Heulin et al. (1990) failed to detect a strong impact of large odonates on cladocerans. By contrast, others have found negative effects of water mites (Kajak et al., 1968) and *Procladius* (Dusoge, 1980).

Refuge Effect in Relation to Structural Complexity and Fish Density

Like zooplankton, small fish may seek refuge in the vegetation to avoid predatory fish and birds (e.g., Carpenter and Lodge, 1986; Gliwicz and Jachner, 1992; Persson and Crowder, this volume, Chapter 1). The prey fish often prefer sparse vegetation (Engels, 1988; Phillips et al., 1996; Jeppesen et al., 1992, 1997; Stansfield et al., 1997), which may reflect that the foraging efficiency of fish decreases with increasing plant density (e.g., Crowder and Cooper, 1982; Savino and Stein, 1982; Anderson, 1984; Diehl, 1988), although there are exceptions to this rule (Winfield, 1986; Persson and Crowder, this volume, Chapter 1). The study in Cromes Broad (Phillips et al., 1996) is an example of aggregation and high predation on zooplankton by planktivorous fish in sparse vegetation (Fig. 5.3). Accordingly,

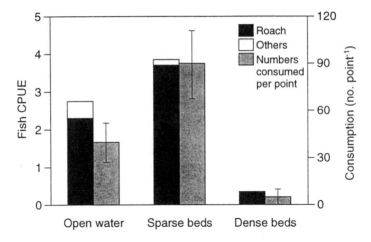

Figure 5.3. Fish community structure in various habitats in Cromes Broad (1993) shown by the mean number of fish captured using point sample electrofishing (left column) and index of point predation (mean ± SE) on zooplankton derived by multiplication of fish density with the mean number of zooplankton in the fish guts. (Data from Phillips et al., 1996.)

zooplankton density was highest in the dense vegetation. Likewise, Lehtovaara and Sarvala (1984) and Kairesalo and Penttilä (1990) found higher fish densities at low density of *Equisetum* than at high density of this plant, with corresponding inverse effects on zooplankton densities.

The presence of predatory fish in the vegetation may further complicate the interaction between planktivorous fish and zooplankton in the littoral zone. On the one hand, the predation pressure on zooplankton may be reduced as the planktivorous fish reduce their activity level (Bean and Winfield, 1995; Jacobsen et al., 1997; L. Jacobsen, unpublished data) or switch to alternative food sources as a consequence of a restricted habitat use (Persson, 1993). As an example, Persson (1993) found that roach in the absence of predators fed on mainly *Bosmina* sp. but switched to detritus/algae in the presence of piscivorous perch. This, in turn, led to an increase in *Bosmina* density. On the other hand, a high predation risk may drive the prey fish to move into even the most dense vegetation (Savino and Stein, 1982; Werner et al., 1983; Persson et al., 1991; Persson and Eklöv, 1995) and accordingly a loss of refuge for large-sized zooplankton may occur (Jeppesen et al., in preparation).

The role of fish density was studied by Schriver et al. (1995), who conducted experiments in 100-m² enclosures with varying PVI and density of 0^+ and 1^+ planktivorous fish (three-spined sticklebacks [*Gasterosteus aculeatus*] and roach [*Rutilus rutilus*]) in Lake Stigsholm, Denmark. They showed a refuge effect when PVI was >15–20% if fish density was below ca. 2/m² (Fig. 5.4). At lower PVI, the zooplankton were dominated by cyclopoid copepods. At fish densities of 2–4/m² and PVI at >15–20%, zooplankton shifted to small cladocerans, and at even higher

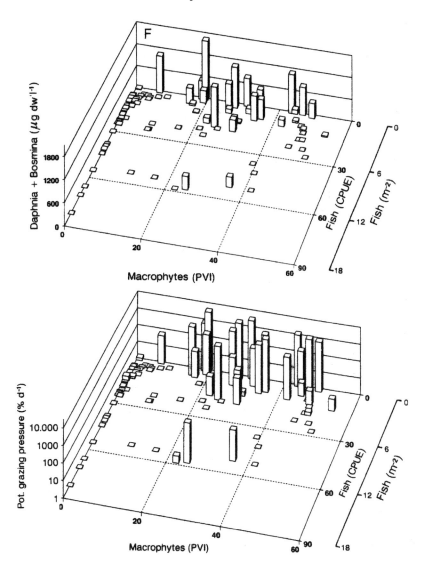

Figure 5.4. Biomass of the dominant pelagic cladocerans (*Bosmina + Daphnia;* upper panel) and their potential grazing pressure on phytoplankton (estimated 24-hour ingestion by *Bosmina + Daphnia* in % of phytoplankton biomass; lower panel) versus the abundance of 0^+ and 1^+ roach and three-spined sticklebacks (catch per unit effort [CPUE] in traps) and macrophyte PVI (%) in Lake Stigsholm enclosure experiments involving manipulation of plants (mainly *Potamogeton* species) and fish density. (From Schriver et al., 1995, published with kind permission of Blackwell Science Ltd.)

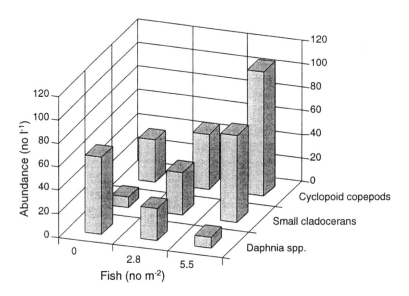

Figure 5.5. Average concentration of various zooplankton in enclosure experiments with high plant biomass (87–165 g/DW/m²) and contrasting fish densities (0 to 5.5 10–12-cm perch m²). The experiment was conducted in hypertrophic Little Mere, England. (Data from Table 2, Beklioglu and Moss, 1996.)

fish densities, to cyclopoid copepods and rotifers. Thus, submerged macrophytes acted as refuge for zooplankton against fish predation if PVI was >15–20%, but the refuge effect almost disappeared—even for small-sized cladocerans—if fish density exceeded a certain threshold level, in this example ca. 4 fish/m². Low refuge effect and dominance by small zooplankton were also observed in the same lake in another experiment with a mixed community of 0⁺ perch and roach (3/m²) and a PVI of 24% (Jeppesen et al., in preparation). A partial loss of refuge for large-sized zooplankton within the vegetation has been observed in other studies. In enclosure experiments with a plant density of 87–165 g dry weight (DW)/m², Beklioglu and Moss (1996) found a significant reduction in *D. hyalina* and *Daphnia cuculata* at densities of 2.8 perch/m² (10–12 cm in length) and a further reduction of daphnids at a fish density of 5.5/m² while the densities of small cladocerans (*Ceriodaphnia* spp., *B. longirostris, Chydorus ovalis*) and cyclopoid copepods increased (Fig. 5.5). No data on total zooplankton biomass are available from this study, but a marked increase in chlorophyll *a* as fish densities increased indicates reduced zooplankton control of phytoplankton. A further evidence of loss of refuge effect for zooplankton is offered by Persson and Eklöv (1995), who used artificial plants with a stem density of 280 m² in enclosures and a fish density of 2.1 m² 0⁺ perch and 1⁺ roach (1:1). During a 48-day experiment, crustacean zooplankton were reduced 40–100-fold to densities <1.6 ind/L, with large *D. longispina,* as well as large plant-associated cladocerans such as *Eurycercus* sp.

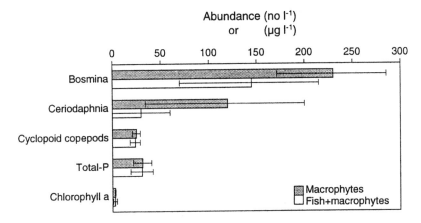

Figure 5.6. Abundance of *Bosmina* spp., *Ceriodaphnia* spp., cyclopoid copepods, total phosphorus, and chlorophyll *a* in macrophyte-rich enclosures (*Elodea canadensis,* 216 g DW m²) without and with fish (3 m² 0⁺ perch). The results stem from the last sampling date of the 21-day-long experiment. (Data from Kairesalo et al., in press.)

reduced to near-zero. In these experiments, no differences were observed whether plants were present or not. Effects of fish presence were also demonstrated by Kairesalo et al. (in press), who found that addition of 3/m² of 0⁺ perch to enclosures with dense beds (216 g DW/m²) of *Elodea canadensis* resulted in a major reduction in the density of *Bosmina* spp. (mainly *B. longispina*) and especially *Ceriodaphnia* spp. (mainly *C. pulchella*) compared with controls without fish, but densities of cyclopoid copepods did not differ (Fig. 5.6).

These last few examples all show that the refuge effect for *Daphnia* spp.—and occasionally also for small-sized cladocerans—even at relatively high plant densities may be partly or totally lost if the density of potentially planktivorous fish exceeds <2–5/m². It is, however, yet to be demonstrated if the same threshold levels also occur in nature. The enclosure size of these experiments was highly variable, ranging from 0.7 m² (Beklioglu and Moss, 1996) to 100 m² (Schriver et al., 1995), but common to all these studies is that they do not allow discernible diel migration of zooplankton and fish between the littoral zone and open water. Lake Stigsholm experiments indicate that such a migration may reduce the strength of the interactions between fish and zooplankton in the vegetation (Jeppesen et al., unpublished data). Both roach and perch fry migrate to open water at night, which in itself may reduce the predation pressure in the littoral zone. This is strengthened by the fact that, for instance, perch feed especially at dusk and dawn outside the littoral zone and are less active during the day (Gliwicz and Jachner, 1992). Thus, in practice, the threshold for loss of refuge effect at high plant density may be higher than indicated by the experiments described above. In sparse beds the fish threshold for loss of refuge effect may be substantially lower (Schriver et al., 1995; Stansfield et al., 1997). The study by Stansfield et al. (1997) suggests, for example,

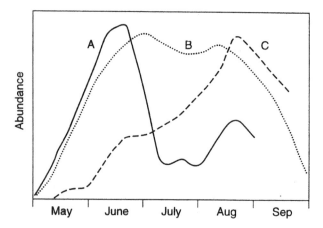

Figure 5.7. Diagrammatic curves that show abundance of littoral zooplankton during the summer months in some northern temperate lakes. Curve A, in which there is a midsummer decline, is the most frequently reported. (From Whiteside, 1988, published with kind permission of E. Schweizerbartsche Verlagsbuchhandlung, Stuttgart.)

a loss of refuge for *Daphnia* spp. at a density of 0.25 0^+ fish/m^2. As densities of 0^+ fish are often higher than 0.25–5/m^2 during the summer (Lehtovaara and Sarvala, 1984; Chick and McIvor, 1994), it is likely that the fish may often have a major structuring impact on zooplankton in the littoral zone. This is supported by the studies of the seasonal patterns of microcrustaceans in the littoral zone. Based on several studies, Whiteside (1988) identified three seasonal patterns (Fig. 5.7): A, a presummer peak followed by low densities during July and by a minor peak in August; B, an almost unimodal pattern with a midsummer maximum; and C, only an autumn peak. Pattern A is very similar to the one observed in the pelagic zone of lakes highly influenced by fish predation (Jeppesen et al., 1997). Whiteside (1988) claimed that predation is responsible for the summer and autumn declines, as there were no arguments in favor of deteriorated food conditions. The bimodal curve (A) is most frequently observed (Whiteside, 1988), indicating that the littoral refuge effect often tends to disappear during midsummer. Subsequent investigations by others support Whiteside's suggestions. Vuille (1991) found a major decline in zooplankton and plant-associated cladocerans in a year with high littoral fish densities and found minor reductions in the preceding year when fish density was low.

Fish-mediated loss of refuge in the littoral zone does not necessarily mean that the predation pressure on zooplankton in the total lake per se is high, because unless the density of older planktivorous fish is high, a predator-mediated aggregation of young fish in the vegetation may lead to a major reduction in the predation on zooplankton in open water. Thus, Boikova (1986) found a reverse migration of crustaceans, with daytime densities being 10–100-fold higher in the pelagic zone than in a dense *Elodea canadensis* plant bed, whereas the crustaceans were more evenly distributed at night. The frequently observed avoidance of the littoral zone (Hutchinson, 1967; Siebeck, 1969, 1980) also indicates that predation pressure in the littoral zone is often higher than in the pelagic zone (Gliwicz and Rybak, 1976; Evans et al., 1980).

Implications for the Lower Trophic Levels

Due to the aggregation of small fish in the littoral zone and the frequently low depth, presumably the overall predation pressure on zooplankton in the plant-free littoral zone may be higher than in the pelagial. Accordingly, the grazing pressure of zooplankton on phytoplankton, protozoans, and bacterioplankton is likely to be lower than in the pelagial. Conversely, if plant density is high and fish are not forced into the vegetation, the cascading effects of grazers may be high due to low fish predation and the daytime aggregation of pelagic zooplankton in the plant beds. The spatial differences in grazing pressure of zooplankton between dense plant beds and open water in the littoral zone are, therefore, expected to be particularly large. This is supported by a few investigations undertaken so far. Jeppesen et al. (in preparation) found that in shallow eutrophic Lake Stigsholm, the daily clearance rates of phytoplankton ranged from very low values of 2% in the littoral zone outside the plant beds to 3.2% in sparse vegetation (PVI = 24%) and to values as high as 300% in dense vegetation (PVI = 50%), whereas the corresponding figures for bacterioplankton clearance were 2.5, 4, and 219%, respectively (Fig. 5.8). The clearance rates were thus 144-fold (phytoplankton) and 88-fold (bacterioplankton) higher in the dense beds than in the macrophyte-free littoral. Such high clearance rates in dense vegetation had significant cascading effects on the trophic structure within the bed. For example, the densities of ciliates, phytoplankton, flagellates, and bacterioplankton were 75-, 4.4-, 4-, and 3-fold, respectively, lower inside than outside the beds (Fig. 5.8) (see also Jeppesen et al., submitted; Søndergaard and Moss, this volume, Chapter 6; Søndergaard et al., this volume, Chapter 15). The significant role played by zooplankton in the determination of these differences was confirmed by size fractionation experiments (Jürgens and Jeppesen, this volume, Chapter 16). Earlier empirical data from experiments involving dense beds of submerged macrophytes and low fish densities have also shown high zooplankton:phytoplankton biomass ratios, suggesting a great cascading impact on the lower trophic levels (Irvine et al., 1989; Moss et al., 1994; Schriver et al., 1995; see also Søndergaard and Moss, this volume, Chapter 6), whereas high fish densities resulted in low ratios inside and especially outside the littoral vegetation (Schriver et al., 1995). The high cascading effect on the lower trophic level within the dense plant beds may not necessarily reflect the role of aggregating pelagic zooplankton. The beds host plant-associated filter feeders such as *Sida,* which may have a potentially high grazing impact (Stansfield et al., 1997). The plant beds also host several filter-feeding macroinvertebrates including mussels with high filtering capacity (Ogilvie and Mitchell, 1995). In addition, the plants and epiphytes may help control the phytoplankton by shading (Straskraba and Pieczynska, 1970) or cause nutrient

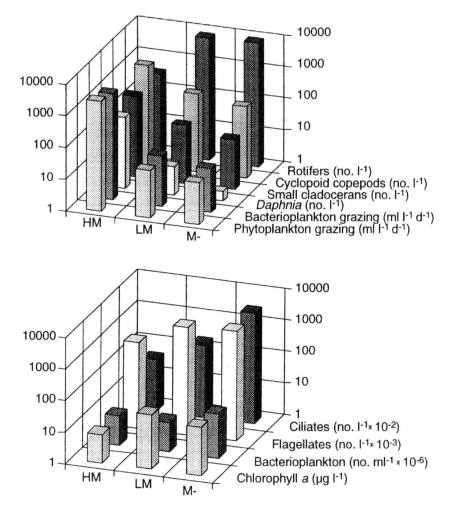

Figure 5.8. Abundance (no. L^{-1}) of various zooplankton and microbial components, chlorophyll a (μg L^{-1}), and zooplankton (>140 μm) clearance (ml/L/day) of phytoplankton and bacterioplankton in Lake Stigsholm enclosures with contrasting densities of submerged macrophytes (HM, plant volume infested (PVI) = 50%; LM, PVI = 24%; M-, without macrophytes). Note the logarithmic scale. (Data from Jeppesen et al., submitted.)

limitation (Kairesalo et al., in press). Multiple factors may thus contribute to the relatively high zooplankton:phytoplankton biomass ratio and the stronger top-down control of phytoplankton in the dense plant beds with low fish densities.

Although aggregation of zooplankton in the vegetation has a substantial impact on both the trophic structure and phytoplankton biomass within the plant bed, we do not know the extent to which night-time migration to the pelagic zone will influence open water phytoplankton. Lauridsen et al. (1996) estimated that a 3% coverage of the lake surface area with 2-m diameter patches of dense *Potamogeton pectinatus* beds is sufficient for a night-time doubling of the density of *Ceriodaphnia* spp. and *B. longirostris* in Lake Stigsholm with a low natural plant coverage, and this must be assumed to have a great impact on the phytoplankton grazing pressure. It does, however, require that the zooplankton are widely spread throughout the pelagic zone during night. There are still no measurements of how far zooplankton migrate horizontally. DVM studies have shown night-time migration of >30 m (e.g., Geller et al., 1992) and a mean migration velocity of 0.2 cm/sec for large-sized *Daphnia* species and at maximum speed as much as .5–2 cm/sec (S.I. Dodson, personal communication). If these figures can be transferred to DHM, the zooplankton may be able to exploit the entire pelagic zone in many shallow lakes. An additional consideration is that plants in shallow lakes are often not restricted to the nearshore area but can be found in patches in large parts or in somewhat deeper areas of the lake, further increasing the possibilities of night-time exploitation of the pelagic zone. We, therefore, predict that in many shallow lakes the possibility of seeking refuge in the vegetation during the day increases grazer control of the phytoplankton in open water and thus contributes to maintaining those lakes with comprehensive and dense macrophyte coverage in the clearwater state. The establishment of macrophyte refuges protected from waterfowl grazing has been proposed as a restoration measure to complement nutrient-loading reductions in shallow lakes (Moss, 1990; Jeppesen et al., 1991). Establishment of numerous small and dense refuges should therefore result in much higher densities of migrating cladocerans than a few large refuges. The higher density of cladocerans will ensure a greater filtering capacity within the beds during the day and in open water during the night. Per unit area, small and dense macrophyte refuges may be better able to promote a shift to a clearwater stage than larger ones with low macrophyte density (Lauridsen et al., 1996; Jeppesen et al., 1997).

Changes in Interactions Along a Nutrient Gradient

Changes in nutrient levels affect both the abundance and composition of macrophytes (Wetzel, 1975) and fish (Persson et al., 1991; Jeppesen et al., 1997), which will alter the zooplankton refuge efficiency of the plants against fish predation. Empirical data are scarce, but we propose various hypotheses that may help initiate future discussions and tests. We restrict ourselves to northern European lakes.

With increasing nutrient levels, the depth limit of submerged macrophytes decreases (Chambers and Kalff, 1985), but at the same time the biomass per unit of area and stem density of plants increase (Wetzel, 1975). This supposedly leads to a reduction of the total refuge area for zooplankton, but in the remaining plant-filled areas the refuge effect will increase. Whether increased density compensates for lower area coverage is open to debate, but with the step-like increases observed by Schriver et al. (1995) in the refuge effect at high plant density, it may be presumed that this is indeed the case. At the highest nutrient levels, submerged macrophytes most often totally disappear due to light limitation caused by phytoplankton (but see Moss et al., 1997), leaving only floating-leaved plants and reed belts, which often have a comparatively low refuge effect (Gliwicz and Rybak, 1976; Winfield, 1986; Venugopal and Winfield, 1993) due to low stem density. Presumably, therefore, the refuge effect at a fixed density of planktivorous fish is greatest in slightly eutrophic lakes in which plant density is high and the area covered not yet severely reduced.

Simultaneously with the changes in density and distribution of plants, there are also plant composition changes. In northern temperate lakes, the succession is often from characeans to elodeids in hardwater lakes and from isotids to elodeids in softwater lakes. Isoetids are small dense rosette plants that may act as an efficient refuge among the leaves. However, due to small stature, their general effect as a refuge is probably poor. Characeans often form dense beds with a high areal biomass. Potentially, they may therefore act as an efficient refuge against predation by fish: Diehl (1988) has shown that high density of characeans results in high density of macroinvertebrates. We do not, however, know if this is true for zooplankton. Elodeids may also appear in high densities, but the biomass per unit volume is often smaller than that of characeans (Diehl, 1988), indicating that characeans potentially may act as a better refuge than elodeids.

The picture is further complicated by changes in the density and the relative contribution of some fish species along the nutrient gradient in lakes (Persson et al., 1988; Jeppesen et al., 1990). In eutrophic and hypereutrophic lakes, planktibenthivorous fish such as roach and bream dominate, being the most efficient fouragers in the pelagic habitat. The preference for the pelagial may be further strengthened by the fact that plant-associated pike (*Esox lucius* L.) is often the dominant piscivore in such lakes (Grimm and Backx, 1990), making it less favorable for prey fish to forage in the vegetation. The pattern will be different if the dominant species is the pelagic forager zander (*Stizostedion lucioperca*), but this species is often not abundant in lakes with extensive growth of submerged macrophytes. Thus, the efficiency of plant beds as a refuge for zooplankton will often be high in eutrophic lakes. The aggregation of zooplankton is further strengthened by the high risk of predation in the pelagic zone.

In less eutrophic to mesotrophic lakes, the impact of predatory fish increases, perch become more important, and the foraging conditions for predatory fish improve due to, for instance, increased transparency. Prey fish thus seek refuge in the vegetation. As planktivorous fish density is relatively high, it is expected that aggregation of young planktivorous fish in the vegetation will be particularly high

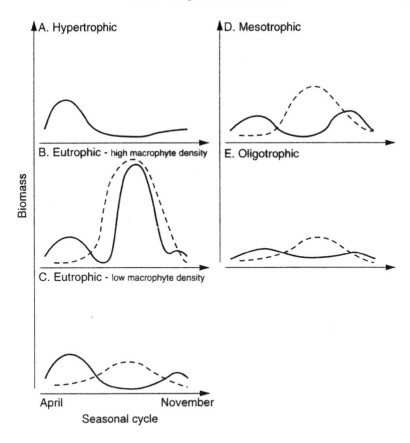

Figure 5.9. Conceptual model showing how the seasonal dynamics of microcrustacean biomass (solid line) in the littoral zone is expected to vary under different nutrient conditions. The broken line represents the average biomass of submerged macrophytes in plant-covered areas.

and the refuge effect correspondingly low. In mesotrophic-oligotrophic lakes, the predator control of planktivorous fish increases. The aggregation is expected to remain high due to a larger share of predatory fish, but the lower abundance of planktivores and lower fish density counteract the effect. The question is if the refuge effect at a given plant density will be higher or lower than in slightly eutrophic lakes. In oligotrophic lakes, the refuge effect for prey fish due to low plant density and plant height will probably be poor, so that the refuge effect for zooplankton may increase.

It is difficult to combine these multiple and complex interactions into a common conceptual model, but we tentatively predict (Fig. 5.9) that

- In hypertrophic lakes without submerged macrophytes, the refuge effect for zooplankton in the littoral zone dominated by reeds is poor throughout the summer. The density of zooplankton and to a lesser extent plant-associated crustaceans is low and dominated by small forms (Fig 5.9A).

- In eutrophic lakes with high areal coverage of submerged macrophytes and high PVI, the refuge effect will be high throughout the summer period, during which planktivorous fish stay in the pelagic zone or in the sparse vegetation. The density of macrocrustaceans in the plant bed will be high, apart from a period during early summer before macrophyte density has become high. (Fig. 5.9B).
- In eutrophic lakes with low PVI, the pattern approaches the one suggested for hypereutrophic lakes (Fig. 5.9C).
- In meso-slightly eutrophic lakes, 0^+ fish seek refuge in the vegetation during mid-summer, and Whiteside's type A (bimodal) or C (unimodal with autumn maximum) response of microcrustacean density (see Fig. 5.7) can be observed, depending on whether the abundance of planktivorous fish (≥ 1 year) is low or high. Very high plant density (e.g., of characeans) may result in a response resembling that of eutrophic lakes. (Fig. 5.9D).
- In oligotrophic lakes, the refuge effect will often be low due to low plant height (Fig. 5.9E).

The patterns described cannot be applied to brackish lakes in northern Europe that deviate substantially in trophic structure and dynamics from freshwater lakes (Leah et al., 1978; Jeppesen et al., 1994, this volume, Chapter 31). Several factors indicate that the refuge effect for zooplankton in nutrient-rich brackish lakes is poor due to aggregation of both sticklebacks and *Neomysis* in the plant beds (Jeppesen et al., this volume, Chapter 31).

Future Research Needs

Although much new information has appeared during recent years about the impact of macrophytes on fish–zooplankton interactions, there are still several unclarified questions. We know that DHM does occur in lakes, but little is known about the distances covered by the zooplankton. This is interesting from a theoretical point of view, but it also has practical implications. As mentioned, the use of macrophyte implantations has been suggested as a restoration tool in lakes. To ensure optimum placement of these macrophyte enclosures to obtain the highest night-time grazing effect of zooplankton in the pelagic zone, more information on the potential migration distances is needed. Also, we know little about how the zooplankton find their way back from the pelagic to the plant-covered areas during the daytime. Moreover, it is unknown what the migration pattern would look like if zooplankton are influenced by both fish and invertebrate predators inhabiting spatially segregated areas (e.g., fish in the pelagic zone and odonates, water mites, etc., in the littoral zone). In addition, cost–benefit analyses are required. When will horizontal migration be cost-efficient and how do such estimates alter along a gradient in food supply and predation risk in the vegetation and the pelagic zone?

What influence does the size of the plant-filled areas have? To answer these questions, we suggest both intensive laboratory and field studies, including detailed studies of zooplankton population dynamics and the use of a modeling approach.

Most of the former studies of the interactions between fish, macrophytes, and zooplankton were undertaken with only one or two species of prey fish and typically one species of predatory fish. Because interactions between a given fish species and zooplankton may change in the presence of other species (Persson and Eklöv, 1995), there is a great need for multispecies experiments. A better insight into the interactions between fish and zooplankton also requires a more thorough knowledge of the feeding behavior of the various zooplanktivorous fish. This is crucial because it is not possible from data on distribution alone to determine where and when the interactions between fish and zooplankton are particularly strong.

It is necessary to undertake investigations under more natural conditions than hitherto has been the case. This means on a scale that allows horizontal migration for both zooplankton and fish (i.e., a large scale and whole-lake basis). Empirical studies of seasonal and diel variations in fish and zooplankton in the pelagic and littoral zones, with contrasting nutrient levels, fish communities, and macrophyte abundance and composition, may contribute to the understanding of natural interactions. Also, the inclusion of quantitative paleoecological investigations, including reconstruction of fish and submerged macrophytes (Jeppesen et al., 1996), will add to our understanding, and it may increase our knowledge about long-time perturbations, factors that are poorly elucidated from the existing short monitoring series and short-term enclosure and whole-lake experiments (Anderson, 1995). We need to know more about the role of plant-associated microcrustaceans and their interactions with zooplankton staying temporarily or permanently in the open water among the plants. Stable isotope (^{13}C and ^{15}N) analyses and grazing measurements on radio-labeled periphyton may be useful methods in such studies. Finally, there is a great need for studies of how density and composition of submerged macrophytes affect the interaction between fish, *Neomysis,* and zooplankton in brackish lakes.

Acknowledgments. The assistance of the technical staff of the National Environmental Research Institute, Silkeborg, is gratefully acknowledged. Technical assistance was provided by K. Møgelvang and A.M. Poulsen. The study was supported by the Centre for Freshwater Environmental Research. We thank Ramesh Gulati, John Iwan Jones, Brian Moss, Greg Cronin, and Martin Søndergaard for valuable comments on the manuscript.

References

Anderson, N.J. Temporal scale, phytoplankton ecology and palaeoecology. Freshwat. Biol. 34:367–378; 1995.

Anderson, O. Optimal foraging by largemouth bass in structured environments. Ecology 65:851–861; 1984.

Bean, C.W.; Winfield, I.J. Habitat use and activity patterns of roach (*Rutilus rutilus*), rudd (*Scardinius erythrophtalmus* (L.)), perch (*Perca fluviatilis*) and pike (*Esox lucius*) in the laboratory: the role of predation threat and structural complexity. Ecol. Freshwat Fish 4:37–46; 1995.

Beklioglu, M.; Moss, B. Mesocosm experiments on the interaction of sediment influence, fish predation and aquatic plants with the structure of phytoplankton and zooplankton communities. Freshwat. Biol. 36:315–325; 1996.

Benndorf, J. Food web manipulation without nutrient control: a useful strategy in lake restoration? Schweiz. Z. Hydrol. 49:237–248; 1987.

Blois-Heulin, C.; Crowley, P.H.; Arrington, M.; Johnson, D.M. Direct and indirect effects of predators on the dominant invertebrates of two freshwater littoral communities. Oecologia 84:295–306; 1990.

Boikova, O.S. Horizontal distribution of crustaceans in Lake Glubokoe. Hydrobiologia 141:113–123; 1986.

Brooks, J.L.; Dodson, S.I. Predation, body size and composition of plankton. Science 150:28–35; 1965.

Carpenter, S.R.; Kitchell, J.F., eds. The trophic cascade in lakes. Cambridge: Cambridge University Press; 1993.

Carpenter, S.R.; Lodge, D.M. Effects of submersed macrophytes on ecosystem processes. Aquat. Bot. 26:341–370; 1986.

Carpenter, S.R.; Kitchell, J.F.; Hodgson, J.R. Cascading trophic interactions and lake productivity. BioScience 35:635–639; 1985.

Chambers, P.A.; Kalff, J. Depth distribution and biomass of submersed aquatic macrophyte communities in relation to Secchi depth. Can. J. Fish. Aquat. Sci. 47:1929–1936; 1985.

Chick, J.H.; McIvor, C.C. Patterns in the abundance and composition of fishes among beds of different macrophytes: viewing a littoral zone as a landscape. Can. J. Fish. Aquat. Sci. 51:2873–2882; 1994.

Christoffersen, K.; Riemann, B.; Klysner, A.; Søndergaard, M. Potential role of zooplankton in structuring a plant community in eutrophic lake water. Limnol. Oceanogr. 38:561–573; 1993.

Crowder, L.B.; Cooper, W.E. Habitat structural complexity and the interaction between bluegill and their prey. Ecology 63:1802–1813; 1982.

Cryer, M.; Townsend, C.R. Spatial distribution of zooplankton in a shallow eutrophic lake, with a discussion of its relation to fish predation. J. Plankton Res. 10:487–501; 1988.

Cryer, M.; Pierson, G.; Townsend, C.R. Reciprocal interactions between roach *Rutilus rutilus,* and zooplankton in a small lake: prey dynamics and fish growth and recruitment. Limnol. Oceanogr. 31:1022–1038; 1986.

Cyr, H.; Downing, J.A. Empirical relationships of phytomacrofaunal abundance to plant biomass and macrophyte bed characteristics. Can. J. Fish. Aquat. Sci. 45:976–984; 1988.

Davies, J. Evidence for a diurnal horizontal migration in *Daphnia hyalina lacustris* Sars. Hydrobiologia 120:103–105; 1985.

De Bernardi, R.; Guisanni, G. Are blue-green algae suitable food for zooplankton? An overview. Hydrobiologia 200/201:29–41; 1990.

De Meester, L.; Maas, S.; Dierckens, K.; Dumont, H.J. Habitat selection of patchiness in *Scapholeberis:* horizontal distribution and migration of *S. mucronata* in a small pond. J. Plankton Res. 15:1129–1139; 1993.

DeMott, W.R. The role of taste in food selection by freshwater zooplankton. Oecologia 69:224–240; 1986.

De Stasio, B.T., Jr. Diel vertical and horizontal migration by zooplankton: population budgets and the diurnal deficit. Bull. Mar. Sci. 53:44–64; 1993.

Diehl, S. Foraging efficiency of three freshwater fishes: effects of structural complexity and light. Oikos 53:207–214; 1988.

Dorgelo, J.; Heykoop, M. Avoidance of macrophytes by *Daphnia longispina.* Verh. Int. Verein. Limnol. 22:3369–3372; 1985.

Downing, J.A.; Plante, C.; Lalonde, S. Fish production correlated with primary productivity and the morphoedaphic index. Can. J. Fish. Aquat. Sci. 47:1929–1936; 1990.

Dusoge, K. The occurrence and role of the predatory larvae of Procladius Skuse (Chironomidae, Diptera) in the benthos of Lake Sniardy. Ekol. Pol. 28:155–186; 1980.

Engels, S. The role and interactions of submersed macrophytes in a shallow Wisconsin lake. J. Freshwat. Ecol. 4:329–340; 1988.

Evans, M.S.; Bethany, E.H.; Sell, D.W. Seasonal features of zooplankton assemblages in the nearshore area of southeastern Lake Michigan. J. Great Lakes Res. 6:275–289; 1980.

Geller, W.; Pinto-Coelho, R.; Pauli, H.-R. The vertical distribution of zooplankton (Crustacea, Rotatoria, Ciliata) and their grazing over the diurnal and seasonal cycles in Lake Constance. Arch. Hydrobiol. Beih. Ergebn. Limnol. 35:79–85; 1992.

Gliwicz, Z.M.; Jachner, A. Diel migration of juvenile fish: a ghost of predation past or present. Arch. Hydrobiol. 124:385–410; 1992.

Gliwicz, Z.M.; Rybak, J.I. Zooplankton. In: Pieczynska, E., ed. Selected problems of lake littoral ecology. Warsaw: University of Warsaw; 1976:69–96.

Grimm, M.P.; Backx, J.J.G.M. The restoration of shallow eutrophic lakes, and the role of northern pike, aquatic vegetation and nutrient concentration. Hydrobiologia 200/201: 557–566; 1990.

Gulati, R.D.; Lammens, E.H.R.R.; Meijer, M.-L.; van Donk, E. Biomanipulation tool for water management. Hydrobiologia 200/201:1–628; 1990.

Hanson, J.R.; Leggett, W.C. Empirical prediction of fish biomass and weight. Can. J. Fish. Aquat. Sci. 39:257–263; 1982.

Hasler, A.D.; Jones, E. Demonstration of the antagonistic action of large aquatic plants on algae and rotifers. Ecology 30:359–364; 1949.

He, X.; Wright, R. An experimental study of piscivore–planktivore interactions: population and community responses to predation. Can. J. Fish. Aquat. Sci. 49:1176–1185; 1992.

Hrbaček, J.; Dvorakova, M.; Korinek, M.; Prochazkova, L. Demonstration of the effect of the fish stock on the species composition of zooplankton and the intensity of metabolism of the whole plankton association. Verh. Int. Verein. Limnol. 14:192–195; 1961.

Hutchinson, G.E. A treatise on limnology II. New York: J. Wiley & Sons; 1967.

Irvine, K.; Moss, B.; Balls, H. The loss of submerged plants with eutrophication II. Relationships between fish and zooplankton in a set of experimental ponds, and conclusions. Freshwat. Biol. 22:89–107; 1989.

Jacobsen, L.; Perrow, M.R.; Landkildehus, F.; Hjørne, M.; Lauridsen, T.L.; Berg, S. Interactions between piscivores, zooplanktivores and zooplankton in submerged macrophytes: preliminary observations from enclosure and pond experiments. Hydrobiologia 342/343:197–205; 1997.

Jakobsen, P.J.; Johnsen, G.H. The influence of predation on horizontal distribution of zooplankton species. Freshwat. Biol. 17:501–507; 1987.

Jeppesen, E.; Jensen, J.P.; Kristensen, P.; Søndergaard, M.; Mortensen, E.; Sortkjær, O.; Olrik, K. Fish manipulation as a lake restoration tool in shallow, eutrophic, temperate lakes. 2: Threshold levels, long-term stability and conclusions. Hydrobiologia 200/201: 219–227; 1990.

Jeppesen, E.; Kristensen, P.; Jensen, J.P.; Søndergaard, M.; Mortensen, E.; Lauridsen, T.L. Recovery resilience following a reduction in external phosphorus loading of shallow, eutrophic Danish lakes: duration, regulating factors and methods for overcoming resilience. Mem. Ist. Ital. Idrobiol. 48:127–148; 1991.

Jeppesen, E.; Sortkjær, O.; Søndergaard, M.; Erlandsen, M. Impact of a trophic cascade on heterotrophic bacterioplankton production and biomass in two shallow fish-manipulated lakes. Arch. Hydrobiol. Ergebn. Limnol. 37:219–231; 1992.

Jeppesen, E.; Søndergaard, M.; Kanstrup, E.; Petersen, B.; Henriksen, R.B.; Hammershøj, M.; Mortensen, E.; Jensen, J.P.; Have, A. Does the impact of nutrients on the biological structure and function of brackish and freshwater lakes differ? Hydrobiologia 275/276: 15–30; 1994.

Jeppesen, E.; Agerbo Madsen, E.; Jensen, J.P.; Anderson, N.J. Reconstructing the past density of planktivorous fish and trophic structure from sedimentary zooplankton fossils: a surface sediment calibration data set from shallow lakes. Freshwat. Biol. 36:115–127; 1996.

Jeppesen, E.; Jensen, J.P.; Søndergaard, M.; Lauridsen, T.; Pedersen, L.J.; Jensen, L. Top-down control in freshwater lakes: the role of nutrient state, submerged macrophytes and water depth. Hydrobiologia 342/343:151–164; 1997.

Johnson, D.M.; Pierce, C.L.; Martin, T.H.; Watson, C.N.; Bohanan, R.E.; Crowley, P.H. Prey depletion by odonate larvae: combining evidence from multiple field experiments. Ecology 68:1459–1465; 1987.

Jürgens, K. Impact of *Daphnia* on planktonic microbial food webs—a review. Microb. Food Webs 8:295–324; 1994.

Kairesalo, T. Diurnal fluctuations within a littoral plankton community in oligotrophic Lake Pääjärvi, southern Finland. Freshwat. Biol. 10:533–537; 1980.

Kairesalo, T.; Penttilä, S. Effect of light and water flow on the spatial distribution of littoral *Bosmina longispina* Leydig (Cladocera). Verh. Int. Verein. Limnol. 24:682–687; 1990.

Kairesalo, T.; Tátrai, I.; Luokkanen, E. Impacts of waterweed (*Elodea canadensis* Michx) on plankton-fish interactions in lake littoral. Verh. Int. Verein. Limnol. (in press).

Kajak, Z.; Dusoge, K.; Stanczykowska, A. Influence of mutual relations of organisms, especially Chironomidae, in natural benthic communities, on their abundance. Ann. Zool. Fenn. 5:49–56; 1968.

Kleiven, O.T.; Larssen, P.; Hobæk, A. Direct distributional response in *Daphnia pulex* to a predator kairomone. J. Plankton Res. 18:1341–1348; 1996.

Kvam, O.V.; Kleiven, O.T. Diel horizontal migration and swarm formation in *Daphnia* in response to *Chaoborus*. Hydrobiologia 307:177–184; 1995.

Lampert, W. Ultimate causes of diel vertical migration of zooplankton: new evidence for the predator-avoidance hypothesis. Arch. Hydrobiol. Beih. Ergebn. Limnol. 39:79–88; 1993.

Lauridsen, T.L.; Buenk, I. Diel changes in the horizontal distribution of zooplankton in the littoral zone of two shallow eutrophic lakes. Arch. Hydrobiol. 137:161–176; 1996.

Lauridsen, T.L.; Lodge, D. Avoidance by *Daphnia magna* fish and macrophytes: chemical cues and predator-mediated use of macrophyte habitat. Limnol. Oceanogr. 41:794–798; 1996.

Lauridsen, T.L.; Pedersen, L.J.; Jeppesen, E.; Søndergaard, M. The importance of macrophyte bed size for cladoceran composition and horizontal migration in a shallow lake. J. Plankton Res. 18:2283–2294; 1996.

Leah, R.T.; Moss, B.; Forrest, D.E. Experiments with large snails in a fertile, shallow, brackish lake. Hickling Broad, United Kingdom. Int. Rev. Ges. Hydrobiol. 63:291–310; 1978.

Lehtovaara, A.; Sarvala, J. Seasonal dynamics of total biomass and species composition of zooplankton in the littoral of an oligotrophic lake. Verh. Int. Verein. Limnol. 22:805–810; 1984.

Leibold, M.A. Resource edibility and the effects of predators and productivity on the outcome of trophic interactions. Am. Nat. 134:922–949; 1990.

Luecke, C.; Vanni, M.J.; Magnuson, J.J.; Kitchell, J.F.; Jacobson, P.J. Seasonal regulation of *Daphnia* populations by planktivorous fish: implications for the clearwater phase. Limnol. Oceanogr. 35:1718–1733, 1990.

McQueen, D.J.; Post, J.R.; Mills, E.L. Trophic relationships in freshwater pelagic ecosystems. Can. J. Fish. Aquat. Sci. 43:1571–1581; 1986.

Meyers, D.G. Diurnal vertical migration in aquatic microcrustaceans: light and oxygen responses of littoral zooplankton. In: Kerfoot, W.C., ed. Evolution and ecology of zooplankton communities. Hanover, NH: University Press of New England; 1980: 80–90.

Mills, E.L.; Forney, J.L.; Wagner, K.J. Fish predation and its cascading effect on the Oneida Lake food chain. In: Kerfoot, W.C.; Sih, A., eds. Predation: direct and indirect effects on aquatic communities. Hanover, NH: University Press of New England; 1987:118–131.

Moss, B. Engineering and biological approaches to the restoration from eutrophication of shallow lakes in which aquatic plant communities are important components. Hydrobiologia 200/201:367–378; 1990.

Moss, B.; McGowan, S.; Carvalho, L. Determination of phytoplankton crops by top-down and bottom-up mechanisms in a group of English lakes, the West Midland Meres. Limnol. Oceanogr. 39:1020–1029; 1994.

Moss, B.; Beklioglu, M.; Carvalho, L.; Kilinc, S.; McGowan, S.; Stephen, D. Vertically challenged limnology: contrasts between deep and shallow lakes. Hydrobiologia 342/343:257–267; 1997.

Ogilvie, S.H.; Mitchell, S.F. A model of mussel filtration in a shallow New Zealand lake, with reference to eutrophication control. Arch. Hydrobiol. 133:471–482; 1995.

Paterson, M. The distribution of microcrustacea in the littoral zone of a freshwater lake. Hydrobiologia 263:173–183; 1993.

Paterson, M. Invertebrate predation and seasonal dynamics of microcrustaceans in the littoral zone of a fishless lake. Arch. Hydrobiol. 99:1–36; 1994.

Pennak, R.W. Structure of zooplankton populations in the littoral macrophyte zone of some Colorado lakes. Trans. Am. Microsc. Soc. 85:329–349; 1966.

Pennak, R.W. Some evidence for aquatic macrophytes as repellents for a limnetic species of *Daphnia*. Int. Rev. Ges. Hydrobiol. 58:569–576; 1973.

Persson, L. Predator-mediated competition in prey refuges: the importance of habitat dependent prey resources. Oikos 68:12–22; 1993.

Persson, L.; Eklöv, P. Prey refuges affecting interactions between piscivorous perch and juvenile perch and roach. Ecology 76:763–784; 1995.

Persson, L.; Anderson, G.; Hamrin, S.F.; Johansson, L. Predation regulation and primary production along the productivity gradient of temperature lake ecosystems. In: Carpenter, S.R., ed. Complex interactions in lake communities. New York: Springer-Verlag; 1988:45–65.

Persson, L.; Diehl, S.; Johanson, L.; Andersson, G.; Hamrin, S.F. Shifts in fish communities along the productivity gradient of temperate lake-patterns and the importance of size-structured populations. J. Fish Biol. 38:281–293; 1991.

Phillips, G.L.; Perrow, M.R.; Stansfield, J. Manipulating the fish–zooplankton interaction in shallow lakes: a tool for restoration. In: Greenstreet, S.P.R.; Tasker, M.L., eds. Aquatic predators and their prey. Oxford: Blackwell Scientific Publications; 1996:174–183.

Ringelberg, J. Enhancement of the phototactic reaction in *Daphnia hyalina* by a chemical mediated by juvenile perch (*Perca fluviatilis*). J. Plankton Res. 13:17–25; 1991.

Sarnelle, O. Nutrient enrichment and grazer effects of phytoplankton in lakes. Ecology 73:551–560; 1992.

Savino, J.; Stein, R.A. Predator–prey interaction between largemouth bass and bluegills as influenced by simulated, submersed vegetation. Trans. Am. Fish. Soc. 111:255–266; 1982.

Schriver, P.; Bøgestrand, J.; Jeppesen, E.; Søndergaard, M. Impact of submerged macrophytes on the interactions between fish, zooplankton and phytoplankton: large-scale enclosure experiments in a shallow lake. Freshwat. Biol. 33:255–270; 1995.

Shapiro, J.; Lamarra, V.; Lynch, M. Biomanipulation: an ecosystem approach to lake restoration. In: Brezonik, P.L.; Fox, J.F., eds. Water quality management through biological control. Gainesville, FL: University of Florida; 1975:85–96.

Siebeck, O. Spatial orientation of planktonic crustaceans. 1. The swimming behaviour on a horizontal plane. Verh. Int. Verein. Limnol. 17:831–840; 1969.

Siebeck, O. Optical orientation of pelagic crustaceans and its consequence in the pelagic and the littoral zones. In: Kerfoot, W.C., ed. Evolution and ecology of zooplankton communities. Hanover, NH: University Press of New England; 1980:28–38.

Søndergaard, M.; Jeppesen, E.; Berg, S. Pike (*Esox lucius* L.) stocking as a biomanipulation tool. 2. Effects on lower trophic levels in lake Lyng (Denmark). Hydrobiologia 342/343:319–325; 1997.

Stansfield, J.H.; Perrow, M.R.; Tench, L.D.; Jowitt, A.J.D.; Taylor, A.A.L. Submerged macrophytes as refuges for grazing *Cladocera* against fish predation: observations on seasonal changes in relation to macrophyte cover and predation pressure. Hydrobiologia 342/343:229–240; 1997.

Straskraba, M.; Pieczynska, E. Field experiments on shading effect by emergents on littoral phytoplankton and periphyton production. Rozpravy Ceskoslovenské Akademie Vêd. Iada Matematickych a Přírodnich Vêd. 80:7–30; 1970.

Szlauer, L. Diurnal migrations of minute invertebrates inhabiting the zone of submerged hydrophytes in a lake. Schweiz. A. Hydrol. 25:56–64; 1963.

Timms, R.M.; Moss, B. Prevention of growth of potentially dense phytoplankton populations by zooplankton grazing in the presence of zooplanktivorous fish, in a shallow wetland ecosystem. Limnol. Oceanogr. 29:472–486; 1984.

Timson, S.; Laybourn-Parry, J. The behavioural responses and tolerance of freshwater benthic cyclopoid copepods to hypoxia and anoxia. Hydrobiologia 127:257–263; 1985.

Venugopal, M.N.; Winfield, I.J. The distribution of juvenile fishes in a hypereutrophic pond: can macrophytes potentially offer a refuge for zooplankton? J. Freshwat. Ecol. 8:389–396; 1993.

von Elert, E.; Loose, C.J. Predator-induced diel vertical migration in *Daphnia:* enrichment and preliminary chemical characterization of a kairomone exuded by fish. J. Chem. Ecol. 22:885–895; 1996.

Vuille, T. Abundance, standing crop and production of microcrustacean populations (*Cladocera, Copepoda*) in the littoral zone of Lake Biel, Switzerland. Arch. Hydrobiol. 123:165–185; 1991.

Vuille, T.; Maurer, V. Body mass of crustacean zooplankton in Lake Biel: a comparison between pelagic and littoral communities. Verh. Int. Verein. Limnol. 24:938–942; 1991.

Walls, M.; Rajasilta, M.; Sarvala, J.; Salo, J. Diel changes in horizontal microdistribution of littoral cladocera. Limnologica 20:253–258; 1990.

Werner, E.E.; Gilliam, J.F.; Hall, D.J.; Mittelbach, G.G. An experimental test of the effects of predation risk on habitat use in fish. Ecology 64:1540–1548; 1983.

Wetzel, R.G. Limnology. Philadelphia: W.B. Saunders; 1975.

Whiteside, M.C. Chydorid (*Cladocera*) ecology: seasonal patterns and abundance of populations in Elk Lake, Minnesota. Ecology 55:538–550; 1974.

Whiteside, M.C. 0^+ fish as major factors affecting abundance patterns of littoral zooplankton. Verh. Int. Verein. Limnol. 23:1710–1714; 1988.

Winfield, I.J. The influence of simulated aquatic macrophytes on the zooplankton consumption rate of juvenile roach, *Rutilus rutilus,* rudd, *Scardinius erythrophthalmus,* and perch, *Perca fluviatilis.* J. Fish Biol. 29:37–48; 1986.

6. Impact of Submerged Macrophytes on Phytoplankton in Shallow Freshwater Lakes

Martin Søndergaard and Brian Moss

Introduction

The ubiquity of phytoplankton, and its fundamental importance as a primary producer and mediator of many major biological processes in lakes, has led to comprehensive research on its biology. Its importance for water quality and its increasing predominance as the main primary producer at the expense of submerged macrophytes in shallow lakes follow from increased nutrient loading. Often, macrophytes have completely disappeared, and nutrients are so abundant that it is difficult to conceive much bottom-up control of phytoplankton through especially phosphorus. Restriction of the nutrient loading to reduce the amount of phytoplankton and to increase water clarity and restore a more diverse biological structure has now started to reverse this process in many areas. Simultaneously, we can anticipate a renewed importance of submerged macrophytes in many lakes.

Most research has, however, been on phytoplankton in the pelagic environment, and few studies have focused on phytoplankton in relation to the presence of submerged macrophytes. Thus, the existence and relative importance of top-down control of phytoplankton by zooplankton and/or bottom-up control through nutrients, which is well known from the pelagic, is poorly documented from macrophyte beds. It is largely unknown whether the mechanisms in the macrophyte beds are similar or to what extent different densities of macrophytes may influence a trophic cascade. By their structuring effect, macrophytes create an environment that is fundamentally different from that of the open water and that potentially may have great impact on the interactions between the different trophic levels (Jeppesen et al., this volume, Chapter 5).

In this chapter, we describe how submerged macrophytes may affect phytoplankton biomass and community structure. We focus on temperate shallow lakes with the potential for an extensive macrophyte cover. It is not our intention to give a comprehensive review but to highlight some of the clearest and, in our opinion, most important relationships. We supplement these with some of our own experience and results, particularly from large-scale enclosure experiments in the shallow and eutrophic Lake Stigsholm, Denmark, and Little Mere in the United Kingdom, where macrophyte and fish densities have been manipulated. We also try to create a framework that defines hypotheses about which phytoplankton biomass and species differences to expect along nutrient, zooplankton, and macrophyte gradients.

Effect of Macrophytes on Phytoplankton Biomass

Assuming a given external flux of nutrients and fish-mediated grazing pressure from zooplankton, several mechanisms may be responsible for the general observation that a lower phytoplankton biomass is found in the presence compared with the absence of macrophytes. The most conspicuous, and probably one of the best documented mechanisms, is that of enhanced grazing pressure from pelagic zooplankton.

Increased Grazing

Large-sized pelagic zooplankters generally occur in greater numbers inside or around the edges of macrophyte beds than outside, although this effect can be modified by fish (see Jeppesen et al., this volume, Chapter 5, for a comprehensive discussion). They appear to use the macrophyte beds as a daytime refuge against fish predation. The concept of this mechanism was first introduced by Timms and Moss (1984) and has been confirmed or indicated by several other studies (De Meester et al., 1993; Stansfield et al., 1995; Lauridsen et al., this volume, Chapter 13). It is discussed in detail by Jeppesen et al. (this volume, Chapter 5), who also suggested that there is much higher grazing pressure on phytoplankton inside than outside the macrophyte beds. An additional consequence of macrophyte presence for decreasing phytoplankton biomass is that their effects seem to extend beyond the border of the macrophyte beds. A horizontal diurnal migration of pelagic zooplankton from dense plant beds covering only 3% of the lake is enough to increase the grazing potential of zooplankton in the open water by 100% (Lauridsen et al., 1996).

Grazing by pelagic zooplankton may often be supplemented by grazing from plant-associated zooplankton species (Irvine et al., 1989; see also Jeppesen et al., this volume, Chapter 5) and other invertebrate filter and suspension feeders (e.g., insect larvae, mollusks). For obvious reasons, plant-associated zooplankters occur in much higher densities in macrophyte beds than outside. For example, in experiments conducted in Little Mere in 1993, the number of grazers in mesocosms containing plants was much greater than in those without, with consequent reductions in chlorophyll *a* concentrations (Beklioglu and Moss, 1996). In addition, Schriver et al. (1995), using 100-m^2 enclosures in Lake Stigsholm, recorded a decline in phytoplankton biomass with increasing macrophyte density in terms of percentage volume infestation (PVI) even when planktonic cladocerans were absent. They argued that plant-associated zooplankton such as chydorids, *Pleuroxus* and *Eurycercus,* which usually were found in tenfold higher densities during both day and night in the vegetation, could be of significance. Although the effects on phytoplankton have been rarely documented directly, there are thus reasons to believe that plant-associated filter feeders are important in decreasing the phytoplankton biomass within macrophyte beds.

Changes in Nutrient Cycling

The presence of macrophytes may in various ways influence nutrient cycling (see Barko and James, this volume, Chapter 10) and thus potentially also phytoplankton biomass and the growth and competition among different phytoplankters. Decreased availability of nitrogen and sometimes phosphorus may be expected in most cases.

Macrophytes, and their associated epiphytes, take up and store nutrients, which are then not available for the phytoplankton (Kufel and Ozimek, 1994; O'Dell et al., 1995). A rapid growth of *Elodea* in Lake Zwemlust was followed by limitation of nitrogen during summer (Ozimek et al., 1990), and it was calculated that 64% of the total lake nitrogen and 61% of that of phosphorus (excluding the sediment pool) were found in the macrophyte compartments (van Donk and Gulati, 1995). Indirect effects are also likely to explain the generally low nitrogen concentrations often seen in the presence of macrophytes, even when the uptake by macrophytes themselves is considered low. The structured environment inside macrophyte beds, which may include gradients leading to anoxic conditions, may increase denitrification (Weisner et al., 1994; Jeppesen et al., submitted a). Changes in nitrogen competition between phytoplankton and sediment-associated bacteria may also be important: when phytoplankton biomass is low (in the presence of macrophytes), the nitrate uptake by the bacteria in the sediment and the subsequent denitrification to N_2 is relatively high compared with a situation with a high phytoplankton biomass (when macrophytes are absent) in which more nitrate is taken up and stored as organic nitrogen by the phytoplankters. The overall nitrogen removal is therefore higher in the presence of macrophytes. The general importance of these mechanisms, however, still needs to be verified, although it is indicated by several whole-lake studies where the macrophyte density has changed markedly (van Donk et al., 1990; Jeppesen et al., 1997; Moss et al., 1997).

When phosphorus availability is concerned, the mechanisms are probably more diverse and the outcome more complicated. Brammer (1979) explained an exclusion of phytoplankton in the proximity of a dense stand of *Stratiotes aloides* to be caused by competition for nutrients together with the co-precipitation of phosphorus with calcium carbonate on the leaf surfaces of the submerged plants. Very dense macrophyte beds, however, may also increase the availability of phosphorus (Stephen et al., 1997). Canopy-forming types, poorly rooted species such as *Elodea* and dense *Chara* beds, which create dense shade, may have more or less anoxic conditions in their bottom layers (Pokorny et al., 1984; Moss et al., 1986), thereby increasing the potential for phosphorus release from the sediment, particularly where the iron-bound and redox-sensitive phosphorus fractions constitute a significant part of the total P in the sediment. This is consistent with observations in Cockshoot Broad, United Kingdom, in which a recovery of macrophytes was associated with an increased release of phosphorus from the sediment (Moss et al., 1996). An analogous mechanism is an increased phosphorus release from the sediment due to enhanced pH, caused by high photosynthetic activity, which has been shown to affect the phosphorus sorption mechanisms in the surface sediment (Søndergaard, 1988). Plant beds frequently have high pH in the surface layers but lower values in the darker, deeper layers toward the sediment (Frodge et al., 1990). The ultimate effects of pH variation in plant beds are thus complex. Finally, release of organic carbon compounds from macrophytes may also stimulate bacterial production (Wetzel and Søndergaard, this volume, Chapter 7), and thus indirectly increase bacterial phosphorus uptake at the expense of phytoplankton.

Less dense macrophyte beds and macrophytes having a more developed root system may, however, affect sediment release rate in a converse way by oxidizing the surface sediment (Andersen and Olsen, 1994; Flessa, 1994; Christensen and Andersen, 1996), thereby reducing the redox-sensitive phosphorus release. The final outcome for phosphorus availability in macrophyte beds may therefore depend on both macrophyte species and density. In general, however, submerged macrophytes may be regarded as phosphorus sinks during their active growth and only as a potential phosphorus source during the relatively short periods of their senescence (Carpenter and Lodge, 1986; Rørslett et al., 1986; Jones, 1990).

Although change in nutrient concentrations is probably related to macrophyte density, it is often difficult to show its effects on phytoplankton biomass because of masking by other mechanisms acting simultaneously. In their enclosure experiments, Schriver et al. (1995) noted that the concentrations of orthophosphate and ammonium were strongly and positively correlated with the potential grazing pressure from zooplankton. They suggested that the shift to low phytoplankton biomass at high grazing pressure from zooplankton should be attributed to enhanced grazing rather than nutrient control of algal growth. In that particular case, nutrient changes did not seem to be important. In enclosure experiments in Little Mere, phosphate and ammonium concentrations were also unaffected by the presence or absence of macrophytes but strongly influenced by the numbers of zooplankters, responding to changes in fish predation (Beklioglu and Moss, 1996). The concentrations of chlorophyll *a* in the phytoplankton, however, were significantly and positively related to the number of fish both in the presence or absence of macrophytes.

Allelopathy

Release of organic compounds from macrophytes, in particular from *Chara,* with allelopathic effects on phytoplankton has also been suggested to reduce phyto-plankton production and biomass (Wium-Andersen, 1982; Gross, 1995; van Donk and Gulati, 1995; Gross et al., 1996) or cause changes in the phytoplankton community structure (Jasser, 1995). However, in field experiments using whole plants and in laboratory experiments using macrophyte extracts, Jasser (1995) only recorded a minor effect on phytoplankton biomass in the presence of macro-phytes or their extracts, although major effects were seen in the community structure. Similarly, Forsberg et al. (1990) did not find any allelopathic effects of *Chara* on chlorophyll *a* concentrations when comparing phosphorus/chlorophyll *a* relationships in *Chara*-dominated and non-*Chara* lakes. Furthermore, they also noticed a dense epiphytic growth on the *Chara* plants, which they interpreted as an absence of allelopathy by *Chara* in situ. The importance of allelopathic substan-ces, although undoubtedly they exist and can be shown to act in laboratory situations, remains still to be fully documented at the ecosystem level.

Increased Sedimentation

Increased sedimentation and reduced resuspension, due to the less turbulent and more quiescent waters among the macrophytes, is yet another factor that may negatively influence the phytoplankton community (see also Barko and James, this volume, Chapter 10). In Chesapeake Bay, with semidiurnal tides, Kemp et al. (1984) recorded a lower phytoplankton biomass in a littoral weedbed area com-pared with nearby open water. They ascribed this to increased sedimentation within the weedbed. Similarly, Jones (1990) recorded a consistently lower phyto-plankton biomass in the vegetated areas of the freshwater Potomac River than in the adjacent unvegetated reaches. Van den Berg et al. (this volume, Chapter 25) recorded a 30-fold lower phytoplankton density inside dense *Chara* beds than outside and attributed it to stiller conditions. In other situations, for example Balls et al. (1989), increased sedimentation of phytoplankton in macrophyte beds in ponds was not found to be important because the phytoplankton community became dominated by flagellates. The importance of this mechanism may also be questioned because of the very high phytoplankton biomass found in nutrient-rich brackish lakes, even when macrophyte coverage is substantial (Baks et al., 1993; Jeppesen et al., 1994; Jeppesen et al., this volume, Chapter 28).

Reduced Light Conditions

Finally, there may be shading effects of macrophytes on phytoplankton (Sand-Jen-sen, 1989; Ozimek et al., 1990). The shading effect is highly related to macrophyte biomass, but macrophyte surface area also depends on plant species (Sher-Kaul et al., 1995), and light extinction coefficients of macrophyte canopies show a four-fold variation among canopy species (review Carpenter and Lodge, 1986). Dense beds are undoubtedly self-shading (Nielsen and Sand-Jensen, 1989; Frodge et al.,

1990). However, quite dense periphyton communities may persist within them, and epipelic algae can be found on the bottom sediment among the plants. The strategies of buoyancy and motility are available to the phytoplankton community to overcome this potential problem.

Macrophyte Threshold Effects

In most of the examples mentioned above, the conclusions have been drawn from comparisons between macrophyte-free and macrophyte-rich areas. An important aspect, not least from a management point of view, is whether macrophyte density must attain a certain threshold before they have a significant effect on the phytoplankton. Several studies indicate that this is the case. Canfield et al. (1984) noted that water transparency was generally high in lakes with a macrophyte PVI exceeding 30%, and Schriver et al. (1995) found a threshold level of 15–20% PVI (*Potamogeton pectinatus* and *Potamogeton pusillus*), above which a sudden decrease from 30 mm^3/L to 10 mm^3/L in phytoplankton biovolume was recorded (Fig. 6.1A). Later experiments in Lake Stigsholm suggested a relatively low refuge effect for zooplankton at a PVI below 20–24%, but a significant effect at a PVI of 50% (Jeppesen et al., this volume, Chapter 5). The presence of a threshold and a nonlinear relation with macrophytes is also proposed with regard to the impact of mollusks (Brönmark and Vermaat, this volume, Chapter 3) and the refuge efficacy for zooplankton (Jeppesen et al., this volume, Chapter 5).

Consensus

Although it may be difficult or even futile to attempt to determine which of the mechanisms mentioned above is decisive, for several may act together, it is becoming increasingly indisputable that strong antagonistic forces exist between macrophytes and phytoplankton and that the presence of macrophytes generally is tantamount to low phytoplankton biomass in freshwater lakes. This statement is supported by numerous studies comparing phytoplankton in the presence and absence of macrophytes, including examples from this book (Blindow et al., this volume, Chapter 26; Faafeng and Mjelde, this volume, Chapter 27; Jeppesen et al., this volume, Chapter 28; Van den Berg et al., this volume, Chapter 25). Balls et al. (1989) also concluded that in experimental ponds where the submerged plants remained intact, plants were able to buffer strongly the effects of added nutrients. Despite large loadings, nutrient concentrations in the presence of plants could not build up sufficiently to support large algal growths. By contrast, where the plants were artificially removed, in a set of replicate but cleared ponds, sufficiently large concentrations did establish and support large algal crops. In 20-m^2 experimental enclosures in the Danish Lake Stigsholm, a fourfold higher chlorophyll *a* concentration and a 10–25-fold higher phytoplankton biomass were recorded in macrophyte-free enclosures compared with enclosures with a 50% PVI (Table 6.1). The statement is also supported by empirical studies of more than 100 Danish lakes, in which Secchi disk transparency was, in general, much higher in

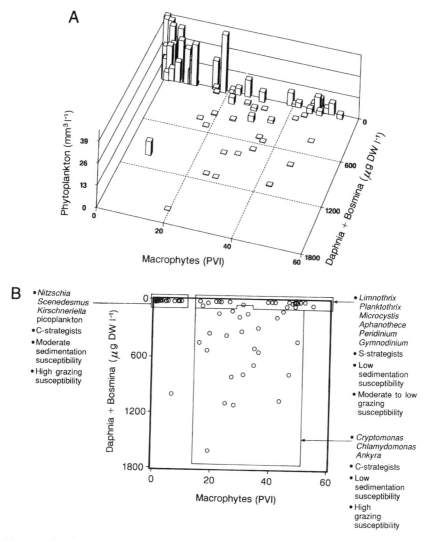

Figure 6.1. Phytoplankton biovolume (A) and community structure (B) in relation to macrophyte PVI (percentage volume infestation) and *Daphnia + Bosmina* biomass. (Modified from Schriver et al., 1995.)

lakes with a high macrophyte coverage compared with those without plants at comparable nutrient concentrations (Jeppesen et al., 1990).

Despite a number of reservations and huge differences among lakes and environmental conditions, we have attempted to summarize the effects and make suggestions concerning the relative importance of different factors associated with macrophytes on phytoplankton biomass in Figure 6.2.

Table 6.1. Chlorophyll *a* Concentrations (Mean of 1 Daytime and 1 Night-Time Sampling in 3 Enclosures, ± SD) and Phytoplankton Biomass (Mean of 3 Night-Time and 3 Daytime Samplings in 3 Enclosures ± SD) in Macrophyte-Free ($n = 3$) and Macrophyte-Rich (50% PVI, $n = 3$) 20-m³ Enclosures from July 25 to 27, 1994, in Lake Stigsholm

	With macrophytes ($n = 3$)		Without macrophytes ($n = 3$)	
	Day	Night	Day	Night
Chlorophyll *a* (μg/L)	17 (19)	8.3 (4.0)	64 (10)	37 (7.9)
Phytoplankton biomass (mm³/L)	0.80 (0.78)	0.31 (0.21)	8.1 (1.4)	7.6 (2.1)

All observations comparing enclosures with and without macrophytes are significantly diferent ($P <$ .01, *t* test unequal variance). Only night and day observations of chlorophyll *a* in macrophyte-free enclosures are significantly different ($P = .02$, *t* test unequal variance).

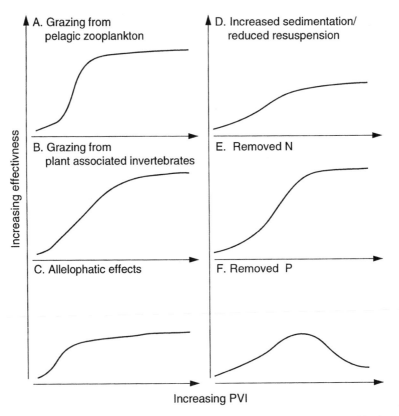

Figure 6.2. The relative importance of different mechanisms potentially affecting the phytoplankton biomass at increasing macrophyte PVI (percentage volume infestation).

The importance of grazing by pelagic zooplankton species is probably not linearly proportional to macrophyte density because of a need for macrophytes to attain a threshold density before they have a refuge effect (Fig. 6.2A). Furthermore, the macrophyte density required depends on fish density, the refuge effect being diminished at increasing fish density (Jeppesen et al., submitted b; Jeppesen et al., this volume, Chapter 5). It is also likely that the importance of grazing by pelagic zooplankton diminishes in very dense macrophyte beds. *Daphnia*, in particular, are frequently most abundant at the edges of macrophyte beds rather than in their interiors (Lauridsen et al., 1996). As beds cover the lake, edge habitat is diminished. We think the importance of grazing overall is likely to be high in most situations, however.

Having exceeded a certain minimum macrophyte density, grazing from plant-associated invertebrates is expected to increase linearly with macrophyte density (Fig. 6.2B). The effect, however, again may depend on fish predation, although in mesocosm experiments in Little Mere (Beklioglu and Moss, 1996), there was no effect of increasing perch (*Perca fluviatilis*) numbers on densities of plant-associated *Cladocera* (*Chydorus, Eurycercus, Simocephalus, Sida*), although severe reductions of *Daphnia* numbers, despite the presence of plant effects. At present, it is difficult to assess the relative importance of plant-associated grazers as there are few quantitative data and conditions in which such grazing is likely to be high (in dense plant beds) are also those in which nutrient effects may be strong.

Allelopathic effects, if present (Fig 6.2C), may be predicted to be proportionately more important at low plant densities, where the plants are growing vigorously before self-shading and perhaps nitrogen limitation sets in, and to reach a plateau of effectiveness at moderate macrophyte densities. Bacterial decomposition of the substances is likely to increase as the beds become dense. The effectiveness of allelopathy is also likely to be dependent on macrophyte species. Based on the present evidence, we suggest that allelopathy is generally of lesser importance in the ecosystem context than other factors.

Effects of stilling of the water and increased sedimentation and/or reduced resuspension are expected to increase proportionally with macrophyte density (Fig. 6.2D), but their overall importance is not well documented. We consider their importance moderate compared with other factors under most circumstances.

Reduction of nitrogen availability is likely to follow a pattern of increased effect after attainment of a threshold macrophyte density sufficient to account for considerable uptake and to create conditions appropriate to denitrification (Fig. 6.2E). This effect is probably an important one overall, especially in dense macrophyte beds and in lakes having a relatively low nitrogen loading and consequently the highest possibility of nitrogen limitation of phytoplankton.

The importance of effects on phosphorus will depend more on macrophyte type and density than most of the other factors. If the macrophyte beds are not too dense, the effect is likely to be one of reduction in concentration; but if macrophytes become very dense, the effect may be the opposite, with the risk of an enhanced release from the sediment (Fig. 6.2F). We think that in terms of mechanisms limiting the growth of phytoplankton in weedy, eutrophic lakes, phosphorus effects are unlikely to be very important.

Fish- and Nutrient-Mediated Effects of Macrophytes on Phytoplankton Biomass

We have not yet fully considered the influence of fish and nutrients. There are reasons to believe, however, that fish modify the influence of macrophytes on phytoplankton. In Chapter 5, it was shown that the behavior of fish and zooplankton and their interactions were highly dependent on macrophyte density. We must therefore anticipate consequential effects on phytoplankton. Based on the threshold level of macrophyte density of 15–20% PVI influencing fish-mediated impact on zooplankton in the Lake Stigsholm experiments (Schriver et al., 1995) and on other results, we have summarized, in Table 6.2, phytoplankton changes along a nutrient, macrophyte, and fish gradient. Schriver et al. (1995) argued also for the existence of a fish threshold. They concluded that the effects on phytoplankton biomass of a high macrophyte coverage were only important as long as the fish density did not exceed approximately 2 fry/m^2. If it was higher, the refuge effect seemed low and even a high macrophyte density would be unable to have any major zooplankton-mediated effects on the phytoplankton. These generalizations are probably modified by the actual nutrient level, and it may, for instance, be possible that nitrogen limitation effects may become more important under such circumstances because high fish densities can frequently be found in association with dense plant beds.

In Table 6.2, we also suggest how some phytoplankton variables may change along nutrient, macrophyte, and fish gradients. Such a generalization is complicated by the fact that the macrophyte community itself will also change along a nutrient gradient. Small vascular plants and *Charophyte*-dominated communities are most common at low nutrient concentrations (say, <50 µg total P/L). Small, shallow, and very fertile waters (say, >200 µg total P/L) tend to become dominated by nymphaeids (lilies), with their partly emergent foliage, perhaps because periphyton infestation inhibits maintenance of submerged leaves. Intermediately fertile waters have mixed communities of tall submerged vascular plants, although charophytes and small vascular plants may still be present and lilies usually are. Clear water is maintained at all combinations of nutrient concentrations and macrophyte densities, except low macrophyte density/high nutrients and high fish densities, by the mechanisms discussed above. Grazing, either by daphnids or by plant-associated animals, is likely to be important in all circumstances in which plants persist, although the balance of these two grazer groups will differ. Daphnids may be less important in very dense submerged macrophyte stands.

The relative importance of the availability of nitrogen and phosphorus is likely to vary along the gradients (Table 6.2), with phosphorus becoming less limiting and nitrogen more so as both nutrient loading and macrophyte density increase.

Table 6.2. Generalized Matrix Showing Proposed Effects of Nutrient Concentration, Fish Density, and Macrophyte Density on Phytoplankton Biomass, Phytoplankton Community Structure, Phytoplankton Cell Size, Transparency, N/P Limitation, and Dominant Macrophtyes[a]

	Low nutrient		High nutrient	
	Low fish	High fish	Low fish	High fish
Phytoplankton biomass				
Low Macrophytes	Low	Low/medium	Medium/high	High
High macrophytes	Low	Low	Low/medium	Medium/high
Phytoplankton community				
Low macrophytes	Flagellates, zyg	Flagellates, zyg	Diatoms, chlor Flagellates	Diatoms, cyano, chlor
High macrophytes	Flagellates, zyg	Flagellates, zyg		Cyano, flagellates
Phytoplankton cell size				
Low macrophytes	Small/large	Small/large	Medium/small	Large/medium
High macrophytes	Small/large	Small/large	Small	Small/large
Transparency				
Low macrophytes	High	High/medium	Medium/low	Low
High macrophytes	High	High	High/medium	Low/medium
N/P limitation				
Low macrophytes	P	P	N/P	N/P
High macrophytes	N/P	N/P	N	N
Dominant macrophytes				
Low macrophytes	s. vas/t. vas	s. vas/t. vas	t. vas/lilies	t. vas/lilies
High macrophytes	s. vas/t. vas	s. vas/t. vas	t. vas/lilies	t. vas/lilies

Abbreviations: cyano, cyanophytes; chlor, chlorococcales; zyg, zygnemetales; s. vas, small vasculars and charophytes; t. vas, tall vasculars.

[a]Approximate definitions used in this context: low nutrient: $P < 25\,\mu g/L$ and N/P high; high nutrient: $P > 100\,\mu g/L$ and N/P low. Low fish density: $<2\,fry/m^2$; high fish density: $>2\,fry/m^2$. Low macrophyte density: PVI < 15%; high macrophyte density: PVI > 30%.

This is because the nitrogen to phosphorus ratios in inflowing waters modified by human activities, especially those concerning sewage and stock effluents, tend to be relatively richer in phosphorus than is required for nutrient sufficiency by algae and because phosphorus is mobilized in dense plant beds, whereas combined nitrogen is denitrified. Finally, the nature of the phytoplankton is likely to change along the gradients, with large cells or colonies being favored by increasing nutrient concentrations, large nitrogen fixers becoming more abundant as nitrogen stress increases, but small and often flagellate cells and cyanophytes being favored at high densities of submerged plants due to sedimentation of large nonbuoyant cells under such conditions.

Indirect Effects of Macrophytes on Phytoplankton Composition and Size

The impact of macrophytes on phytoplankton extends not only to biomass. Macrophytes may also have marked effects on the species and size of individuals forming the community. As with biomass, several mechanisms may be involved in determining species composition. Change in the nutrient regime, as mentioned above, is probably one of the most important. But other factors such as differential tolerance to shading (Sand-Jensen, 1989), susceptibility to sedimentation in the darker and quiescent water inside the macrophyte beds, allelopathic effects (Jasser, 1995), and increased input of particulate dissolved organic carbon from the macrophyte–epiphyte complex, altering the planktonic food chain and thereby also influencing the phytoplankton, may also be important.

One general tendency is that of increased importance of flagellates in the presence of macrophytes and/or large-sized zooplankton. For example, Balls et al. (1989), in fish-free pond experiments, observed no systematic pattern of change with nutrient loading in either plant-dominated or cleared ponds and a great deal of variation within the limits of communities all dominated by small and often flagellated organisms. They attributed this to a stronger selection for small organisms with low sinking rates in the very short water column, compared with any possible selection for large phytoplankters through grazing. Godmaire and Planas (1986) in enclosure experiments also recorded a prevalence of flagellates during the entire season in the presence of *Myriophyllum spicatum* and suggested that neither light nor nutrient conditions seemed responsible for this particular community composition. And in Lake Veluwemeer, Van den Berg et al. (this volume, Chapter 25) recorded a shift in species composition from cyanobacteria to flagellates over a transect from no vegetation to a dense vegetation of *Chara.*

A tendency toward a greater dominance of small flagellates at increasing macrophyte abundance, despite an increase in large-sized daphnids, was also observed by van Donk et al. (1990) in a whole-lake experiment in the biomanipulated Lake Zwemlust (The Netherlands). In this lake, submerged macrophyte density increased from 0 to 80% within 3 years, and apart from a much lower phytoplankton biomass, this was accompanied by a shift toward small edible species. Larger phytoplankton species such as *Aphanizomenon,* which is considered more resistant to zooplankton grazing, did not develop. This was explained by dependence of phytoplankton growth on several factors operating simultaneously or successively. Thus, zooplankton grazing in spring, nitrogen limitation caused by the macrophytes and grazing in summer, and temperature and light availability in winter controlled phytoplankton growth. It was concluded that large-sized inedible phytoplankton taxa were unable to oust the small-sized and edible but fast-growing C strategists, which were more effective in taking up nutrients during nutrient limitation due to their high surface-to-volume ratio.

A shift to edible cryptophyte flagellates has also been observed in the clear open water following a fish kill (Sarnelle, 1993) and in biomanipulated lakes where the fish stock has been reduced (Leah et al., 1980; Reinertsen et al., 1990;

Søndergaard et al., 1990). Thus the effects of herbivory on algal succession were not predictable from the relative susceptibilities of these algal species to grazing mortality.

In a series of 100-m^2 enclosures, Schriver et al. (1995) investigated the importance of zooplankton and macrophyte densities on the phytoplankton composition. Among their findings was that when *Daphnia* and *Bosmina* were abundant, fast-growing small flagellates such as *Cryptomonas* and *Chlamydomonas* dominated. Although cryptophytes (as percentage of biomass) were positively related to both zooplankton biomass and the interaction between macrophyte PVI and zooplankton biomass, *Chlamydomonas* was significantly related only to zooplankton biomass, indicating that *Cryptomonas* is more dependent on the presence of macrophytes. In the absence of planktonic cladocerans but in the presence of macrophytes, cyanophytes (mainly represented by *Planktothrix, Limnothrix, Microcystis,* and *Aphanothece*) and dinoflagellates (mainly *Gymnodinium* and *Peridinium*) were dominant. Cyanophytes were negatively related to the interaction between PVI and zooplankton biomass. In the absence of both planktonic cladocerans and macrophytes, the community was dominated by fast-growing diatoms (mainly *Nitzschia*) and chlorophytes (mainly *Scenedesmus* and *Kirchneriella*). Chlorococccales were therefore negatively related to PVI*zooplankton biomass. Basically, these findings are in agreement with the life history strategies described by Reynolds (1987). Fast-growing algae are moderately susceptible to loss by sedimentation and are dominant in the turbulent environment of the macrophyte-free areas, whereas slow-growing buoyant cyanophytes are dominant in the more quiescent water among the plants, where the risk of sedimentation is higher.

However, flagellates such as *Cryptomonas,* with a high grazing susceptibility, were present at high grazing pressure from zooplankton, and cyanophytes, with low susceptibility, were present at low grazing pressure, contrary to expectation, but also often seen in whole-lake studies. On the basis of these findings, Schriver et al. (1995) argued that the term *grazing susceptible* should maybe be replaced by *grazing resistant* and *grazing tolerant,* the implication being that grazing-resistant algae such as *Microcystis* are only favored until grazing pressure increases to a certain limit whereafter grazing-tolerant algae such as *Cryptomonas* take over (see also Table 6.2). The results of Schriver et al. (1995) have also been summarized in Figure 6.1B, which defines the three groups of phytoplankton.

It may be asked why small flagellates, vulnerable to zooplankton grazing, seem to be such successful competitors in the presence of macrophytes, with the considerable grazer pressure associated with them. A possible explanation (Sommer, 1988) is that flagellates, due to their motility, are better adapted to exploit an environment that is heterogeneous and structured with respect to nutrient distribution, such as that created by macrophytes. Furthermore, many flagellates are mixotrophic and able to use DOC excreted from macrophytes and their associated epiphytic environment. A combination of these advantages and their high growth rates could be a major reason to their success.

There may also be effects of macrophytes on species composition as well as changes in the balance of groups such as diatoms, cyanophytes, or flagellates. In

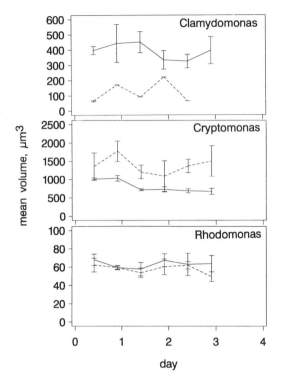

Figure 6.3. Biovolume (μm³) of *Chlamydomonas* sp., *Cryptomonas* spp. (*C. reflexa, C. curvata,* and *C. marssonii*), and *Rhodomonas* sp. in 20-m² enclosure experiments in Lake Stigsholm from July 25 (12 AM) to 27 (12 PM), 1994 (sampling at 12 AM and 12 PM), using macrophyte-free (solid line) and macrophyte-dense (broken line) compartments. Mean values and standard error of three enclosures with and three enclosures without macrophytes. For total phytoplankton biomass, see Table 7.1.

the Little Mere mesocosm experiments, in which the phytoplankton community was studied in relation to different densities of fish (Beklioglu and Moss 1996), the presence or absence of macrophytes had no effect on the total biomass of diatoms but very significant effects on particular diatom species. *Aulacoseira granulata* was much less abundant and *Nitzschia palea* much more abundant when macrophytes were present.

Results from the Lake Stigsholm experiments, furthermore, indicate that macrophytes can also affect the size distribution at the phytoplankton genus/species level (Søndergaard et al., unpubl.). As an example, marked differences in the size of the flagellates *Chlamydomonas* and *Cryptomonas* were recorded between macrophyte-rich and nonmacrophyte areas, whereas no differences were recorded for *Rhodomonas* (Fig. 6.3). Whereas *Chlamydomonas* sp. was much smaller inside the macrophyte beds, the mean size of *Cryptomonas* spp. (three different species) during the whole experimental period was significantly higher in the presence of macrophytes (1,100 to 1,800 μm^3) than in their absence (700–1,000 μm^3). Differences in grazing susceptibility and size-specific mechanisms to avoid grazing among different species within macrophyte beds could reinforce the suggestion that grazing is a major agent affecting the phytoplankton in and around macrophyte beds. The difference in size may also be related to differences in nutrient concentrations, which are known to influence the size of phytoplankton (Watson et al., 1992; Hansson and Carpenter, 1993; Larocque et al., 1996). No matter which mechanism lies behind the results, however, they indicate that macrophytes can have a structuring effect also on phytoplankton size distribution.

Concluding Remarks

The presence of macrophytes clearly has a negative effect on phytoplankton biomass. There are indications that along a macrophyte density gradient, the response is nonlinear, a threshold being recognizable at a macrophyte density of approximately 15–30% PVI, depending on fish abundance. Several mechanisms contribute to maintenance of low phytoplankton crops in association with macrophyte beds. These include direct mechanisms associated with the plants themselves, including creation of a still water environment, poor light climate, and secretion of allelopathic substances as well as mechanisms indirectly linked with the plants, such as provision of refuges or habitat for grazers on algae, and modification of the ambient nutrient regime by the metabolic activity of the plants. In turn, these mechanisms are affected by a second level of influences—predation by fish on the grazers, imposition of nutrient loads from the catchment, and hydrodynamic conditions that determine the degree of water movement through the plant beds.

All these mechanisms interact. The hydrodynamics must determine the extent to which anoxic gradients that promote denitrification can be established; excretion by grazers modifies the nutrient regime. Furthermore, the influence of specific composition of the macrophyte community is likely to be important. There are metabolic as well as structural differences among different species of submerged plants, let alone those between, for example, nymphaeids and charophytes. The degree of grazing on algae will also be affected by the composition of the grazer community, which in turn, is determined by the size structure and composition of the fish community. The macrophyte growth will also be subject to grazing by vertebrates so that the structure and composition of the beds will vary.

Major gaps exist concerning the differential effects of different macrophyte species and of those of mixed fish communities. Most experiments have used monospecific or simple communities because of the need to balance adequate replication against the considerable costs of ecosystem experiments, even in mesocosms. A further gap is that of determining whether allelopathy is important in the complex ecosystem context. All in all, the mutual interaction of all these factors makes it highly unlikely that anything other than a very general model of how the macrophyte and phytoplankton communities interact can be constructed. The importance of different mechanisms and processes will, to some extent, be lake-specific. Much progress has been made in this area through the use of mesocosm experiments, and this should continue. In addition, accumulating experience from a variety of different lakes may also begin to reveal greater generalities than we have been able to discern here. The comparative approach coupled with whole-lake and mesocosm experiments of increasing scope are likely to be the way forward.

Acknowledgments. The assistance of the technical staff of the National Environmental Research Institute, Silkeborg, is gratefully acknowledged. Phytoplankton analysis was conducted by B. Laustsen and L. Nørgaard. Technical assistance was provided by K. Møgelvang, J. Jacobsen, and A.M. Poulsen. The study was supported by the Centre for Freshwater Environmental Research.

References

Andersen, F.Ø.; Olsen, K.J. Nutrient cycling in shallow, oligotrophic Lake Kvie, Denmark. Hydrobiologia 275/276:267–276; 1994.

Baks, M.; Moss, B.; Phillips, G.; Irvine, K.; Stansfield, J. The changing ecosystem of a shallow, brackish lakes, Hickling Broad Norfolk. II. Long-term changes in water chemistry and ecology and their implications for restoration of the lake. Freshwat. Biol. 29:141–165; 1993.

Balls, H.; Moss, B., Irvine, K. The loss of submerged plants with eutrophication. I. Experimental design, water chemistry, aquatic plant and phytoplankton biomass in experiments carried out in ponds in the Norfolk Broad. Freshwat. Biol. 22:71–87; 1989.

Beklioglu, M.; Moss, B. Mesocosm experiments on the interaction of sediment influence, fish predation and aquatic plants with the structure of phytoplankton and zooplankton communities. Freshwat. Biol. 36:315–325; 1996.

Brammer, E.S. Exclusion of phytoplankton in the proximity of dominant water-soldier (*Stratiotes aloides*). Freshwat. Biol. 9:233–249; 1979.

Canfield, D.E., Jr.; Shireman, J.W.; Colle, D.E.; Haller, W.T.; Watkins, C.E., II; Maceina, M.J. Prediction of chlorophyll *a* concentrations in Florida lakes: importance of aquatic macrophytes. Can. J. Fish. Aquat. Sci. 41:497–501; 1984.

Carpenter, S.; Lodge, D. Effects of submersed macrophytes on ecosystem processes. Aquat. Bot. 26:341–370; 1986.

Christensen, K.K.; Andersen, F.Ø. Influence of *Littorella uniflora* on phosphorus retention in sediment supplied with artificial porewater. Aquat. Bot. 55:183–197; 1996.

De Meester, L.; Maas, S.; Dierckens, K.; Dumont, H.J. Habitat selection and patchiness in *Scapholeberis:* horizontal distribution and migration of *S. mucronata* in a small pond. J. Plankton Res. 15:1129–1139; 1993.

Flessa, H. Plant-induced changes in the redox potential of the rhizospheres of the submerged vascular macrophytes *Myriophyllum verticillatum* L. and *Ranunculus circinatus* L. Aquat. Bot. 47:119–129; 1994.

Forsberg, C.; Kleiven, S.; Willén, T. Absence of allelopathic effects of *Chara* on phytoplankton in situ. Aquat. Bot. 38:289–294; 1990.

Frodge, J.D.; Thomas, G.L.; Pauley, G.B. Effects of canopy formation by floating and submergent aquatic macrophytes on the water quality of two shallow Pacific Northwest lakes. Aquat. Bot. 38:231–248; 1990.

Godmaire, H.; Planas, D. Influence of *Myriophyllum spicatum* L. on the species composition, biomass and primary productivity of phytoplankton. Aquat. Bot. 23:299–308; 1986.

Gross, E.M. Allelopathische Interaktionen zwischen Makrophyten und Epiphyten: Die Rolle hydrolysierbarer Polyphenole aus *Myriophyllum spicatum*. Dissertation, Univ. of Kiel. Göttingen: Cuvillier Verlag; 1995.

Gross, E.M.; Meyer, H.; Schilling, G. Release and ecological impact of algicidal hydrolyzable polyphenols in *Myriophyllum spicatum*. Phytochemistry 41:133–138; 1996.

Hansson, L.A.; Carpenter, S.R. Relative importance of nutrient availability and food chain for size and community composition in phytoplankton. Oikos 67:257–263; 1993.

Irvine, K.; Moss, B.; Balls, H. The loss of submerged plants with eutrophication. II. Relationships between fish and zooplankton in a set of experimental ponds, and conclusions. Freshwat. Biol. 22:89–107; 1989.

Jasser, I. The influence of macrophytes on a phytoplankton community in experimental conditions. Hydrobiologia 306:21–32; 1995.

Jeppesen, E.; Jensen, J.P.; Kristensen, P.; Søndergaard, M.; Mortensen, E.; Sortkjær, O.; Olrik, K. Fish manipulation as a lake restoration tool in shallow, eutrophic, temperate lakes 2: threshold levels, long-term stability and conclusions. Hydrobiologia 200/201: 219–227; 1990.

Jeppesen, E.; Søndergaard, M.; Kanstrup, E.; Petersen, B.; Eriksen, R.B.; Hammershøj, M.; Mortensen, E.; Jensen, J.P.; Have, A. Does the impact of nutrients on the biological structure and function of brackish and freshwater lakes differ? Hydrobiologia 275/276: 15–30; 1994.

Jeppesen, E.; Søndergaard, M.; Kronvang, B.; Jensen, J.P.; Svendsen, L.M.; Lauridsen, T.L. Lake and catchment management in Denmark. In: Harper, D.; Brierley, B.; Ferguson, A.; Phillips, G.; Madgwick, J., eds. Ecological basis for lake and reservoir management. London: J. Wiley & Sons (submitted).

Jeppesen, E.; Jensen, J.P.; Søndergaard, M.; Lauridsen, T.; Hald Møller, P.; Sandby, K. Changes in nitrogen retention in shallow eutrophic lakes following a decline in the density of cyprinids. (submitted a).

Jeppesen, E.; Søndergaard, Ma.; Søndergaard, Mo.; Christoffersen, K.; Jürgens, K.; Theil-Nielsen, J.; Schlüter, L. Cascading trophic interactions in the littoral zone of a shallow lake. Limnol. Oceanogr. (submitted b).

Jones, R.C. The effect of submersed aquatic vegetation on phytoplankton and water quality in the tidal freshwater Potomac river. J. Freshwat. Ecol. 5:279–288; 1990.

Kemp, W.M.; Boynton, W.R.; Twilley, J.C.; Ward, L.G. Influence of submersed vascular plants on ecological processes in upper Chesapeake Bay. In: Kennedy, V.S., ed. The estuary as a filter. Orlando: Academic Press; 1984.

Kufel, L.; Ozimek, T. Can *Chara* control phosphorus cycling in Lake Lukajno (Poland)? Hydrobiologia 275/276:277–283; 1994.

Larocque, I.; Mazumder, A.; Prouix, M.; Lean, D.R.S.; Pick, F.R. Sedimentation of algae: relationships with biomass and size distribution. Can. J. Fish. Aquat. Sci. 53:1133–1142; 1996.

Lauridsen, T.; Pedersen, L.J.; Jeppesen, E.; Søndergaard, M. The importance of macrophyte bed size for cladoceran composition and horizontal migration in a shallow lake. J. Plankton Res. 18:2283–2294; 1996.

Leah, R.T.; Moss, B.; Forrest, D. The role of predation in causing major changes in the limnology of a hypereutrophic lake. Int. Rev. Ges. Hydrobiol. 65:223–247; 1980.

Moss, B.; Balls, H.R.; Irvine K.; Stansfield, J. Restoration of two lowland lakes by isolation from nutrient-rich water sources with and without removal of sediment. J. Appl. Ecol. 23:391–414; 1986.

Moss, B.; Stansfield, J.; Irvine, K.; Perrow, M.R.; Phillips, G. Progressive restoration of a shallow lake—a 12-year experiment in isolation, sediment removal and biomanipulation. J. Appl. Ecol. 33:71–86; 1996.

Moss, B.; Beklioglu, M.; Carvalho, L.; Kilinc, S.; McGowan, S.; Stephen, D. Vertically challenged limnology: contrasts between deep and shallow lakes. Hydrobiologia 342/343:257–267; 1997.

Nielsen, S.L.; Sand-Jensen, K. Regulation of photosynthetic rates of submerged rooted macrophytes. Oecologia 81:364–368; 1989.

O'Dell, K.M.; VanArman, J.; Welch, B.H.; Hill, S.D. Changes in water chemistry in a macrophyte-dominated lake before and after herbicide treatment. Lake Reserv. Manage. 11:311–316; 1995.

Ozimek, T.; van Donk, E.; Gulati, R.D. Can macrophytes be useful in the biomanipulation of lakes? The Lake Zwemlust example. Hydrobiologia 200/201:399–409; 1990.

Pokorny, J.; Kvet; Ondok, J.P.; Toul, Z.; Ostry, I. Production-ecological analysis of a plant community dominated by *Elodea canadensis* Michx. Aquat. Bot. 19:263–292; 1984.

Reinertsen, H.; Jensen, A.; Kokksvik, J.I.; Langeland, A.; Olsen, Y. Effects of fish removal on the limnetic ecosystem of a shallow lake. Can. J. Fish. Aquat. Sci. 47:166–173; 1990.

Reynolds, C.S. The response of phytoplankton communities to changing lake environments. Schweiz. Z. Hydrobiol. 49:220–236; 1987.

Rørslett, B.; Berge, D.; Johansen, S.W. Lake enrichment by submersed macrophytes: a Norwegian whole-lake experience with *Elodea canadensis*. Aquat. Bot. 26:325–340; 1986.

Sand-Jensen, K. Environmental variables and their effect on photosynthesis of aquatic plant communities. Aquat. Bot. 32:5–25; 1989.

Sarnelle, O. Herbivore effects on phytoplankton succession in a eutrophic lake. Ecol. Monogr. 63:129–149; 1993.

Schriver, P.; Bøgestrand, J.; Jeppesen, E.; Søndergaard, M. Impact of submerged macrophytes on fish-zooplankton-phytoplankton interactions: large-scale enclosure experiments in a shallow eutropic lake. Freshwat. Biol. 33:255–270; 1995.

Sher-Kaul, S.; Oertli, B.; Castella, E.; Lachavanne, J-B. Relationship between biomass and surface area of six submerged aquatic plant species. Aquat. Bot. 51:147–154; 1995.

Sommer, U. Some size relationships in phytoflagellated motility. Hydrobiologia 161:125–131; 1988.

Søndergaard, M. Seasonal variations in the loosely sorbed phosphorus fraction of the sediment of a shallow and hypereutrophic lake. Environ. Geol. Wat. Sci. 11:115–121; 1988.

Søndergaard, M.; Jeppesen, E.; Mortensen, E.; Dall, E.; Kristensen, P.; Sortkjær, O. Phytoplankton biomass reduction after planktivorous fish reduction in a shallow, eutrophic lake: a combined effect of reduced internal P-loading and increased zooplankton grazing. Hydrobiologia 200/201:229–240; 1990.

Stansfield, J.H.; Perrow, M.R.; Tench, L.D.; Jowitt, A.J.D.; Taylor, A.A.L. Do macrophytes act as refuges for grazing cladocera against fish predation? Wat. Sci. Techn. 32:217–220; 1995.

Stephen, D.; Moss, B.; Phillips, G.L. Do rooted macrophytes increase sediment phosphorus release? Hydrobiologia 342/343:27–34; 1997.

Timms, R.M.; Moss, B. Prevention of growth of potentially dense phytoplankton populations by zooplankton grazing, in the presence of zooplanktivorous fish, in a shallow wetland ecosystem. Limnol. Oceanogr. 29:472–486; 1984.

van Donk, E.; Gulati, R.D. Transition of a lake to turbid state six years after biomanipulation: mechanisms and pathways. Wat. Sci. Techn. 32:197–206; 1995.

van Donk, E.; Grimm, M.P.; Gulati, R.D.; Klein Breteler, J.P.G. Whole-lake food-web manipulation as a means to study community interactions in a small ecosystem. Hydrobiologia 200/201:275–289; 1990.

Watson, S.; McCauley, E.; Downing, J.A. Sigmoid relationships between phosphorus, algal biomass, and algal community structure. Can. J. Fish. Aquat. Sci. 49:2605–2610; 1992.

Weisner, S.; Eriksson, G.; Granéli, W.; Leonardson, L. Influence of macrophytes on nitrate removal in wetlands. Ambio 23:363–366; 1994.

Wium-Andersen, S.; Christophersen, C.; Houen, G. Allelopathic effects on phytoplankton by substances isolated from aquatic macrophytes (Charales). Oikos 39:187–190; 1982.

7. Role of Submerged Macrophytes for the Microbial Community and Dynamics of Dissolved Organic Carbon in Aquatic Ecosystems

Robert G. Wetzel and Morten Søndergaard

Introduction

Examination of the multifaced functions of submerged macrophytes in shallow lakes has once again placed emphasis on the habitat characteristics of the vegetation for fish and motile invertebrate communities. The habitat support functions of the macrophytic community are critical to these animals for refuge and related cryptic behavioral functions. We argue here, however, that much more fundamental structuring of microbial metabolism and biogeochemical cycling of the ecosystem result from the development of submerged macrophytic communities. These metabolic functions not only control biogeochemical cycling within these lake ecosystems but are essential to the success of diverse integrated higher trophic levels.

Plant Growth and Habitat Characteristics

The general productivity gradient across the land–water interface clearly recognizes that the maximum productivity of the biosphere occurs in the zone of emergent macrophytes (e.g., Westlake, 1963; Wetzel, 1990). These plants have several morphological, physiological, growth, and reproductive adaptations that capitalize on the high availability of water and nutrients and the suppression of competitive interactions with many less tolerant plant species.

Macrophytic productivity declines precipitously when submerged. Ample physiological evidence indicates that the primary suppressing mechanisms are the exponential light attenuation with depth and the reduction of gas and nutrient exchange, particularly for inorganic carbon, largely because diffusion processes for gases are some 10^4 times slower than in air (e.g., Raven, 1984). Although, theoretically, the exchange processes should be influenced greatly by boundary layer characteristics, the macrophyte beds themselves change the flow rates and the flow characteristics, both as a function of density and species. In addition, the presence of attached epiphytic microbial communities complicate and increase the diffusive layers and both physically and metabolically separate the macrophyte surfaces from surrounding water (Losee and Wetzel, 1993; Wetzel, 1993a, 1996). Thus, two common conditions emerge. (1) External water flow rates are rapidly dissipated within a few centimeters of the outer plant-bed boundary, even under severe external flow-rate conditions (30 cm/s or more), particularly among submerged macrophytes with large surface areas per unit biovolume (Losee and Wetzel, 1993). As a result, within-bed flows are so low that boundary layers are many millimeters in thickness and totally diffusion dependent. The same reduction is also found in macrophyte beds within moderately flowing streams (Madsen and Warncke, 1983; Sand-Jensen and Mebus, 1996). (2) Chemical conditions within periphyton communities are not only considerably different from the surrounding waters but highly dynamic on short time scales (minutes) (e.g., Carlton and Wetzel, 1987, 1988; Riber and Wetzel, 1987). These epiphytic communities are frequently viewed as resource competitors for the submerged macrophytes (e.g., Phillips et al., 1978), and indeed they can be when excessively developed under very eutrophic conditions. However, *from the standpoint of the ecosystem,* the productivity is effectively shifted from the macrophytes to the epiphytic communities on the submerged macrophytes. The macrophytes, although not profoundly productive, survive and are essential in providing substrata and massively diverse three-dimensional habitats for microbiotal colonization. As discussed from a mechanistic viewpoint in greater detail below, the result is *a shift in productivity from the macrophytes to the very high productivity of the attached microbiota.*

Certain morphological and physiological adaptations occur in submerged angiosperms, such as marked increases in surface area and pigment concentrations, as well as particularly efficient recycling of gases and nutrients within gas lacunae distributed throughout foliage and rooting tissues (cf. Søndergaard and Wetzel, 1980; Wetzel et al., 1984). Two important growth characteristics emerge in regard to submerged macrophytes that are directly germane to their roles in shallow lake ecosystems. (1) First, from the standpoint of the ecosystem, the most successful and productive submerged macrophytes are the herbaceous perennials that develop highly dissected foliage **within** the water column and not the rapidly growing surface canopy-forming species such as *Myriophyllum* and *Hydrilla* under conditions of eutrophy. Productivity of rosette perennials such as the isoetids are very much less productive, as are most of the few annual submerged plants. The most abundant emergent, floating-leaved, and submerged aquatic macrophytes are herbaceous perennial plants. True annual submerged plants, germinating from seed and essentially returning to seed biomass at the end of a brief growing season, constitute less than 10% of aquatic plants (Hutchinson, 1975). Two important ecosystem characteristics emerge from this characteristic. (a) Much of the nutrient pool is significantly recycled by translocation to the rooting tissues as foliage of each cohort completes growth. Nutrient recycling is further enhanced by acquisition of nutrients by the plants largely from interstitial waters of the hydro-soils, and most of it from microbial degradation of residual plant tissue and epiphytic microorganisms. (b) These herbaceous perennials exhibit continuous growth with numerous overlapping cohorts. For example, the emergent rush *Juncus effusus* has five cohorts per year (latitude 33°; Wetzel and Howe, in press) and *Typha latifolia* three cohorts per year (latitude 42°; Dickerman and Wetzel, 1985), and the rooted floating-leaved *Nymphaea odorata* has seven per year (latitude 33°; Carter, 1995). The submerged plant *Scirpus subterminalis* has three cohorts per year (latitude 42°; Rich et al., 1971) and *Ceratophyllum demersum* four to six cohort turnovers per year (latitude 42°; Spencer and Wetzel, 1993). Both of the latter species overwinter under ice cover (ca. 4 months) in viable evergreen, basal metabolic condition, which gives them great advantage in ac-quisition of nutrients during winter periods when nutrients are relatively abundant and rapid growth in the spring before that of competitors. Therefore, the variable but continuous growth results in continuous turnover of organic matter with production of relatively resistant particulate organic matter in senescent tissues and provides massive and constantly regenerating surface areas for photosynthetic and heterotrophic microbial colonization over the entire year. Annual plants do not exhibit such continuous growth and biomass turnover, and some perennials do not overwinter in large accumulations of aboveground biomass. Where examined in detail (e.g., references cited above; also Wetzel, 1983a), however, many of the submerged macrophytes exhibit high rates of turnover and, importantly, regenera-tion of surfaces for epiphytic microflora.

(2) Second, the thin, finely divided, and reticulated foliage increases surface area, which can greatly enhance exchange of gases with those of the water and interception of light (Sculthorpe, 1967). The resulting high ratio of surface area to volume increases substrata available for colonization by epiphytic algae, cyano-bacteria, and other microbes. For example, leaf surface area available for coloniza-tion by epiphytic microbes on the submerged linear-leaved macrophyte *Scirpus subterminalis* averaged 24 m^2/m^2 of lake area in the moderately developed littoral zone of a lake in southwestern Michigan (Burkholder and Wetzel, 1989). Such values are likely conservative and in reality can be much larger with even more finely dissected macrophytes. The extensive area of these submerged macrophyte surfaces provide myriad diverse microhabitats with attendant attached algal/cyanobacterial communities upward into littoral environments of relatively abun-dant light and dissolved gases from photosynthesis (O_2) and decomposition (CO_2). Nutrients within senescing macrophytes and their epiphytic microflora are dis-placed to the sediment surfaces and recycled. Some nutrients diffuse from the sites of decomposition to the interstitial waters of the sediments and detrital particulate matter of macrophytic origins. As a result, the productivity of epiphytic algae/

cyanobacteria and of benthic algae on macrophytic detritus frequently exceeds that of the supporting macrophytes (Wetzel, 1983b, 1996). Very high algal and microbial diversity (e.g., >80% of the freshwater species of algae; Round, 1981) are associated with highly dynamic and productive microhabitats of the surfaces of submerged macrophytes.

Loading and Fates of Dissolved Organic Carbon

Bacterioplankton

A widely held opinion in limnology is a direct quantitative coupling of bacterial productivity to phytoplankton productivity (Cole et al., 1988). Loadings of dissolved organic carbon (DOC) to shallow aquatic ecosystems are manifold, however, from both external and numerous internal sources of photosynthesis and decomposition (discussed further below). Most of the loadings of organic carbon from terrestrial and wetland sources is as DOC, which although certainly not refractory contains many compounds that are recalcitrant to rapid degradation (e.g., Wetzel, 1995). In addition, many other sources of DOC enter the pelagic zone from littoral productivity and from decomposition of imported organic matter. Thus, the dependence of bacterioplankton on DOC released from phytoplankton is highly variable. In every case in which the sources of DOC for bacterioplankton have been examined carefully in lakes, DOC produced by phytoplankton is inadequate to support bacterioplanktonic metabolism and growth by a factor of 3 to as much as 20 in shallow lakes (e.g., Börsheim and Andersen, 1987; Hessen, 1992; Tranvik, 1992; Søndergaard, 1993; del Giordio and Peters, 1994; Coveney and Wetzel, 1995). The rates of nonphytoplanktonic particulate organic carbon being mineralized by bacterioplankton can be modified strongly by grazing interactions of protozoan and metazoan grazing (e.g., Lyche et al., 1996).

Submerged macrophytes function in two major ways in regard to the impact of microheterotrophs and DOC turnover in the water column: (1) as a source of DOC for bacterioplanktonic productivity (discussed below), and (2) as a refuge for mesozooplankton from vertebrate predation, which allows much greater population densities of cladocerans and their grazing on flagellates, ciliates, and other protists. Although the planktonic protists have little direct link to submerged macrophytes, *Daphnia* can be very effective in reducing nanoflagellates and ciliates (e.g., to as much as 70% of planktonic nanoflagellate biovolume; Christoffersen, this volume, Chapter 17). As has been shown often, *Daphnia* populations can be very large as they develop in the food-abundant predation refuge of submerged macrophytes (Jeppesen et al., this volume, Chapter 5). Many of the *Daphnia* migrate laterally from dense to less densely colonized areas of the littoral or into the pelagic zone at night to feed on phytoplankton and protistian communities in the predation-reduced time window. Bacterioplankton abundance is low, but specific productivity can be high under these periods of heavy mesozooplankton grazing (Søndergaard et al., this volume, Chapter 15), but the bacteria can compensate with morphotype changes induced by high grazing pressures (Jürgens and Jeppesen, this volume, Chapter 16). Grazing-resistant morphotype changes are found under high nanoflagellate and ciliate grazing and not when the bacterioplanktonic grazing is dominated by mesozooplankton. When submerged macrophytes are removed or suppressed, fish predation reduce *Daphnia* populations to low levels (e.g., <10/L) and a corresponding increase in nanoflagellates and other protists. The latter microheterotrophs cycle carbon among the bacteria, and relatively little is incorporated into zooplankton and higher trophic levels (Lyche et al., 1996).

Bacterioplanktonic couplings with protistian grazing and their interactions with higher trophic levels can sometimes be effective in regulating bacterial productivity in the pelagic zones during some productive periods of the year. Although the macrophytes are functioning importantly as DOC sources for bacterioplankton and as refuge areas for influencing grazers on phyto- and bacterioplankton, we contend that *with the development of submerged macrophyte communities, bacterial productivity shifts from moderate levels within planktonic communities to a dominance by attached bacterial productivity.* The greatly increased attached bacterial productivity results not only from the large habitat development provided by the macrophyte surfaces but because of the metabolic couplings between attached algae/cyanobacteria and the attached bacteria. As a result, the primary mechanisms of nutrient retention and recycling are shifted from planktonic to sessile microbiota.

Attached Bacteria of Surfaces

In every lake examined in detail with quantitative measures of rate functions (not simply correlations), and particularly in shallow lakes, DOC obtained from phytoplanktonic sources (extracellular release, lysis) is totally inadequate to support the observed rates of even bacterioplankton productivity (e.g., Hessen, 1992; Tranvik, 1992; Coveney and Wetzel, 1995; Søndergaard et al., this volume, Chapter 15) or are not related directly to phytoplankton (Jeppesen et al., 1992; Søndergaard, 1993). The submerged macrophytes have heavily colonized surfaces, and the summation of the bacterial productivity on a three-dimensional basis is very much greater than are those rates in the bacterioplankton. We hypothesize here that *most of the bacterial productivity is associated with surfaces, particularly those of submerged macrophytes and particulate organic detritus* of these plants as they are constantly growing and cohorts are turning over with the detrital mass collecting at the sediment–water interface. Although the conceptual framework of the biofilm–macrophyte–pelagic interactions was proposed many years ago (Wetzel and Allen, 1970) and variously analyzed and verified (e.g., Allen, 1971; Allanson, 1973), good quantitative measurements of attached bacterial productivity are very few, difficult because of methodological problems caused by the large diffusive boundary layer, and inadequate for extrapolation (Table 7.1). A few other examples exist. In Lake Stigsholm, attached bacterial productivity averages ca. 1,000 mg C/m^2/day on leaves of *Potamogeton pectinatus* at 24 m^2/m^2, whereas the bacterioplanktonic

Table 7.1. Bacterial Abundance and Production on Submerged Macrophytes

Species	Abundance (cell/cm^2)	Production (μg C cm^{-2} h^{-1})	Remarks	References
Zostera marina[a]	8×10^7	0.14	TTI, Initial decomposition	Blum and Mills (1991)
Z. marina				
Green leaves	2×10^7	2.8	FDC	Newell (1981)
Brown leaves	4–6×10^7	4–6		
Z. marina	10^7	—		Kirchman et al. (1980)
Z. marina	2–8×10^6	0.4	TTI, average	Kirchman et al. (1984)
Z. capricorni[b]	5.2×10^7	0.005	TTI	Moriarty et al. (1985)
Cymodocea nodosa[a]	1.5×10^7	0.2	From O$_2$ respiration	Peduzzi & Herndl (1994)
Laminaria pallida	10^3–10^7	—	Min-max season average for tip	Mazure & Field (1980)
	10^6			
Ecklonia maximata	10^3–10^7	—	Highest viable counts	Mazure & Field (1980)
Laminaria longicruris	10^4–10^5	—	Seasonal plate counts	Laycock (1974)
Macrocystis integrifolia	2–37×10^6	—	Young to senescing blades	Velji & Albright (1986)
Rhizophora mangle	2.5×10^6	4.6×10^6 cells	TTI, day 1 day 6	Benner et al. (1988)
	2.2×10^8	1.6		
Ranunculus penicillatus	10^5	—	Leaf 1, direct count	Hossell & Baker (1979)
	2.6×10^7		Leaf 6, tropological variation	
Juncus effusus	—	0.06	LEU, average for plant detritus	Thomaz & Wetzel (1995)
		(4–32/g DW)		

[a]Z. marina: 350 cm^2/g DW, same for Cymodocea.
[b]The value applies to sediment area; leaf are index is unknown.
Abbreviations: FDC, frequency of dividing cells; TTI, ^{3}H-thymidine incorporation; LEU, ^{3}H-leucine incorporation.

productivity was ca. 300 mg C/m^2/day (Mo. Søndergaard, unpublished data). In another small Danish lake, Dystrup, heavily colonized by *Ceratophyllum submersum*, estimates in July and August of attached bacterial productivity (leucine to protein methodology) was ca. 68 g C/m^2/48 days versus ca. 21 g C/m^2 by the bacterioplankton over the same time period (Theil-Nielsen and Søndergaard, 1997). Even though the actual rates per unit area are often modest, at least with the conservative methods used, the total area in three-dimensional water column

RELATIVE BACTERIAL
PRODUCTION / AREA

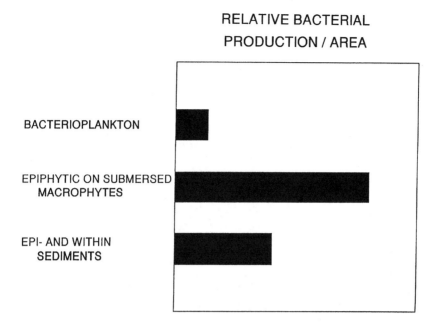

Figure 7.1. Suggested relationships of bacterial production per unit of lake area at surface in three habitats of shallow lakes with submersed macrophytes. (A) Bacterioplankton = production per m^2 of water column per year; (B) epiphytic on submersed macrophytes = production on all plant surfaces within the water column greater than 1 m^2 of sediment per year; (C) epipelic and within sediments = production per m^2 of sediments per year.

is very large. Hence, the collective bacterial productivity per square meter of water column below each square meter of water surface can be very large. A depth-scaling factor for the bacterioplankton is of equal importance, as is the leaf-area scaling factor within the water column for the macrophytes.

Because the productivity of the emergent, floating-leaved, and submerged macrophytes is high and constantly turning over, leachate from both living and senescent aquatic macrophytes is high and supports large communities of microbiota. Much of this leachate is used efficiently (4–45%/day) by the attached microbial community; relatively recalcitrant DOC compounds are released down-gradient and used at slower rates (ca. 1%/day) (Otsuki and Wetzel, 1974; Mickle and Wetzel, 1978, 1979; Søndergaard, 1983, 1990; Moran and Hodson, 1989; Wetzel, 1992; Mann and Wetzel, 1995, 1996). Partial photolytic degradation of recalcitrant dissolved organic compounds from allochthonous and from macrophytic sources occurs by natural ultraviolet (UV-B) of sunlight (cf. Wetzel et al., 1995; Moran and Zepp, in press). These photolytic products from humic and fulvic substances include many low-molecular-weight carbonyl compounds, such as numerous fatty acids, and nutrients including NH_4^+ and PO_4^{3-}, which are readily used and enhance bacterial productivity.

Thus, most bacterial metabolism in shallow lakes with submerged macrophytes is associated with benthic surfaces, as was emphasized long ago (Wetzel et al., 1972; Wetzel, 1983a). We project here that a major attribute of shallow lakes in which submerged macrophytic communities are well developed is the major increase in bacterial productivity associated with macrophytic surfaces (Fig. 7.1). The submerged macrophytes function as a major source of organic carbon, but a primary function is provision of three-dimensional surfaces within the water column. The bacteria are directly coupled to the productivity of the epiphytic algae. One cannot view P:R ratios of the phyto- and bacterioplankton alone; this myopic evaluation must be expanded to include the collective bacterial metabolism within the water column and that associated with surfaces (i.e., an ecosystem evaluation). When this corrected perspective is done, the P:R ratio is always less than 1 because of the importation of large amounts of particulate and especially DOC from allochthonous and land–water interface sources. Most of that respiration is detrital-based and associated with surfaces, particularly benthic surfaces (Wetzel, 1995).

Pivotal Metabolic Roles of Epiphytic Microbiota

For many years, microbiota attached to surfaces of living and detrital organic materials in littoral areas and wetlands have been largely ignored as quantitatively insignificant. Supporting evidence is still sparse but overwhelming that these microbial communities of autotrophic algae and cyanobacteria, and bacteria, fungi, and other heterotrophic microorganisms are not only very productive but can often serve as major regulators of nutrient dynamics in many freshwaters (Wetzel et al., 1972; Wetzel, 1983a,b, 1990). The importance of epiphytic algal productivity of Lawrence Lake, southwestern Michigan, is taken as an example because it has been studied in detail for many years. Submerged macrophytes extend in dense beds from 0 to 5.3 m of depth but constitute less than one-quarter of the annual net primary productivity in the lake of 180 g C/m^2 (Table 7.2). The rest originates from phytoplanktonic (13%) and epiphytic (71%) algae (Burkholder and Wetzel, 1989). These ratios are likely very common among shallow oligo- to moderately eutrophic lakes and certainly are not exceptional. Very few lakes have been studied in the required detail to make such comparisons (Søndergaard and Sand-Jensen, 1979).

The exceptionally high rates of primary productivity of the attached algae and cyanobacteria are only possible because of the intensive internal recycling of nutrients, including carbon, and gases within the attached microcommunities (Wetzel, 1993a). Steep gradients and very large boundary layers (millimeters in thickness) exist between the overlying water and within these attached microbial communities embedded in dense mucopolysaccharide matrices. The encapsulation of polysaccharide matrices increases densities and reduces diffusion rates even further. To maintain the very high ratio of attached algal productivity always measured in these communities, rapid, intensive recycling of inorganic nutrients

Table 7.2. Reconstructed Carbon Budget for Primary Production in Lawrence Lake, Considering Epiphytic Algae from Depths 0–5 m[a]

Component	Mean daily (mg C/m^2/day)	Mean annual (kg C/lake/yr)	Contribution (%)
Phytoplankton	119	2,154	13
Littoral zone algae <1 m	2,001	5,512	34
Littoral zone algae 1–5 m	500	5,968	37
Macrophytes ≤5 cm	241	2,701	16
Total		16,335	100

[a]Epipelic algal productivity was evaluated and constitutes <1% of the total in this lake (Carlton and Wetzel, unpublished).

(Adapted from data of Wetzel et al. [1972], Rich et al. [1971], and Burkholder and Wetzel [1989].)

and particulate and dissolved organic matter must be occurring among the producers and the heterotrophic organisms (bacteria, fungi, and protists) within the periphyton. This mutualistic recycling, demonstrated experimentally (Neely and Wetzel, 1995), allows the communities to maintain themselves and to efficiently sequester external sources of nutrients from the overlying water or from the substrata on which they grow for net growth, export, and reproduction.

Critical to the efficiency of the internal recycling is the coupling between attached microbial photosynthesis and heterotrophic bacterial metabolism. Rapid recycling of nutrients and gases within the periphytic complex is facilitated by the proximate juxtaposition of cells, often in immediate contact with each other. Experimental analyses have demonstrated a tight metabolic coupling among these organisms. For example, suppression of photosynthesis of the attached algae results in an immediate reduction of attached bacterial productivity (Neely and Wetzel, 1995) and certainly also nutrient and gas recycling. As a result, the nutrient retention capabilities of the attached communities is reduced markedly if the photosynthetic interactions within these mutually coupled communities are reduced or removed.

Extracellular mucilaginous material (i.e., exopolysaccharides or exopolymer secretions) occurs as coatings around most individual microbial cells or projections from cells. This material ultimately forms a matrix inhabited by a variety of microorganisms, particularly bacteria, algae, protozoa, and fungi (e.g., Fletcher and Marshall, 1982). Excretion of exopolymer fibrils by bacteria is an important initial phase of attachment of microbes to surfaces (Fletcher and Floodgate, 1973; van Loosdrecht et al., 1990; Fletcher, 1991), and a substantial portion of any attached microbial community will be composed of nonliving mucilaginous materials. Most of this material is mucopolysaccharide, although the exact composition and texture varies with environmental conditions as well as the nature and condition of the organisms that secrete it (e.g., Sutherland, 1985).

Problems of slow nutrient diffusion into attached microbial communities are particularly acute in relatively thick communities, where high productivity is maintained by intensive internal recycling of nutrients and gases among the metabolically interdependent microbiota (Wetzel, 1993a). Oxygen and CO_2 are interchanged from photosynthesis and respiration, DOC, and organic micro-nutrients (e.g., vitamins) released from algae or bacteria—large portions are exchanged rapidly without leaving the confines of the attached communities. Certainly, the metabolism of the supporting macrophyte functions both as a source and recipient of materials of the attached community. For example, some epiphytic algal species growing adnate to submerged macrophyte leaves can obtain more than 60% of their phosphorus from the living macrophytes (Moeller et al., 1988). Similarly, adnate bacteria can obtain significant carbon from the macrophyte (Kirchman et al., 1984).

Abiotic adsorption of particulate and dissolved organic carbon, as well as incorporation of inorganic nutrients and metals into the mucilaginous matrix, can provide a supplemental mechanism to diffusion that results in nutrients/metal inputs into attached microbiota (Lock, 1981, 1990; Wetzel, 1983b; Lock et al., 1984; Roemer et al., 1984; Beveridge and Graham, 1991). This mechanism likely not only stimulates metabolism of the community but, in the case of certain cations (e.g., Ca, Mg, Fe, Mn), affects the physical properties of the mucilage matrix, such as hydrophobicity (e.g., Lemke et al., 1995) and possibly texture. The extracellular matrix also provides sites of attachment for extracellular enzymes (e.g., phosphatases and proteases) that are critical in rendering nutrients and carbon available to microorganisms (Wetzel, 1990, 1991).

Further studies have examined the effects of varying concentrations of natural dissolved organic acids from decomposing aquatic plants on the metabolism and productivity of attached algae and bacteria (Wetzel, 1991, 1992; Kim and Wetzel, 1993; Wetzel et al., 1997). Microscopic examination of periphyton communities during these studies revealed apparent reductions in and alterations of mucilaginous matrices on exposure to various types of natural and purified organic acids from decomposing macrophytes. Direct experiments, using both scanning electron microscopy and energy dispersive X-ray microanalyses of copper-containing dyes that bind to glycosaminoglycans of the mucilage, showed that DOC-treated communities contained 10–57% less mucilage than non–DOC-treated controls (Wetzel et al., 1997). These examples only suggest (1) the importance of the mucopolysaccharide matrices in influencing the rates of adsorption and transport of nutrients and gases into and from the periphytic communities and (2) that many environmental parameters potentially influence the diffusion rates into the communities from the overlying water. These parameters are likely dynamic in their effects on mucopolysaccharide permeability properties on a short time basis.

Conclusions

1. In addition to habitat functions critical for refuge and behavior of fish and invertebrates, we argue that submerged macrophyte-epiphytic microbiotic communities function in fundamental structuring of microbial metabolism and

biogeochemical cycling of entire shallow lake ecosystems. Additional structuring roles encourage development of mesoplankton and depression of phytoplankton and microbial protista. Bacterial productivity is increased by high rates of nutrient recycling and the metabolic couplings within the macrophyte–periphyton complex.

2. The instantaneous primary productivity of submerged macrophytes is moderated because of a number of physical and physiological constraints, particularly reduced light and gaseous/nutrient diffusion in water. These environmental limitations have resulted in (1) a predominance of continuously growing and senescing herbaceous perennial species with a number of multiple, simultaneously growing cohorts at different stages of development and (2) a marked increase in surface areas of living, senescing, and dead tissues.

3. The extremely high surface area of living, senescing, and dead tissues projecting *within and up into* the water column promotes development of a highly mutualistic attached microbial community. Preliminary experimental data suggest that these attached communities require rapid intensive recycling of carbon, phosphorus, nitrogen, and other nutrients and growth factors between producers, particulate and dissolved detritus, and bacteria and protists. The organization of the attached communities in a matrix of mucopolysaccharides in a diffusion controlled environment facilitates internal community recycling and results in a highly efficient reutilization of resources. This high growth can then augment the internal recycling and small losses with external inputs of carbon and nutrients from the surrounding water or from the supporting substrata. As a result, a shift occurs in the ecosystem productivity from submerged macrophytes of modest productivity to very high productivity of the attached microbiota. Attached algal productivity can exceed that of the submerged macrophytes and phytoplankton combined. Composite attached bacterial productivity also can be much greater than that of the bacterioplankton and can dominate ecosystem degradative productivity.

4. DOC from allochthonous sources entering the submerged macrophyte–epiphyte complex can be effectively scavenged by physical and metabolic use. Use of recalcitrant DOC can be markedly enhanced by partial UV-B photolysis by natural sunlight. Rapid use of much of the DOC from this photolytic decomposition and that of senescing submerged macrophytes results in selective increases in the chemical recalcitrance of DOC passing through the submerged macrophyte–detritus–epiphyte complex. The P:R ratios for the lake ecosystems are always less than 1 because of the importation of large amounts of DOC.

5. Productivity and nutrient recycling capacities are maximized within the water by extensive development of submerged macrophytes and associated epiphytic microbiota. Prevention of the development of the submerged macrophyte–epiphytic complexes by any mechanism (e.g., turbidity, excessive nutrient enrichment, epidemic herbivory, toxicity) will usually result in precipitous declines in productivity, nutrient recycling, and nutrient retention capacities.

We recognize that these conclusions are drawn from relatively few data. Much additional supporting information is needed, and we particularly recommend intensive investigation of the following:

- Intensive comparative quantitative measurements of microbial production in attached communities against those of the open water column. These measures are complicated by the slow diffusion rates and rapid recycling that occur within the attached communities, as discussed in this chapter.
- The sources of DOC used by attached and planktonic bacteria from allochthonous, macrophytic, and phytoplanktonic sources. It is particularly important to evaluate the suggestions made in this discourse that most submerged macrophytes do not senesce en mass but rather are constantly turning over with multiple continuous cohorts.
- The relative impacts on bacterial productivity of DOC from submerged macrophytes released rapidly from cells undergoing slow, progressive senescence versus DOC released from attached bacteria and fungi that are slowly degrading the structural tissues of submerged macrophytes.
- If the presence of submerged macrophytes in general promote the development of large populations of mesozooplankton and suppress protistian micrograzers and thereby reduce the effectiveness of microbial loop cycling in the water.

References

Allanson, B.R. The fine structure of the periphyton of *Chara* sp. and *Potamogeton natans* from Wytham Pond, Oxford, and its significance to the macrophyte-periphyton metabolic model of R.G. Wetzel and H.L. Allen. Freshwat. Biol. 3:535–542; 1973.

Allen, H.L. Primary productivity, chemo-organotrophy, and nutritional interactions of epiphytic algae and bacteria on macrophytes in the littoral of a lake. Ecol. Monogr. 41:97–127; 1971.

Benner, R.; Hodson, R.E.; Kirchman, D. Bacterial abundance and production on mangrove leaves during initial stages of leaching and biodegradation. Arch. Hydrobiol. Beih. Ergebn. Limnol. 31:19–26; 1988.

Beveridge, T.J.; Graham, L.L. Surface layers of bacteria. Microbiol. Rev. 55:684–705; 1991.

Blum, L.K.; Mills, A.L. Microbial growth and activity during the initial stages of seagrass decomposition. Mar. Ecol. Progr. Ser. 70:73–82; 1991.

Börsheim, K.Y.; Andersen, S. Grazing and food selection by crustacean zooplankton compared to production of bacteria and phytoplankton in a shallow Norwegian mountain lake. J. Plankton Res. 9:367–379; 1987.

Burkholder, J.M.; Wetzel, R.G. Epiphytic microalgae on natural substrata in a hardwater lake: seasonal dynamics of community structure, biomass and ATP content. Arch. Hydrobiol. Suppl. 83:1–56; 1989.

Carlton, R.G.; Wetzel, R.G. Distributions and fates of oxygen in periphyton communities. Can. J. Bot. 65:1031–1037; 1987.

Carlton, R.G.; Wetzel, R.G. Phosphorus flux from lake sediments: effect of epipelic algal photosynthesis. Limnol. Oceanogr. 33:562–570; 1988.

Carter, S.M. Herbivory by *Donacia rufescens* Lacordaire and *Donacia cincticornis* Newman on the white water-lily, *Nymphaea odorata* Aiton. M.Sc. thesis, Univ. Alabama, Tuscaloosa; 1995.

Cole, J.J.; Findlay, S.; Pace, M.L. Bacterial production in fresh and saltwater ecosystems: A cross-system overview. Mar. Ecol. Progr. Ser. 43:1–10; 1988.

Coveney, M.F.; Wetzel, R.G. Biomass, production, and specific growth rate of bacterioplankton and coupling to phytoplankton in an oligotrophic lake. Limnol. Oceanogr. 40:1187–1200; 1995.

del Giordio, P.A.; Peters, R.H. Patterns in planktonic P:R ratios in lakes: influence of alke trophy and dissolved organic carbon. Limnol. Oceanogr. 39:772–787; 1994.

Dickerman, J.A.; Wetzel, R.G. Clonal growth in *Typha latifolia:* population dynamics and demography of the ramets. J. Ecol. 73:535–552; 1985.

Fletcher, M. The physiological activity of bacteria attached to solid surfaces. Adv. Microb. Physiol. 32:53–85; 1991.

Fletcher, M.; Floodgate, G.D. An electron-microscopic demonstration of an acidic polysaccharide involved in the adhesion of a marine bacterium to solid surfaces. J. Gen. Microbiol. 74:325–334; 1973.

Fletcher, M.; Marshall, K.C. Are solid surfaces of ecological significance to aquatic bacteria? Adv. Microb. Ecol. 6:199–236; 1982.

Hessen, D.O. Dissolved organic carbon in a humic lake: effects on bacterial production and respiration. Hydrobiologia 229:115–123; 1992.

Hossell, J.; Baker, J.H. Epiphytic bacteria of the freshwater plant *Ranunculus penicillatus:* enumeration, distribution and identification. Arch. Hydrobiol. 86:322–337; 1979.

Hutchinson, G.E. A treatise on limnology. III. Limnological botany. New York: John Wiley & Sons; 1975.

Jeppesen, E.; Sortkjaer, O.; Søndergaard, M.; Erlandsen, M. Impact of a trophic cascade on heterotrophic bacterioplankton production in two shallow fish-manipulated lakes. Arch. Hydrobiol. Beih. Ergebn. Limnol. 37:219–231; 1992.

Kim, B.; Wetzel, R.G. The effect of dissolved humic substances on the alkaline phosphatase and the growth of microaigae. Verh. Int. Verein. Limnol. 25:122–128; 1993.

Kirchman, D.L.; Mazzella, L.; Mitchell, R.; Alberte, R.S. Bacterial epiphytes on *Zostera marina* surfaces. Biol. Bull. 159:461–462; 1980.

Kirchman, D.L.; Mazzella, L.; Alberte, R.S.; Mitchell, R. Epiphytic bacterial production on *Zostera marina.* Mar. Ecol. Progr. Ser. 15:117–123; 1984.

Laycock, R.A. The detrital food chain based on seaweeds. I. Bacteria associated with the surface of *Laminaria* fronds. Mar. Biol. 25:223–231; 1974.

Lemke, M.J.; Churchill, P.F.; Wetzel, R.G. The effect of substrate and cell surface hydrophobicity on phosphate uptake in bacteria. Appl. Environ. Microbiol. 61:913–919; 1995.

Lock, M.A. River epilithon—a light and organic energy transducer. In: Lock, M.A.; Williams, D.D., eds. Perspectives in running water ecology. New York: Plenum Press; 1981:3–40.

Lock, M.A. The dynamics of dissolved and particulate organic material over the substratum of water bodies. In: Wotton, R.S., ed. The biology of particles in aquatic systems. Boca Raton, FL: CRC Press; 1990:117–144.

Lock, M.A.; Wallace, R.R.; Costerton, J.W.; Ventullo, R.M.; Charlton, S.E. River epilithon (biofilm): toward a structural functional model. Oikos 42:10–22; 1984.

Losee, R.F.; Wetzel, R.G. Littoral flow rates within and around submersed macrophyte communities. Freshwat. Biol. 29:7–17; 1993.

Lyche, A.; Andersen, T.; Christoffersen, K.; Hessen, D.O.; Hansen, P.H.B.; Klysner, A. Mesocosm tracer studies. 2. The fate of primary production and the role of consumers in the pelagic carbon cycle of a mesotrophic lake. Limnol. Oceanogr. 41:475–487; 1996.

Madsen, T.V.; Warncke, E. Velocities of currents around and within submerged aquatic vegetation. Arch. Hydrobiol. 97:389–394; 1983.

Mann, C.J.; Wetzel, R.G. Dissolved organic carbon and its utilization in a riverine wetland ecosystem. Biogeochemistry 31:99–120; 1995.

Mann, C.J.; Wetzel, R.G. Loading and utilization of dissolved organic carbon from emergent macrophytes. Aquat. Bot. 53:61–72; 1996.

Mazure, H.G.F.; Field, J.G. Density and ecological importance of bacteria on kelp fronds in an upwelling region. J. Exp. Mar. Biol. Ecol. 43:173–182; 1980.

Mickle, A.M.; Wetzel, R.G. Effectiveness of submersed angiosperm-epiphyte complexes on exchange of nutrients and organic carbon in littoral systems. II. Dissolved organic carbon. Aquat. Bot. 4:317–329; 1978.

Mickle, A.M.; Wetzel, R.G. Effectiveness of submersed angiosperm-epiphyte complexes on exchange of nutrients and organic carbon in littoral systems. III. Refractory organic carbon. Aquat. Bot. 6:339–355; 1979.

Moeller, R.E.; Burkholder, J.M.; Wetzel, R.G. Significance of sedimentary phosphorus to a rooted submersed macrophytes (*Najas flexilis* (Willd.) Rostk. and Schmidt) and its algal epiphytes. Aquat. Bot. 32:261–281; 1988.

Moran, M.A.; Hodson, R.E. Formation and bacterial utilization of dissolved organic carbon derived from detrital lignocellulose. Limnol. Oceanogr. 34:1034–1047; 1989.

Moran, M.A.; Zepp, R.G. Role of photolysis in the formation of biologically labile compounds from dissolved organic matter. Limnol. Oceanogr. (in press).

Moriarty, D.J.W.; Boon, P.I.; Hansen, J.A.; Hunt, W.G.; Poiner, I.R. Microbial biomass and productivity in seagrass beds. Geomicrobiol. J. 4:21–51; 1985.

Neely, R.K.; Wetzel, R.G. Simultaneous use of ^{14}C and ^{3}H to determine autotrophic production and bacterial protein production in periphyton. Microb. Ecol. 30:227–237; 1995.

Newell, S.Y. Fungi and bacteria in or on leaves of eelgrass (*Zostera marina* L.) from Chesapeake Bay. Appl. Environ. Microbiol. 41:1219–1224; 1981.

Otsuki, A.; Wetzel, R.G. Release of dissolved organic matter by autolysis of a submerged macrophyte, *Scirpus subterminalis*. Limnol. Oceanogr. 19:842–845; 1974.

Peduzzi, P.; Herndl, G.J. Decomposition and significance of seagrass leaf litter (*Cymodocea nodosa*) for the microbial foodweb in coastal waters (Gulf of Trieste, northern Adriatic Sea). Mar. Ecol. Progr. Ser. 71:163–174; 1994.

Phillips, G.L.; Eminson, D.; Moss, B. A mechanism to account for macrophyte decline in progressively eutrophicated fresh waters. Aquat. Bot. 4:103–126; 1978.

Raven, J.A. Energetics and transport in aquatic plants. New York: Alan R. Liss, Inc.; 1984.

Riber, H.H.; Wetzel, R.G. Boundary-layer and internal diffusion effects on phosphorus fluxes in lake periphyton. Limnol. Oceanogr. 32:1181–1194; 1987.

Rich, P.H.; Wetzel, R.G.; Thuy, N.V. Distribution, production and role of aquatic macrophytes in a southern Michigan marl lake. Freshwat. Biol. 1:3–21; 1971.

Roemer, S.C.; Hoagland, K.D.; Rosowski, J.R. Development of a freshwater periphyton community as influenced by diatom mucilages. Can. J. Bot. 62:1799–1813; 1984.

Round, F.E. The ecology of algae. Cambridge: Cambridge University Press; 1981.

Sand-Jensen, K.; Mebus, J.R. Fine-scale patterns of water velocity within macrophyte patches in streams. Oikos 76:169–180; 1996.

Sculthorpe, C.D. The biology of aquatic vascular plants. New York: St. Martin's Press; 1967.

Søndergaard, M. Heterotrophic utilization and decomposition of extracellular organic carbon (EOC) released by the aquatic angiosperm *Littorella uniflora*. Aquat. Bot. 16:59–73; 1983.

Søndergaard, M. Extracellular organic carbon (EOC) in the genus *Carpophyllum* (Phaeophyceae): Diel release patterns and EOC lability. Mar. Biol. 104:143–151; 1990.

Søndergaard, M. Organic carbon pools in two Danish lakes: flow of carbon to bacterioplankton. Verh. Int. Verein. Limnol. 25:593–598; 1993.

Søndergaard, M.; Sand-Jensen, K. Total autotrophic production in oligotrophic Lake Kalgaard, Denmark. Verh. Int. Verein. Limnol. 20:667–673; 1979.

Søndergaard, M.; Wetzel, R.G. Photorespiration and internal recycling of CO_2 in the submersed angiosperm *Scirpus subterminalis* Torr. Can. J. Bot. 58:591–598; 1980.

Spencer, W.E.; Wetzel, R.G. Acclimation of photosynthesis and dark respiration of a submersed angiosperm beneath the ice in a temperate lake. Plant Physiol. 101:985–991; 1993.

Sutherland, I.W. Biosynthesis and composition of gram-negative bacterial extracellular and wall polysaccharides. Annu. Rev. Microbiol. 39:243–270; 1985.

Thomaz, S.M.; Wetzel, R.G. [^{3}H]leucine incorporation methodology to estimate epiphytic bacterial biomass production. Microb. Ecol. 29:63–70; 1995.

Tranvik, L. Allochthonous dissolved organic matter as an energy source for pelagic bacteria and the concept of the microbial loop. Hydrobiologia 29:107–114; 1992.

van Loosdrecht, M.C.M.; Lyklema, J.; Norde, W.; Zehnder, A.J.B. Influence of interfaces on microbial activity. Microbiol. Rev. 54:75–87; 1990.

Velji, M.I.; Albright, L.J. Microscopic enumeration of attached marine bacteria of seawater, marine sediment, fecal matter, and kelp blade samples following pyrophosphate and ultrasound treatments. Can. J. Microbiol. 32:121–126; 1986.

Westlake, D.F. Comparisons of plant productivity. Biol. Rev. 38:385–425; 1963.

Wetzel, R.G. Limnology. 2nd Ed. Philadelphia: Saunders College Publishing; 1983a.

Wetzel, R.G. Attached algal-substrata interactions: fact or myth, and when and how? In: Wetzel, R.G., ed. Periphyton in freshwater ecosystems. The Hague: Dr. W. Junk Publishers; 1983b:207–215.

Wetzel, R.G. Land–water interfaces: metabolic and limnological regulators. Verh. Int. Verein. Limnol. 24:6–24; 1990.

Wetzel, R.G. Extracellular enzymatic interactions in aquatic ecosystems: storage, redistribution, and interspecific communication. In: Chróst, R.J., ed. Microbial enzymes in aquatic environments. New York: Springer-Verlag; 1991:6–28.

Wetzel, R.G. Gradient-dominated ecosystems: sources and regulatory functions of dissolved organic matter in freshwater ecosystems. Hydrobiologia 229:181–198; 1992.

Wetzel, R.G. Microcommunities and microgradients: linking nutrient regeneration, microbial mutualism, and high sustained aquatic primary production. Netherlands J. Aquat. Ecol. 27:3–9; 1993a.

Wetzel, R.G. Humic compounds from wetlands: complexation, inactivation, and reactivation of surface-bound and extracellular enzymes. Verh. Int. Verein. Limnol. 25:122–128; 1993b.

Wetzel, R.G. Death, detritus, and energy flow in aquatic ecosystems. Freshwat. Biol. 33:83–89; 1995.

Wetzel, R.G. Benthic algae and nutrient cycling in lentic freshwater ecosystems. In: Stevenson, R.J.; Bothwell, M.L.; Lowe, R.L., eds. Algal ecology: freshwater benthic ecosystems. New York: Academic Press; 1996:641–667.

Wetzel, R.G.; Allen, H.L. Functions and interactions of dissolved organic matter and the littoral zone in lake metabolism and eutrophication. In: Kajak, Z.; Hillbricht-Illkowska, A., eds. Productivity problems of fresh waters. Warsaw: PAN Publishers; 1970:333–347.

Wetzel, R.G.; Howe, M.J. Population dynamics and seasonal growth and biomass patterns of an emergent rush (*Juncus effusus* L.) in a subtemperate Alabama wetland. Aquat. Bot. (in press).

Wetzel, R.G.; Rich, P.H.; Miller, M.C.; Allen, H.L. Metabolism of dissolved and particulate detrital carbon in a temperate hard-water lake. Mem. Ist. Ital. Idrobiol. 29(suppl.):185–243; 1972.

Wetzel, R.G.; Brammer, E.S.; Forsberg, C. Photosynthesis of submersed macrophytes in acidified lakes. I. Carbon fluxes and recycling of CO_2 in *Juncus bulbosus* L. Aquat. Bot. 19:329–342; 1984.

Wetzel, R.G.; Hatcher, P.G.; Bianchi, T.S. Natural photolysis by ultraviolet irradiance of recalcitrant dissolved organic matter to simple substrates for rapid bacterial metabolism. Limnol. Oceanogr. 40:1369–1380; 1995.

Wetzel, R.G.; Ward, A.K.; Stock, M. Effects of natural dissolved organic matter on mucilaginous matrices of biofilm communities. Arch. Hydrobiol. 140:1–11; 1997.

8. Impact of Herbivory on Plant Standing Crop: Comparisons Among Biomes, Between Vascular and Nonvascular Plants, and Among Freshwater Herbivore Taxa

David M. Lodge, Greg Cronin, Ellen van Donk, and
Adrienne J. Froelich

Introduction

Two contradictory traditions exist regarding the impact of herbivores on the ecology and evolution of plants. For ecologists studying terrestrial ecosystems, the interaction between plants and their consumers has been a focal point for research in recent decades. Herbivores are widely regarded as an important determinant of plant abundance and species composition and as an important selective force in the evolution of terrestrial plant traits (Rhoades, 1985; Herms and Mattson, 1992; Rosenthal and Berenbaum, 1992). Similarly, the abundance of many marine plants is often reduced by herbivores, and many seaweed traits are thought to have evolved in response to herbivory (Lubchenco and Gaines, 1981; Gaines and Lubchenco, 1982; Estes and Steinberg, 1988; Hay, 1991). By contrast, for decades the paradigm in limnology has been that live freshwater macrophytes are too tough for the mouthparts of aquatic herbivores, are of low nutritional quality, and are rarely consumed by herbivores (Lodge, 1991; Newman, 1991).

The idea that freshwater herbivores are unimportant has recently come under more careful scrutiny. Data contradict the assumption of low nitrogen content (Duarte, 1992) and suggest that live macrophytes may be consumed far more than is often appreciated (Lodge, 1991). Whereas many past freshwater studies documented large reductions of periphytic algae by herbivores (Feminella and Hawkins, 1995; Lamberti, 1996; Steinman, 1996), an increasing number of direct tests of the impact of freshwater herbivores on macrophyte abundance are now being conducted.

If freshwater herbivores reduce aquatic macrophyte biomass or change the species composition of plants in lakes, they could modify the many important biotic, abiotic, and biogeochemical processes in which macrophytes are involved (Carpenter and Lodge, 1986; many chapters in this volume). Given this potentially important role of freshwater herbivores in community and ecosystem function, it becomes essential to better understand freshwater macrophyte–herbivore interactions.

In this chapter, we use recent studies to evaluate the impact of herbivores on freshwater macrophytes. First, we compare the magnitude of impact of herbivores on plant standing crop in freshwater, marine, and terrestrial habitats. If the magnitude of herbivore impact on freshwater macrophytes is at least as great as that in other biomes, then there is no a priori reason to believe its evolutionary or ecological importance is less than that of terrestrial herbivores.

Second, for freshwater and marine habitats, we compare the herbivore impact on nonvascular with that on vascular plants. Interbiome comparisons could be partly confounded by the different taxonomic affinities of the plants common to different biomes. Marine studies focus on macroalgae (because few marine vascular species exist), terrestrial studies focus on vascular plants, and freshwater studies have historically focused on microalgae (but are increasingly addressing macroalgae and vascular macrophytes). Perhaps algae are generally more susceptible to herbivory than vascular macrophytes because most algae contain less structural materials (e.g., cellulose, lignin, cuticles), which render some vascular plants very tough and/or indigestible (Grubb, 1986; Rosenthal and Berenbaum, 1992), and because the lack of vascular tissues mean algae cannot efficiently transport defensive compounds (Cronin and Hay, 1996b). Or the typically higher growth rates of algae may simply allow them to better tolerate herbivory (Lubchenco and Gaines, 1981). Our comparison of herbivore impact will test whether algae are more reduced by herbivores than vascular plants.

Third, narrowing our focus to the freshwater habitat, we compare the impact on both algae and macrophytes of different taxa of herbivores. Most freshwater studies of herbivory have focused on insects and snails. It could be that other groups of herbivores, including crustaceans, fishes, turtles, mammals, and aquatic birds, have an equal or larger impact on freshwater plants that has simply not been measured because the traditional training of limnologists has not included work with these taxa.

Fourth, because our analysis suggests that some of these other taxa may, in fact, often reduce macrophyte standing stock, we briefly review literature suggesting that the spatial and temporal dynamics of some of these taxa are complex and require much more thorough study by limnologists. We address the hypothesis that the abundance and impact of mammalian and avian herbivores have been lower during the past century than during most of the evolutionary history of freshwater plants because of anthropogenic reductions in their populations. We also present data that suggest this trend has reversed, at least in North America and northern

Europe. Thus, we might expect herbivory by mammals and birds to be increasingly important.

Fifth, we use the literature to assess the degree of diet specialization of different herbivore taxa. Available data suggest that many of the least studied taxa (which may also be increasing in importance) are generalists (i.e., consume many macrophyte species in several plant families). To understand and predict their impact on lake macrophyte communities, we propose a conceptual model of feeding selectivity by generalist herbivores.

Sixth and finally, we evaluate how the impact of herbivores on macrophyte abundance and species composition may affect the shifts of lakes between the clear water and turbid states.

Does Herbivore Impact Differ Among Terrestrial, Marine, and Freshwater Biomes, and Between Vascular and Nonvascular Plants?

Using previously published primary and secondary literature sources, we compared the reduction in standing crop of vascular and nonvascular plants (excluding phytoplankton) caused by herbivores in terrestrial, marine, and freshwater habitats (Fig. 8.1). With respect to the herbivores, we excluded studies on livestock and on herbivores introduced specifically to control plants. With respect to methods, we used both experimental studies (in which herbivores were directly excluded and/or included with cages) and comparisons of otherwise similar unmanipulated sites with and without herbivores. From all studies, we calculated the percentage reduction of plant abundance (A) as $A = [(A_{-herbivore} - A_{+herbivore})/A_{-herbivore}]100$. We used whatever index of abundance that the original author used (e.g., biomass, numbers, leaf area damaged).

This or similar indices have been used to estimate grazer consumption and the impact of herbivores on plant productivity (Cyr and Pace, 1993; Cattaneo and Mousseau, 1995), but such uses can lead to errors of large magnitude because of the timing or duration of experiments, density-dependent plant growth, and uncontrolled herbivore feedbacks (Jacobsen and Sand-Jensen, 1994; Mitchell and Wass, 1996a; Wass and Mitchell, this volume, Chapter 18; Mitchell and Perrow, this volume, Chapter 9). Thus, with our comparisons, we do not imply anything about consumption rates of herbivores or about the impact of herbivores on plant production. Rather, we use it solely as an index of the differences in plant standing stock that result from the entangled set of complex interactions that differ in the presence and absence of herbivores. Yet even this application requires considerable caution in interpretation because plant abundance A is often very dependent on the timing and duration of a study relative to the growth cycle of the plants (Mitchell and Wass, 1996a; Wass and Mitchell, this volume, Chapter 18; Mitchell and Perrow, this volume, Chapter 9). Nevertheless, A is the statistic that can be calculated from the greatest number of studies, and few studies provide the time course of data necessary to evaluate the time dependence of A. In our literature survey, the spatial and temporal scale of studies ranged widely and no doubt

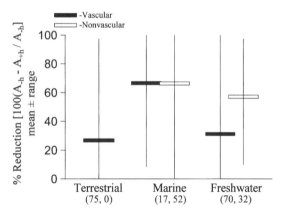

Figure 8.1. Mean and range of percentage reduction of standing crop of vascular and nonvascular plants (excluding phytoplankton) in terrestrial, marine, and freshwater habitats. Sample numbers (numbers of separate experiments or comparisons) are in parentheses below the habitat name (vascular, nonvascular). The few negative values were not plotted but are listed here: −2, −20, −38 for terrestrial; −5, −5, −50 for marine; −24, −47 for freshwater. (Data for freshwater habitat from Abrahamsson, 1966; Dean, 1969; Pelkian et al., 1971; Soska, 1975; Urban, 1975; Anderson and Low, 1976; Jupp and Spence, 1977; Verhoeven, 1978; Kiorboe, 1980; van der Velde, 1982; Prejs, 1984; Carter and Rybicki, 1985; Smith and Kadlec, 1985; Wallace and O'Hop, 1985; Hansson et al., 1987; Lodge and Lorman, 1987; Scott and Haskins, 1987; Korshgen et al., 1988; Painter and McCabe, 1988; Esler, 1989; Feminella and Resh, 1989; Crutchfield et al., 1992; Hart, 1992; Jacobsen and Sand-Jensen, 1992, 1994, 1995; Julien et al., 1992; Maceina et al., 1992; Martin et al., 1992; Matthews and Reynolds, 1992; Shaffer et al., 1992; Underwood et al., 1992; Wootton and Oemke, 1992; Sarnelle et al., 1993; Urbanc and Blejec, 1993; Brönmark, 1994; Chikwenhere, 1994; Conover and Kania, 1994; Creed, 1994; Hoyer and Canfield, 1994; Kornijów, 1994, 1996; Lodge et al., 1994; Schutten et al., 1994; Taylor et al., 1994; van Donk et al., 1994; Vermaat, 1994; Feminella and Hawkins, 1995; Richardson et al., 1995; Taylor and Grace, 1995; van Donk and Gulati, 1995; Søndergaard et al., 1996. Data for marine habitat from Castenholz, 1961; Randall, 1961, 1965; Sammarco et al., 1974; Wanders, 1977; Ogden et al., 1979; Vance, 1979; Duggins, 1980; Sammarco, 1980, 1982; Carpenter, 1981, 1986; Hatcher, 1981; Hay, 1981, 1984; Gaines and Lubchenco, 1982; Hatcher and Larkum, 1983; Hay and Goertemiller, 1983; Hay et al., 1983; Himmelman et al., 1983; Bertness, 1984; Cargill and Jefferies, 1984, 1986; Hay and Taylor, 1985; Bazely and Jeffries, 1986; Lewis, 1986; Fletcher, 1987; Foster, 1987; Lewis et al., 1987; Russ, 1987; van Tamelan, 1987; Morrison, 1988; Santelices and Martinez, 1988; Buschmann, 1990; Keats et al., 1990; Geller, 1991; Valentine and Heck, 1991; Jones, 1992; Polunin and Klumpp, 1992; Andrew and Underwood, 1993; Benedetti-Cecchi and Cinelli, 1993; Mitchell et al., 1994; Estes and Duggins, 1995; Heck and Valentine, 1995; Prince, 1995; Steinberg et al., 1995; Miller and Hay, 1996. Data on terrestrial habitats from Witkowski, 1980; Ydenberg and Prins, 1981; Coppock et al., 1983; McNaughton, 1985; Hughes et al., 1987; Huntley, 1987; Owen-Smith and Cooper, 1987; Risley and Crossley, 1988; Whicker and Detling, 1988; Coley and Aide, 1991; Marquis and Whelan, 1994; Bonser and Reader, 1995; Hulme, 1996; Schreiner et al., 1996.)

affected results of individual studies. We made no attempt to account for the scale at which studies were performed. We suspect that most investigators select the experimental duration that maximizes the differences in standing stock between enclosures and exclosures, thus maximizing our A. For example, in temperate zone habitats where winter foliar standing stock is typically near zero regardless of herbivores ($A = 0$), investigators end their experiment before seasonal senescence of plants begins so that at least the potential exists for A to exceed zero. Thus although most results reported in the literature may produce impact indices that are more maximal than minimal, we are not aware of any systematic methodological differences that would bias our comparisons across habitat types and plant types.

For estimating some ecosystem impacts (e.g., nutrient cycling), knowledge of plant production and how it is affected by herbivory might be important (Mitchell and Wass, 1996a), but plant standing stock per se is an important determinant in community and ecosystem interactions involving, for example, predation refuge, microclimate effects, and provision of surface areas for epiphytic organisms. Thus, our focus on differences in standing stock during the growing season has great applicability to these latter interactions.

For the terrestrial habitat, we relied primarily on secondary sources (e.g., Coley and Aide, 1991, on tree leaves), but added some recent primary literature on terrestrial vegetation types not well represented in the secondary sources (e.g., studies by McNaughton and colleagues on grasslands). Terrestrial nonvascular plants are not included because few relevant data exist. For the marine habitat, we relied completely on primary sources because no reviews have been published. For vascular freshwater plants, we added more recent sources to those found in Lodge (1991) and Newman (1991). For nonvascular freshwater plants, we relied heavily on the review by Feminella and Hawkins (1995). Complete information on sources is provided (see Fig. 8.1 legend). Using the results of our literature survey, we make qualitative comparisons only, because in some cases we did not have access to the original data (e.g., from review articles), which would be necessary to calculate anything other than a mean and range. Statistical comparisons were not possible for these data.

Results suggest that herbivores in all three habitats may often reduce plant abundance substantially (Fig 8.1). Herbivore impact on nonvascular plants overlaps broadly with that on vascular plants, but in freshwater, mean impact on nonvascular plants is greater. Freshwater nonvascular plants might be more susceptible to a wide variety of herbivores because of their smaller size, lack of tough structural material, and possibly lower levels of defensive compounds. Herbivory on nonvascular plants in marine habitats is comparable with that in freshwater habitats, but to some extent the plant groups being compared are different. Most marine studies focus on macroalgal seaweeds (predominantly Phaeophyta, Rhodophyta, and Chlorophyta), for which there are almost no similar-sized counterparts in freshwater habitats. Most freshwater studies focus on microalgae (predominantly Chlorophyta, Chrysophyta, and Cyanobacteria), which also occur in marine habitats (including on the surfaces of macroalgae) but have attracted less study in marine habitats. The implications of these interhabitat differences in algal taxa and anatomy for the plant–herbivore interaction are largely unknown.

For vascular plants, herbivore impact is similar in terrestrial and freshwater habitats and considerably higher in marine habitats. We suspect that the means for both terrestrial and freshwater habitats are biased low because both terrestrial (Dirzo and Miranda, 1991) and freshwater (see next section) plant–herbivore ecologists have been preoccupied with insects, which tend to have lower impact than many other herbivores on vascular plants. However, even the current data (Fig. 8.1) suggest that freshwater herbivores reduce macrophyte abundance substantially, thus confirming earlier suggestions (Lodge, 1991) that herbivory is at least as important in freshwater as it is in terrestrial habitats. It is therefore reasonable to hypothesize that freshwater herbivores may often exert substantial selection pressure on plant characteristics and alter the role of macrophytes in aquatic ecosystem function. It is important to understand in more detail what sort of freshwater herbivores most reduce macrophytes and what role herbivory plays in altering the ecosystem impact of macrophytes. For the remainder of this chapter, we narrow our focus to the plant–herbivore interaction in freshwaters.

Does Herbivore Impact Differ Among Freshwater Herbivore Taxa?

To compare the impact of different taxa of freshwater herbivores, we replotted the freshwater data from Figure 8.1 by herbivore type (Fig. 8.2). The absence of studies addressing the impact on periphyton of herbivores with larger body size— mammals and aquatic birds—reflects the widely held belief (apparently supported by abundant observations) that these herbivores rarely intentionally consume periphyton (although counter examples exist, such as black swan studies in Mitchell et al., 1988; Mitchell and Perrow, this volume, Chapter 9), but may often consume epiphytic periphyton incidentally with macrophytes. Intermediate-sized herbivores—fishes and crayfishes—reduce macrophytes and periphyton similarly, with the literature suggesting that the periphyton consumed by these species is primarily filamentous macroalgae that may be intentionally ingested by the herbivores. By contrast, the smallest herbivores—snails and insects—have little (insects) or no impact (snails) on macrophytes but do substantially reduce periphyton, especially microalgae (Fig. 8.2).

For periphyton, the largest mean percentage reductions were by fishes and crayfishes. In some U.S. midwestern streams, herbivorous fishes (Cyprinidae), whose distribution and abundance among stream pools is determined by piscivorous fishes (Centrarchidae), determine whether pools are green with filamentous algae or grazed bare (Matthews et al., 1987; Power et al., 1988). Similarly, crayfishes can exert control over the abundance of stream *Cladophora* (Hart, 1992; Creed, 1994). By contrast, the most studied herbivore taxa, snails and insects (see sample numbers on Fig. 8.2), had lower but still large mean reductions of periphyton.

For vascular plants, it is clear from comparing the number of studies on different taxa (Fig. 8.2) that past bias in studying insects to the exclusion of other

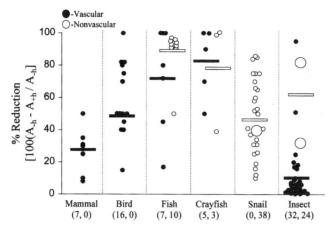

Figure 8.2. Data points and means (bars) for the percentage reduction of standing crop of freshwater vascular and nonvascular plants (excluding phytoplankton) by different taxa of herbivores. Herbivore taxa are listed in rough order of declining body size from left to right. Large dots are mean values from several studies summarized in Feminella and Hawkins (1995). Small dots represent one experiment or comparison. Sheldon (1987), which seemed to show large reductions of macrophytes by snails, was excluded because later communication suggested that snails had not consumed the macrophytes and probably were not responsible for the macrophyte reductions (Brönmark, 1990; Sheldon, 1990). Sample numbers (numbers of separate experiments or comparisons) are given below the name of the herbivore taxon (vascular, nonvascular). Values of −24 and −47 were not plotted for birds. (Data on birds from Anderson and Low, 1976; Jupp and Spence, 1977; Verhoeven, 1978; Kiorboe, 1980; Carter and Rybicki, 1985; Korshgen et al., 1988; Esler, 1989; Urbanc and Blefec, 1993; Conover and Kania, 1994; Hoyer and Canfield, 1994; Schutten et al., 1994; van Donk et al., 1994; van Donk and Gulati, 1995; Søndergaard et al., 1996. Data on crayfish from Abrahamsson, 1966; Dean, 1969; Lodge and Lorman, 1987; Feminella and Resh, 1989; Hart, 1992; Matthews and Reynolds, 1992; Lodge et al., 1994. Data on fish from Prejs, 1984; Hansson et al., 1987; Crutchfield et al., 1992; Maceina et al., 1992; Wootton and Oemke, 1992; Urbanc and Blejec, 1993; Creed, 1994; Feminella and Hawkins, 1995; Richardson et al., 1995. Data on insects from Soska, 1975; Urban, 1975; van der Velde, 1982; Wallace and O'Hop, 1985; Scott and Haskins, 1987; Painter and McCabe, 1988; Jacobsen and Sand-Jensen, 1992, 1994, 1995; Underwood et al., 1992; Chikwenhere, 1994; Kornijów, 1994, 1996; Feminella and Hawkins, 1995. Data on mammals from Pelikan et al., 1971; Smith and Kadlec, 1985; Julien et al., 1992; Shaffer et al., 1992; Taylor et al., 1994; Taylor and Grace, 1995. Data on snails from Martin et al., 1992; Sarnelle et al., 1993; Brönmark, 1994; Vermaat, 1994.)

herbivores has distorted our view of the role of herbivory and contributed to what is probably an artifactually low mean for the freshwater habitat (Fig. 8.1). About 48% of all the studies on herbivory of macrophytes have been done on insects. Yet they have the smallest impact on macrophytes of any herbivore group (Fig. 8.2). The largest reductions of macrophyte biomass were caused by the lesser-studied herbivores (i.e., crayfishes, fishes, and aquatic birds). Mammals, primarily muskrat

(*Ondatra zibethicus*) and nutria (*Myocastor coypus*), typically caused reductions of macrophyte standing crop intermediate between insects and other vertebrates (Fig. 8.2).

An additional reason for caution in interpreting these patterns is that most experiments targeted one species or guild of herbivores. In many habitats, other herbivores were probably present, but their impact and any direct or indirect effect on their impact by the manipulation of the target herbivore were usually not quantified.

Some important biogeographical differences in the occurrences of macrophyte-eating herbivores are obscured by these data summaries. For example, there are no fishes native to North America for which macrophytes are an important dietary component. Macrophyte-eating fishes are, however, common in Europe (e.g., *Rutilus rutilus, Scardinius erythrophthalmus, Cyprinus carpio*) and contribute to reductions in macrophytes that may destabilize the clearwater state of shallow northern European lakes (van Donk and Otte, 1996). Macrophyte-eating fishes are also common in Asia (e.g., *Ctenopharyngodon idella*), Africa, and South America (many species of Cichlidae and Characidae). Such biogeographical patterns interact with other natural and anthropogenic spatial and temporal patterns in herbivore impact that operate on much shorter time scales. In the next section, we narrow our focus further to vascular plants only. We consider how short-term spatial and temporal variation in herbivore abundance may differ among taxa and how that variation may affect the plant–herbivore interaction.

Spatial and Temporal Dynamics of Herbivore Abundance

One difficulty in interpreting the patterns of herbivore impact from the literature (Figs. 8.1 and 8.2) is knowing what spatial and temporal scales the documented reductions of macrophytes by herbivores represent. Evidence suggests that in the absence of major anthropogenic changes in the lake environment (e.g., addition or removal of herbivores or exotic plant species, eutrophication, acidification), macrophyte abundance and species composition are remarkably constant over decades, at least in oligotrophic-mesotrophic lakes (Carpenter and Titus, 1984; Lodge et al., 1989). Thus impact of changes in herbivory can be distinguished from background variation in macrophyte abundance. It then becomes essential to consider the temporal and spatial patterns of variation in the abundance and impact of herbivores.

Invertebrates, Fishes, and Turtles

The natural abundance of many taxa of herbivores and their impact on macrophytes may be relatively similar over wide geographical regions and relatively constant from year to year. Natural interannual variation certainly exists for many invertebrates and is well documented for many fishes. However, natural disappearance or appearance of herbivore species on an interannual time scale is probably rare, such that macrophytes are relatively constantly subject to herbivory exerted by many invertebrates, fishes, and turtles (Clark and Gibbons, 1969; Carter and Rybicki, 1985).

By contrast, anthropogenic interannual variation induced by harvesting, other management practices, and eutrophication can be large and is especially well documented for fishes. For example, eutrophication leads to increases in zooplanktivorous and benthivorous fishes in northern Europe (Persson et al., 1991), with consequences for water clarity and macrophyte abundance that are beginning to be understood from subsequent removal of zooplanktivorous fishes (e.g., *Rutilus rutilus* removal in biomanipulation).

Other anthropogenically driven changes important for macrophyte abundance and lake management involve the spread of exotic species of herbivorous invertebrates and fishes. Probably the most important North American examples include three cyprinid fishes: the goldfish (*Carassius auratus*), introduced in the late 1600s; the common carp (*Cyprinus carpio*), introduced in 1831; and the grass carp (*Ctenopharygodon idella*), introduced in the early 1960s. By the late 19th century, the common carp was already extremely abundant throughout eastern North America (Laird and Page, 1996), and large impacts on macrophytes, invertebrates, and other fishes are well documented (Taylor et al., 1984). Grass carp were imported specifically to reduce nuisance macrophytes, which they have done very effectively (Shireman, 1984).

The potential role of crayfishes as herbivores has been highlighted by reductions in macrophytes as crayfishes have been introduced outside their native range (Lodge and Hill, 1994). The best studied example is the rusty crayfish (*Orconectes rusticus*), a native of Indiana, that has been widely introduced elsewhere in eastern North America. Its establishment has caused large declines in submerged macrophytes and snails in lakes (Lodge et al., 1994). Introductions of exotics are increasing globally and include both aquatic plants and herbivores; thus, new plant–herbivore interactions can be expected in lakes (Lodge et al., in press), with potentially large consequences for macrophyte abundance and ecosystem function.

Mammals

The reduction or elimination by humans of large mammalian herbivores in terrestrial ecosystems has been well documented (Wilson, 1992). Although less studied, the pattern has been similar for freshwater mammalian herbivores that were dramatically reduced in earlier centuries by hunting, trapping, and other human activities (e.g., beaver [*Castor canadensis*], manatees [*Trichechus manatus*]) (Henderson, 1960; Campbell and Irvine, 1977; Whitaker, 1980; Jones and Birney, 1988; Lacki et al., 1990; Wilsey and Chabreck, 1991; Shaffer et al., 1992; Doucet and Fryxell, 1993). Other freshwater mammalian herbivores have, however, increased their geographical range, even as populations have fluctuated in response to changing hunting and fur-trading pressure. Introductions of muskrat (*Ondatra zibethicus*) into Europe (Hengeveld, 1989) and of nutria (*Myocastor coypus*) into North America (Wilsey and Chabreck, 1991; Shaffer et al., 1992) are striking examples. Limnologists need to consider more carefully the impact of such herbivores on lake macrophytes as their ranges and abundance change. Because many of the most dramatic changes in abundance of these herbivores happened so long ago, it is difficult to know what the impact of their removal or addition was. We should not, however, simply assume that the impact was negligible.

Aquatic Birds

Both the evolution and past ecology of macrophytes may have been strongly affected by waterfowl. Because limnologists have not often considered this possibility, we devote disproportionate space to waterfowl (relative to better studied freshwater herbivores) in this chapter (also see Mitchell and Perrow, this volume, Chapter 9). North American fossils (Bickart, 1990), northern European rock etchings (Maringer and Bandi, 1953), and other evidence indicate that waterfowl, including swans, which can consume up to 9 kg per capita of aquatic vegetation daily (Bellrose, 1980), have been feeding from northern hemisphere lakes for about a million years. However, since the 19th century, abundance of aquatic birds has been very low relative to prehistorical levels.

Unregulated hunting in North America before 1916, the loss of 53% of all wetlands in the continental United States (Dahl, 1990), and the degradation of remaining wetlands caused a significant decline in waterfowl populations (Bellrose, 1980; Baldassare and Bolen, 1994). Although population estimates for most waterfowl species do not exist for the time before European colonization, data for trumpeter swan (*Cygnus buccinator*) may be representative of the trend for other aquatic birds. Before European colonization, there were probably 100,000 trumpeter swans in the Great Lakes region of North America; yet, by 1900 there were none (Gillette and Shea, 1995). Clearly, not only the absolute numbers of swans (and probably other waterfowl) declined precipitously, but their numbers per unit area of lake and wetland habitat have also been very low in the past century; the decline in waterfowl numbers has been substantially greater than the decline in wetland habitat area. Therefore, studies of herbivory on macrophytes and the general impressions of limnologists about the importance of herbivory have been shaped during what has probably been the period of lowest densities of herbivorous aquatic birds in the past many thousands of years.

Although quality and quantity of habitat continue to decline in both North America and Europe, population trends for most herbivorous waterfowl have reversed in recent decades in North America (Fig. 8.3) and Europe. Over the past 40 years in North America, diving duck populations have remained roughly constant, but dabbling ducks have increased about 35%, geese about 225%, and swans about 80% (Fig. 8.3). Over the past 25 years in northern and central Europe, populations of three of the four most important herbivorous species have also increased. While coot (*Fulica atra*) numbers have been relatively constant or declining slightly, mute swans (*Cygnus olor*) have increased about 15%, Bewick's

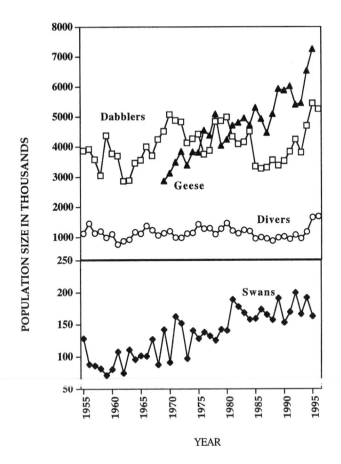

Figure 8.3. Abundance of herbivorous aquatic birds in North America during the latter half of the 20th century. Data have been pooled across species with similar eating habits: dabblers (*Anas americana, A. strepera*), which typically eat seeds and other parts of macrophytes in water <0.5 m deep; divers (*Aythya americana, A. valisineria*), which commonly eat macrophytes at 1–2-m depth; geese (*Branta canadensis, Chen caerulescens*), which eat terrestrial plants and/or emergent macrophytes; and swans (*Cygnus buccinator, C. columbianus, C. olor*), which feed on emergent and submerged macrophytes at water depths down to 1 m. Data for coots (*Fulica americana*), the only major herbivorous aquatic bird not included, were unavailable. All groups include both aboveground and belowground feeders. Data are estimates of continentwide populations except for *C. olor* (for which only Atlantic Flyway data are available). Data for geese before 1970 are not included because census methods differed from later data. A few missing data points for some swan species were estimated by interpolation. (Data from Caithamer and Dubovsky, 1995, for all ducks; Caithamer and Dubovsky, 1995 (1970–1995) and courtesy of U.S. Fish and Wildlife Service (1955–1970) for geese and tundra swans; and Allin et al., 1987; Gillette and Shea, 1995; Allin, personal communication, for other swans.)

swans (*C. bewickii*) have increased about 30%, and red-crested pochards (*Netta rufina*) have increased about 15,000% (Rose, 1994).

The impacts that these higher densities of waterfowl may have on macrophytes in wetlands, shallow lakes, and estuaries have been best studied for snow geese (*Chen caerulescens*) in North America. Higher populations of snow goose have resulted in reductions in a variety of terrestrial and emergent plant variables: cover, biodiversity, aboveground standing crop, productivity, nitrogen content, and belowground biomass (Smith and Odum, 1981; Cargill and Jefferies, 1984; Bazely and Jefferies, 1986; Giroux and Bedard, 1987; Rockwell et al., 1996). Snow geese predominantly affect emergent and wetland plants, but large reductions of submerged macrophytes and macroalgae by coots in Europe (van Donk et al., 1994; Søndergaard et al., 1996; van Donk and Otte, 1996; van Donk, this volume, Chapter 19), black swans in New Zealand (Mitchell et al., 1988; Mitchell and Wass, 1996b), and diving ducks in Texas (Mitchell et al., 1994) have also been well documented in recent years (see Fig. 8.2). Clearly, limnologists need to examine more carefully when, where, and by how much aquatic birds reduce macrophyte abundance. Because of the strong seasonal migratory patterns of many aquatic birds and the longer-term population trends, judging how widely results of one study may apply is difficult.

Spatial and Temporal Patterns of Aquatic Bird Abundance

On large spatial scales (lake-to-lake) and long time scales (year-to-year), a positive relationship exists between abundance of herbivorous waterfowl and macrophyte abundance (McAtee, 1911; Wilson and Atkinson, 1995; Mitchell and Wass, 1996b). Herbivorous waterfowl choose lakes with higher macrophyte biomass and preferred species composition (Lovvorn, 1989; Squires, 1991; Baldassare and Bolen, 1994). The high mobility of aquatic birds relative to their resources makes this positive relationship between consumer and resource unsurprising (Sih, 1984) and suggests that birds may usually move before eliminating plants (Reinecke et al., 1989). However, ample evidence suggests that birds often reduce macrophyte abundance during the periods they inhabit a lake. What remains almost untested (with the exception of the work of Jefferies and colleagues cited above), is the long-term impact of seasonal plant depletion, especially if the consumption is not constant year to year. In this section, we can only begin to suggest the issues that limnologists need to address before any general conclusions on the impact of waterfowl on macrophytes are reached (see Mitchell and Perrow, this volume, Chapter 9).

Waterfowl that are territorial on the breeding grounds (e.g., most dabblers, divers, and swans except black swans) may have a low impact on macrophytes because bird population densities are low. In North America, maximum densities of adult breeding birds are about 5.4/ha and 1.6/ha, respectively, for dabbling and diving ducks (Kantrud and Stewart, 1977), whereas the ranges of densities are about 0.03–0.07/ha for trumpeter swan (*Cygnus buccinator;* Banko, 1960), and 0.05–0.1/ha for mute swan (*C. olor;* Wood and Gelston, 1972). In addition, breeding females and young are primarily carnivorous (Baldassare and Bolen, 1994). By contrast, colonially nesting species, like the snow goose, may have dramatic local effects on their resources and habitat—creating barren "eat-outs"— during the breeding season (Rockwell et al., 1996). Coots also may reduce macrophyte standing crop during the growing season, but like many other aquatic birds, the most obvious impacts appear during autumn and winter aggregations (Søndergaard et al., 1996; van Donk and Otte, 1996, van Donk, this volume, Chapter 19).

During migrations and on the wintering grounds, almost all species are gregarious. These large and diverse assemblages of waterfowl could have a large impact on both aboveground biomass and overwintering structures of macrophytes (Lovvorn, 1989; Mitchell et al., 1994). However, at this point in the growing season of macrophytes, most aboveground tissue is senescent or soon to be senescent. Consumption of belowground parts (including nutrient storage and overwintering structures) might have a greater long-term impact on the macrophytes, but few studies have measured consumption of belowground parts (Korshgen et al., 1988; Lovvorn, 1989; Michot and Chadwick, 1994), and few, if any, have looked at the impact on the following year's growth. It is critical to assess the importance of this common manifestation of herbivory by aquatic birds.

Superimposed on the rebounding populations of many waterfowl, herbivorous mute swans (*Cygnus olor*) have been introduced to North America. Mute swans have spread throughout eastern North America in recent decades, where they may reduce submerged vegetation through direct consumption, wasteful feeding, and ~~~~~~ emergent vegetation for nest construction, although the area affected is typically smaller (Willey and Halla, 1972).

Thus, the temporal dynamics of aquatic bird populations—from seasonal migrations to decades-long population trends—make it difficult to generalize about impact on macrophytes in the past or the present. Even infrequent, but intense, bouts of macrophyte feeding by aquatic birds on a lake might restart succession of macrophytes, causing a long-term impact the cause of which would be easy to miss. Enough examples exist, however, of large reductions of macrophytes by aquatic birds that limnologists can no longer ignore them as potential determinants of macrophyte abundance and species composition. Much work remains to determine whether the impact of waterfowl is usually small or whether limnologists must often consider waterfowl in understanding dynamics of macrophytes.

Diet Specialization and a Model of Plant Selection

The past bias of limnologists in focusing herbivory studies on insects rather than other herbivores has been misleading for at least two reasons. First, insects usually cause much less damage than other herbivorous taxa (Fig. 8.2). Second, most terrestrial (Strong et al., 1984; Bernays, 1989) and freshwater (Newman, 1991) insects found on vascular plants are oligophagous. Although freshwater macrophyte-eating insects are less oligophagous on average than terrestrial insects

Table 8.1. Degree of Specialization Among Freshwater Taxa That Are Primarily Herbivorous on Macrophytes[a]

	Plants as % of diet	Plants eaten per herbivore species	
		Families	Genera
Insects[b]	?–100%	1–3	<3
Mammals[c]	NA	4–7	NA
Birds[d]	43–99%	7–19	9–32
Turtles[e]	27–89%	NA	4–8
Fishes[f]	20–95%	9	NA
Crayfishes[g]	12–80%	>11	NA

[a]For birds, data are for taxa in Figure 8.3. NA, not available.
[b]Data from Newman, 1991.
[c]Data from Henderson, 1960; Whitaker, 1980; Jones and Birney, 1988; Lacki et al., 1990; Wilsey and Chabreck, 1991; Shaffer et al., 1992; Doucet and Fryxell, 1993.
[d]Data from Fasset, 1957; Willey and Halla, 1972; Palmer, 1976; Mitchell, 1994.
[e]Data from Carr, 1952; Conant, 1958; Clark and Gibbons, 1969; Minton, 1972; Mount, 1975; Behler and King, 1979; Parmenter, 1980; Ernst, 1983; Parmenter and Avery, 1990.
[f]Data from Nichols, 1991; Bain, 1993.
[g]Data from Lodge and Hill, 1994; Lodge and Cronin, unpublished data.

(Newman, 1991), most aquatic insects are still far more specialized than are other aquatic herbivore taxa (Table 8.1). Because (1) the herbivore taxa that cause the most damage (Fig. 8.2) are generalists (Table 8.1), (2) generalist herbivores often change the relative abundance of macrophyte species (Lodge, 1991), and (3) macrophyte community composition, in addition to macrophyte abundance, affects the ecosystem role of macrophytes (Carpenter and Lodge, 1986; van Donk, this volume, Chapter 19), it becomes important to understand which plants will be most affected by herbivores and how macrophyte communities will change under the influence of herbivory.

Model of Plant Selection by Generalist Herbivores

Although few direct tests of the basis of plant selection by freshwater herbivores have been conducted, enough evidence has accumulated to suggest a conceptual model to guide further work (Fig. 8.4). To be preferentially consumed, a macrophyte must (1) have a structure (morphology, toughness, and surface features) that makes it possible for an herbivore to take a bite; (2) lack chemical deterrents; and (3) be nutritious. Evidence for each element of this model is described briefly below.

Plant Structure

For crayfishes, the freshwater herbivore for which the most experiments addressing plant selection have been conducted, emergent plants are typically much less reduced than submerged plants (Chambers et al., 1991; Lodge, 1991; Cronin, this

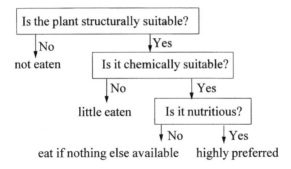

Figure 8.4. Conceptual model of how diet composition is determined in freshwater herbivores.

volume, Chapter 21). This is also true for *Limnephilus* caddisfly larvae (Lodge, unpublished data). For both taxa, low consumption of emergent plants seems to result from the structure of the plants (i.e., toughness and/or the mismatch between small mouthparts and broad, flat plant surfaces). When the same plants are freeze-dried, ground, and reconstituted in an alginate gel, they often become highly preferred, showing clearly the deterrent quality of plant structure (Cronin, this volume, Chapter 21).

Trichomes, thick cuticles, and other surface structures that defend many terrestrial plants from herbivores are largely absent from submerged aquatic plants (Levin, 1973; Grubb, 1986). This probably results from selection pressure to minimize boundary layers to increase gas diffusion across the submerged plant surface. Thus, submerged macrophytes may require alternative deterrents.

Chemical Deterrents and Stimulants

Many authors have suggested that chemical deterrents and attractants play a role in the macrophyte–herbivore interaction (Otto, 1983; Sterry et al., 1983; Haynes and Taylor, 1984; Suren and Lake, 1989; Jefferies, 1990; Center and Wright, 1991; Lodge, 1991; Newman, 1991). Nevertheless, only one example exists of an identified chemical defense (glucosinolate-myrosinase system) that deters herbivores from eating an aquatic plant, watercress (*Nasturtium officinale*) (Newman et al., 1992). However, in many freshwater macrophytes, many classes of compounds exist that are known to be deterrent to many terrestrial herbivores (Lodge, 1991; Rosenthal and Berenbaum, 1992). More important, examples are mounting of unidentified plant compounds that deter freshwater herbivores (Buchsbaum et al., 1984; Cronin, this volume, Chapter 21).

Nutritional Value

Herbivores are, in general, nitrogen limited, and thus many terrestrial herbivores preferentially consume high-nitrogen plants (Mattson, 1980). Although supporting data for freshwater habitats are still few and primarily correlative, the same pattern appears to be true for freshwater herbivores. A snail, three species of caddisflies, and an amphipod prefer high-nitrogen green watercress tissue over low-nitrogen yellowed watercress, as long as the chemical defense mechanism in the green tissue is inoperative (Newman et al., 1996). The crayfish *Procambarus* also prefers high-nitrogen plants among undefended species (Cronin, this volume, Chapter 21). Species selection by the crayfish *Orconectes* can be reversed by reversing the nitrogen content of different submerged macrophyte species (Lodge et al., unpublished data). These examples are all consistent with the model suggesting that among plants that are neither structurally nor chemically defended, nitrogen content may often determine plant selection (Fig. 8.4).

Implications for Herbivore Impact on Macrophyte Species Composition

The specific predictions of this model will differ to some extent among herbivore taxa because, for example, a plant that is structurally defended against caddisflies may be handled easily by swans. In addition, other factors important to herbivores may differ among plants and affect herbivore impact (e.g., predation refuge offered by the plant for the herbivore [Duffy and Hay, 1994], satiation of the herbivore [Cronin and Hay, 1996a], and feeding history of the herbivore [Provenza, 1995]). Nevertheless, the model (Fig. 8.4) is consistent with the data known to us and provides a useful framework for additional experimental and observational work on the impact of herbivores on macrophyte communities.

Conclusions and Role of Herbivores in Stabilizing–Destabilizing Alternate Lake States

From our summary of the literature on herbivory, it is clear that herbivores often reduce plant standing stock in freshwater habitats, as they do in terrestrial and marine habitats. To the limited extent that limnologists have studied herbivory on macrophytes, they have given undue attention to insects and insufficient attention to the role of larger herbivores—especially birds, fishes, and crayfishes—in determining macrophyte abundance in lakes. Increasing abundance of aquatic birds and mammals in both North America and Europe makes them especially deserving of studies of their long-term impact on macrophyte abundance. In addition, theory predicts that the generalist nature of plant selection characteristic of the understudied herbivores makes them potentially more powerful suppressers of overall macrophyte abundance than more specialized insect herbivores (Murdoch and Bence, 1987).

Any herbivore that substantially reduces macrophyte standing stock in shallow lakes could play a role in destabilizing any clearwater macrophyte-dominated state (Scheffer et al., 1993). Ample evidence already exists of the importance of European coots in suppressing the recovery of macrophytes (Lake Stigsholm; Søndergaard et al., 1996) or reducing macrophyte abundance (Lake Zwemlust; van Donk and Otte, 1996; van Donk, this volume, Chapter 19) and thereby resisting the establishment of the macrophyte-dominated state or the reversal of the phyto-plankton-dominated state. In addition, coots were apparently responsible for a shift in macrophyte community composition from *Elodea* (the evergreen nature of which enhanced the clearwater state) to *Potamogeton* (the seasonal nature of which was less effective in maintaining the clearwater state) (van Donk, this volume, Chapter 19). Thus, if large generalist herbivores such as aquatic birds continue to increase, they may become an impediment to managing lakes for the clearwater state.

An additional reason that herbivory may be more important now than in the past in reducing macrophytes and destabilizing the clearwater state is that eutrophication has both decreased the occurrence of nutrient conditions under which the clearwater state can prevail and increased the abundance of fishes that help maintain the turbid water state in Europe (Persson et al., 1991) and North America (Laird and Page, 1996). If the boundary between alternative lake states approximates a threshold (Scheffer and Jeppesen, this volume, Chapter 31), a small increase in herbivory by fishes or birds (or any other herbivore) could tip the balance toward the turbid phytoplankton-dominated state. Thus, more rigorous experiments testing the impact of different herbivores and more observational work on the abundance, feeding rates, and plant selection of a variety of fresh-water herbivores will produce insight on the plant-herbivore interaction that has immediate implications for lake management.

Acknowledgments. We thank Laura Eidietis for assistance with the literature review and analysis and for preparation of the bibliography. NSFDEB 94-08452 (D.M.L.) has supported the recent research on herbivory by D.M.L., G.C, and A.F. A.F. has benefited from a NSF Graduate Research Traineeship (NSFGER 94-5265-001). Dee Butler (USFWS) gave A.F. much assistance in locating waterfowl data sources. G.C. and D.M.L. thank Mark Hay for many stimulating discussions about plant–animal interactions. For many helpful suggestions on the manuscript, we thank Mark Hoyer, Robert McIntosh, and Stuart Mitchell.

References

Abrahamsson, S.A.A. Dynamics of an isolated population of the crayfish *Astacus astacus* Linne. Oikos 17:96–107; 1966.

Allin, C.C.; Chasko, G.C.; Husband, T.P. Mute swans in the Atlantic flyway: a review of the history, population growth and management needs. Trans. N. E. Sec. Wildl. Soc. 44:32–47; 1987.

Anderson, M.G.; Low, J.P. Use of sago pondweed by waterfowl on the Delta Marsh, Manitoba. J. Wildl. Manage. 40:233–242; 1976.

Andrew, N.L.; Underwood, A.J. Density-dependent foraging in the sea urchin *Centrostephanus rodgersii* on shallow subtidal reefs in New South Wales, Australia. Mar. Ecol. Prog. Ser. 99:89–98; 1993.

Bain, M.B. Assessing impacts of introduced aquatic species: grass carp in large systems. Environ. Manage. 17:211–224; 1993.

Baldassarre, G.A.; Bolen, E.G. Waterfowl ecology and management. New York: John Wiley & Sons; 1994.

Banko, W.E. The trumpeter swan: its history, habits and population in the United States. North American Fauna 63. Washington, DC: Bur. Sport Fish and Wildl.; 1960.

Bazely, D.R.; Jeffries, R.L. Changes in the composition and standing crop of salt-marsh communities in response to the removal of a grazer. J. Ecol. 74:693–706; 1986.

Behler, J.L.; King, F.W. The Audubon Society field guide to North American reptiles and amphibians. New York: Knopf; 1979.

Bellrose, F.C. Ducks, geese & swans of North America. Harrisburg, PA: Stackpole Books; 1980.

Benedetti-Cecchi, L.; Cinelli, F. Early patterns of algal succession in a midlittoral community of the Mediterranean sea: a multifactorial experiment. J. Exp. Mar. Biol. Ecol. 169:15–31; 1993.

Bernays, E.A. Host range in phytophagous insects: the potential role of generalist predators. Evol. Ecol. 3:299–311; 1989.

Bertness, M.D. Habitat and community modification by an introduced herbivorous snail. Ecology. 65:370–381; 1984.

Bickart, K.J. Recent advances in the study of neogene fossil birds I. The birds of the late Miocene-early Pliocene Big Sandy Formation, Mohave County, Arizona. Ornithol. Monogr. 44:1–72; 1990.

Bonser, S.P.; Reader, R.J. Plant competition and herbivory in relation to vegetation biomass. Ecology 76:2176–2183; 1995.

Brönmark, C. Effects of tench and perch on interactions in a freshwater, benthic food chain. Ecology 75:1818–1828; 1994.

Brönmark, C. How do herbivorous freshwater snails affect macrophytes?—a comment. Ecology 71:1212–1215; 1990.

Buchsbaum, R.; Valiela, I.; Swain, T. The role of phenolic compounds and other plant constitutents in feeding by Canada geese in a coastal marsh. Oecologia 63:343–349; 1984.

Buschmann, A.H. The role of herbivory and desiccation on early successional patterns of intertidal macroalgae in southern Chile. J. Exp. Mar. Biol. Ecol. 139:221–230; 1990.

Caithamer, D.F.; Dubovsky, J.A. Waterfowl population status, 1995. Laurel, MD: U.S. Fish and Wildl. Ser.; 1995.

Campbell, H.W.; Irvine, A.B. Feeding ecology of the West Indian manatee *Trichechus manatus* Linnaeus. Aquaculture 12:249–251; 1977.

Cargill, S.M.; Jefferies, R.L. The effects of grazing by lesser snow geese on the vegetation of a sub-arctic salt marsh. J. Aquat. Ecol. 21:669–686; 1984.

Cargill, S.M.; Jefferies, R.L. Changes in the composition and standing crop of salt-marsh communities in response to the removal of a grazer. J. Ecol. 74:693–706; 1986.

Carpenter, R.C. Grazing by *Diadema antillarum* (Philippi) and its effects on the benthic algal community. J. Mar. Res. 39:749–765; 1981.

Carpenter, R.C. Partitioning herbivory and its effects on coral reef algal communities. Ecol. Monogr. 56:345–363; 1986.

Carpenter, S.R.; Lodge, D.M. Effects of submersed macrophytes on ecosystem processes. Aquat. Bot. 26:341–370; 1986.

Carpenter, S.R.; Titus, J.E. Composition and spatial heterogeneity of submersed vegetation in a softwater lake in Wisconsin. Vegetatio 57:153–165; 1984.

Carr, A. Handbook of turtles. Ithaca, NY: Comstock Publishing; 1952.

Carter, V.; Rybicki, N.B. The effects of grazers and light penetration on the survival of transplants of *Vallisneria americana* Michx. in the tidal Potomac River, Maryland. Aquat. Bot. 23:197–213; 1985.

Castenholz, R.W. The effect of grazing on marine littoral diatom populations. Ecology 42:783–794; 1961.

Cattaneo, A.; Mousseau, B. Empirical analysis of the removal rate of periphyton by grazers. Oecologia 103:249–254; 1995.

Center, T.D.; Wright, A.D. Age and phytochemical composition of waterhyacinth (Pontederiaceae) leaves determine their acceptability to *Neochetina eichhorniae* (Coleoptera: Curculionidae). Environ. Entomol. 20(1):323–334; 1991.

Chambers, P.A.; Hanson, J.M.; Prepas, E.E. The effect of aquatic plant chemistry and morphology on feeding selectivity by the crayfish, *Orconectes virilis*. Freshwat. Biol. 25:339–348; 1991.

Chikwenhere, G.P. Biological control of water lettuce in various impoundments in Zimbabwe. J. Aquat. Plant Manage. 32:27–29; 1994.

Clark, D.B.; Gibbons, J.W. Dietary shift in the turtle *Pseudemys scripta* (Schoepff) from youth to maturity. Copeia 4:704–706; 1969.

Coley, P.D.; Aide, T.M. Comparison of herbivory and plant defenses in temparate and tropical broad-leaved forests. In: Price, P.W.; Lewinsohn, T.M.; Fernandes, G.W.; Benson, W.W., eds. Plant–animal interactions: evolutionary ecology in tropical and temparate regions. New York: John Wiley & Sons; 1991:25–49.

Conant, R. A field guide to reptiles and amphibians. Boston: Houghton Mifflin; 1958.

Conover, M.R.; Kania, G.S. Impact of interspecific aggression and herbivory by mute swans on native waterfowl and aquatic vegetation in New England. Auk 3:744–748; 1994.

Coppock, D.L.; Detling, J.K.; Ellis, J.E.; Dyer, M.I. Plant–herbivore interactions in a North American mixed-grass prairie. I. Effects of black-tailed prairie dogs on intraseasonal aboveground plant biomass and nutrient dynamics and plant species diversity. Oecologia 56:1–9; 1983.

Creed, R.P. Direct and indirect effects of crayfish grazing in a stream community. Ecology 75:2091–2103; 1994.

Cronin, G.; Hay, M. Seaweed–herbivore interactions depend on recent history of both the plant and animal. Ecology 77:1531–1543; 1996a.

Cronin, G.; Hay, M. Within-plant variation in seaweed palatability and chemical defenses: optimal defense theory versus the growth differentiation balance hypothesis. Oecologia 105:361–368; 1996b.

Crutchfield, J.U.; Schiller, D.H.; Herlong, D.D.; Mallin, M.A. Establishment and impact of redbelly tilapia in a vegetated cooling resevoir. J. Aquat. Plant. Manage. 30:28–35; 1992.

Cyr, H.; Pace, M.L. Magnitude and patterns of herbivory in aquatic and terresrial ecosystems. Nature 361:148–150; 1993.

Dahl, T.E. Wetlands losses in the United States, 1780's to 1980's. Washington, DC: U.S. Fish and Wildl. Serv.; 1990.

Dean, J.L. Biology of the crayfish *Orconectes causeyi* and its control of aquatic weeds in Trout Lake. Tech. Paper, Bureau Sport Fish Wildl. 24:1–15; 1969.

Dirzo, R.; Miranda, A. Altered patterns of herbivory and diversity in the forest understory: a case study of the possible consequences of contemporary defaunation. In: Price, P.W.; Lewinsohn, T.M.; Fernandes, G.W.; Benson, W.W., eds. Plant–animal interactions: evolutionary ecology in tropical and temparate regions. New York: John Wiley & Sons; 1991:273–287.

Doucet, C.M.; Fryxell, J.M. The effect of nutritional quality on forage preference by beavers. Oikos 67:201–208; 1993.

Duarte, C.M. Nutrient concentration of aquatic plants: patterns across species. Limnol. Oceanogr. 37:882–889; 1992.

Duffy, J.E.; Hay, M.E. Herbivore resistance to seaweed chemical defense: the roles of mobility and predator risk. Ecology 75:1304–1319; 1994.

Duggins, D.O. Kelp beds and sea otters: an experimental approach. Ecology 61:447–453; 1980.

Ernst, C. Turtles of the United States. Washington, DC: Smithsonian Institute Press; 1972.

Esler, D. An assessment of American coot herbivory of hydrilla. J. Wildl. Manage. 53: 1147–1149; 1989.

Estes, J.A.; Duggins, D.O. Sea otters and kelp forests in Alaska: generality and variation in a community ecological paradigm. Ecol. Monogr. 65:75–100; 1995.

Estes, J.A.; Steinberg, P.D. Predation, herbivory and kelp evolution. Paleobiology 14:19–36; 1988.

Fassett, N.C. A manual of aquatic plants. Madison, WI: University of Wisconsin Press; 1957.

Feminella, J.W.; Hawkins, C.A. Interactions between stream herbivores and periphyton: a quantitative analysis of past experiments. J. North Am. Benth. Soc. 14:465–509; 1995.

Feminella, J.W.; Resh, V.H. Submersed macrophytes and grazing crayfish: an experimental study of herbivory in a California freshwater marsh. Holarct. Ecol. 12:1–8; 1989.

Fletcher, W.J. Interactions among subtidal Australian sea urchins, gastropods, and algae: effects of experimental removal. Ecol. Monogr. 57:89–109; 1987.

Foster, S.A. The relative impacts of grazing by Caribbean coral reef fishes and *Diadema*: effects of habitat and surge. J. Exp. Mar. Biol. Ecol. 15:1–20; 1987.

Gaines, S.D.; Lubchenco, J. A unified approach to marine plant–herbivore interactions. II. Biogeography. Annu. Rev. Ecol. Syst. 13:111–138; 1982.

Geller, J.B. Gastropod grazers and algal colonization on a rocky shore in northern California: the importance of body size of grazers. J. Exp. Mar. Biol. Ecol. 150:1–7; 1991.

Gillette, L.N.; Shea, R. An evaluation of trumpeter swan management today and a vision for the future. Trans. 60th North Am. Wildl. & Natur. Resour. Conf. Washington, DC: Wildlife Management Institute; 1995:258–265.

Giroux, J.F.; Bedard, J. The effects of grazing by greater snow geese on the vegetation of tidal marshes in the St. Lawrence Estuary. J. Appl. Ecol. 24:773–788; 1987.

Grubb, P.J. Sclerophylls, pachyphylls, and pycnophylls: the nature and significance of hard leaf surfaces. In: Juniper, B.E.; Southwood, T.R.E., eds. Insects and the plant surface. London: Edwin Arnold; 1986:137–150.

Hansson, L.; Johansson, L.; Persson, L. Effects of fish grazing on nutrient release and succession of primary producers. Limnol. Oceanogr. 32:723–729; 1987.

Hart, D.D. Community organization in streams: the importance of species interactions, physical factors, and chance. Oecologia 91:220–228; 1992.

Hatcher, B.G. The interaction between grazing organisms and the epilithic algal community of a coral reef: a quantitative assessment. Proc. 4th Int. Coral Reef Congr. 2; 1981:515–524.

Hatcher, B.G.; Larkum, A.W.D. An experimental analysis of factors controlling the standing crop of the epilithic algal community on a coral reef. J. Exp. Mar. Biol. Ecol. 69:61–84; 1983.

Hay, M.E. Herbivory, algal distribution, and the maintenance of between-habitat diversity on a tropical fringing reef. Am. Nat. 118:520–540; 1981.

Hay, M.E. Predictable spatial escapes from herbivory: how do these affect the evolution of herbivore resistance in tropical marine communities? Oecologia 64:396–407; 1984.

Hay, M.E. Fish-seaweed interactions on coral reefs: effects of herbivorous fishes and adaptations of their prey. In: Sale, P.F., ed. The ecology of fishes on coral reefs. San Diego, CA: Academic Press; 1991:96–119.

Hay, M.E.; Goertemiller, T. Between-habitat differences in herbivore impact on Caribbean coral reefs. In: Reake, M.L., ed. The ecology of deep and shallow coral reefs. Symposia Series for Undersea Research. Vol. 1. Rockville, MD: Office of Undersea Research, NOAA; 1983:97–102.

Hay, M.E.; Taylor, P.R. Competition between herbivorous fishes and urchins on Caribbean reefs. Oecologia 65:591–598; 1985.

Hay, M.E.; Colburn, T.; Downing, D. Spatial and temporal patterns in herbivory on a Caribbean fringing reef: the effects on plant distribution. Oecologia 58:299–308; 1983.

Haynes, A.; Taylor, B.J.R. Food finding and food preference in *Potamopyrgus jenkinsi* (E.A. Smith) (Gastropoda: Prosobranchia). Arch. Hydrobiol. 100:479–491; 1984.

Heck, K.L.; Valentine, J.F. Sea urchin herbivory: evidence for long-lasting effects in subtropical seagrass meadows. J. Exp. Mar. Biol. Ecol. 189:205–217; 1995.

Henderson, R.F. Beaver in Kansas. Lawrence, KS: State Biological Survey and Museum of Natural History; 1960.

Hengeveld, R. Dynamics of biological invasions. New York: Chapman & Hall; 1989.

Herms, D.A.; Mattson, W.J. The dilemma of plants: to grow or defend. Q. Rev. Biol. 67:283–335; 1992.

Himmelman, J.H.; Cardinal, A.; Bourget, E. Community development following removal of urchins, *Strongylocentrotus droebachiensis,* from the rocky subtidal zone of the St. Lawrence Estuary, Eastern Canada. Oecologia 59:27–39; 1983.

Hoyer, M.V.; Canfield, D.E. Bird abundance and species richness on Florida lakes: influence of trophic status, lake morphology, and aquatic macrophytes. Hydrobiologia 279/280:107–119; 1994.

Hughes, T.P.; Reed, D.C.; Boyle, M.J. Herbivory on coral reefs: community structure following mass mortalities of sea urchins. J. Exp. Mar. Biol. Ecol. 113:39–59; 1987.

Hulme, P.E. Herbivores and the performance of grassland plants: a comparison of arthropod, mollusc and rodent herbivory. J. Ecol. 84:43–51; 1996.

Huntley, N.J. Influence of refuging consumers (Pikas: *Ochotona princeps*) on subalpine meadow vegetation. Ecology 68:274–283; 1987.

Jacobsen, D.; Sand-Jensen, K. Herbivory of invertebrates on submerged macrophytes from Danish freshwaters. Freshwat. Biol. 28:301–308; 1992.

Jacobsen, D.; Sand-Jensen, K. Invertebrate herbivory on the submerged macrophyte *Potamogeton perfoliatus* in a Danish stream. Freshwat. Biol. 31:43–52; 1994.

Jacobsen, D.; Sand-Jensen, K. Variability of invertebrate herbivory on the submerged macrophyte *Potamogeton perfoliatus.* Freshwat. Biol. 34:357–365; 1995.

Jefferies, M. Evidence of induced plant defences in a pondweed. Freshwat. Biol. 23:265–269; 1990.

Jones, G.P. Interactions between herbivorous fishes and macroalgae on a temperate rocky reef. J. Exp. Mar. Biol. Ecol. 159:217–235; 1992.

Jones, J.K.; Birney, E.C. Handbook of mammals of the north central states. Minneapolis, MN: University of Minnesota Press; 1988.

Julien, M.H.; Bourne, A.S.; Low, V.H.K. Growth of the weed *Alternanthera philoxeroides* (Martius) Grisebach, (alligator weed) in aquatic and terrestrial habitats. Aust. Plant Protection Q. 7:102–108; 1992.

Jupp, B.P.; Spence, D.H.N. Limitations of macrophytes in an eutrophic lake, Loch Leven. J. Ecol. 65:431–446; 1977.

Kantrud, H.A.; Stewart, R.E. Use of natural basin wetlands by breeding waterfowl in North Dakota. J. Wildl. Manage. 41:243–253; 1977.

Keats, D.W.; South, G.R.; Steele, D.H. Effects of an experimental reduction in grazing by green sea urchins on a macroalgal community in eastern Newfoundland. Mar. Ecol. Prog. Ser. 68:181–193; 1990.

Kiorboe, T. Distribution and production of submerged macrophytes in Tipper Grund (Ringkobing Fjord, Denmark), and the impact of waterfowl grazing. J. Appl. Ecol. 17:675–687; 1980.

Kornijów, R. The importance of invertebrates as consumers of freshwater macrophytes. Wiad. Ekol. 40:181–195; 1994.

Kornijów, R. Cumulative consumption of the lake macrophyte *Elodea* by abundant generalist invertebrate herbivores. Hydrobiologia 319:185–190; 1996.

Korshgen, C.E.; George, L.S.; Green, W.L. Feeding ecology of canvasbacks staging on pool 7 of the upper Mississippi River in waterfowl in winter. In: Weller, M.W., ed. Minneapolis, MN: University of Minnesota Press; 1988:237–249.

Lacki, M.J.; Peneston, W.T.; Adams, K.B.; Vogt, F.D.; Houppert, J.C. Summer foraging patterns and diet selection of muskrats inhabiting a fen wetland. Can. J. Zool. 68:1163–1167; 1990.

Laird, C.A.; Page, L.M. Non-native fishes inhabiting the streams and lakes of Illinois. Ill. Nat. History Surv. Bull. 35:1–51; 1996.

Lamberti, G.A. The role of periphyton in benthic food webs. In: Stevenson, R.J.; Bothwell, M.L.; Lowe, R.L., eds. Algal ecology: freshwater benthic ecosystems. New York: Academic Press; 1996:533–573.

Levin, D.A. The role of trichomes in plant defense. Q. Rev. Biol. 48:3–15; 1973.

Lewis, S.M. The role of herbivorous fishes in the organization of a Caribbean reef community. Ecol. Monogr. 56:183–200; 1986.

Lewis, S.M.; Norris, J.N.; Searles, R.B. The regulation of morphological plasticity in tropical reef algae by herbivory. Ecology 68:636–641; 1987.

Lodge, D.M. Herbivory on freshwater macrophytes. Aquat. Bot. 41:195–224; 1991.

Lodge, D.M.; Hill, A.M. Factors governing species composition, population size, and productivity of cool-water crayfishes. Nordic J. Freshwat. Res. 69:111–136; 1994.

Lodge, D.M.; Lorman, J.G. Reductions in submersed macrophyte biomass and species richness by the crayfish *Orconectes rusticus*. Can. J. Fish. Aquat. Sci. 44:591–597; 1987.

Lodge, D.M.; Krabbenhoft, D.P.; Striegl, R.G. A positive relationship between groundwater velocity and submersed macrophyte biomass in Sparkling Lake, Wisconsin. Limnol. Oceanogr. 34:235–239; 1989.

Lodge, D.M.; Kershner, M.W.; Aloi, J.E.; Covich, A.P. Effects of an omnivorous crayfish (*Orconectes ructicus*) on a freshwater littoral food web. Ecology 75:1265–1281; 1994.

Lodge, D.M.; Stein, R.A.; Brown, K.M.; Covich, A.P.; Brönmark, C.; Garvey, J.E; Klosiewski, S.P. Predicting impact of freshwater exotic species on native biodiversity: challenges in spatial scaling. Aust. J. Ecol. (in press).

Lovvorn, J.R. Distribution responses of canvasback ducks to weather and habitat change. J. Appl. Ecol. 26:113–130; 1989.

Lubchenco, J.; Gaines, S.D. A unified approach to marine plant–herbivore interactions. I. Populations and communities. Annu. Rev. Ecol. Syst. 12:405–437; 1981.

Maceina, M.J.; Cichra, M.F.; Betsill, R.K.; Bettoli, P.W. Limnological changes in a large reservoir following vegetation removal by grass carp. J. Freshwat. Ecol. 7:81–95; 1992.

Maringer, J.; Bandi, H.G. Art in the ice age. New York: Frederick A. Praeger; 1953.

Marquis, R.J.; Whelan, C.J. Insectivorous birds increase growth of white oak through consumption of leaf-chewing insects. Ecology 75:2007–2014; 1994.

Martin, T.H.; Crowder, L.B.; Dumas, C.F.; Burkholder, J.M. Indirect effects of fish on macrophytes in Bays Mountain Lake: evidence for a littoral trophic cascade. Oecologia 89:476–481; 1992.

Matthews, M.; Reynolds, J.D. Ecological impact of crayfish plague in Ireland. Hydrobiologia 234:1–6; 1992.

Matthews, W.J.; Stewart, A.J.; Power, M.E. Grazing fishes as components of North American stream ecosystems: Effects of *Campostoma anomalum*. In: Matthews, W.J.; Heins, D.C., eds. Community and evolutionary ecology of North American stream fishes. Norman, OK: University of Oklahoma Press; 1987:128–135.

Mattson, W.J., Jr. Herbivory in relation to plant nitrogen content. Annu. Rev. Ecol. Syst. 11:119–161; 1980.

McAtee, W.L. Three important wild duck foods. U.S. Bur. Biol. Surv. Circ. 81; 1911.

McNaughton, S.J. Ecology of a grazing ecosystem: the Serengeti. Ecol. Monogr. 55:259–294; 1985.

Michot, T.C.; Chadwick, P.C. Winter biomass and nutrient values of three seagrass species as potential foods for redheads (*Aythya americana* Eyton) in Chandeleur Sound, Louisiana. Wetlands 14:276–283; 1994.

Miller, M.W.; Hay, M.E. Coral-seaweed-grazer-nutrient interactions on temperate reefs. Ecol. Monogr. 66:323–344; 1996.

Minton, S.A. Amphibians and reptiles of Indiana. Indianapolis, IN: Indiana Academy of Science; 1972.

Mitchell, C.A.; Thomas, W.C.; Zwank, P.J. Herbivory on shoalgrass by wintering redheads in Texas. J. Wildl. Manage. 58:131–141; 1994.

Mitchell, C.D. Trumpeter swan (*Cygnus buccinator*). In: Poole, A.; Gill, F., eds. The birds of North America, No. 105. Philadelphia, PA: The Academy of Natural Sciences; Washington, DC: The American Ornithologists' Union; 1994.

Mitchell, S.F.; Wass, R.T. Quantifying herbivory: grazing consumption and interaction strength. Oikos 77:1–4; 1996a.

Mitchell, S.F.; Wass, R.T. Grazing by black swans (*Cygnus atratus* Lathram), physical factors, and the growth and loss of aquatic vegetation in a shallow lake. Aquat. Bot. 55:205–215; 1996b.

Mitchell, S.F.; Hamilton, D.P.; Macgibbon, W.S.; Nayer, P.K.B.; Reynolds, R.N. Interrelations between phytoplankton, submerged macrophytes, black swans (*Cygnus atratus*) and zooplankton in a shallow New Zealand lake. Int. Rev. Ges. Hydrobiol. 73:145–170; 1988.

Morrison, D. Comparing fish and urchin grazing in shallow and deeper coral reef algal communities. Ecology 69:1367–1382; 1988.

Mount, R.H. Reptiles and amphibians of Alabama. Auburn, AL: Auburn Printing Company; 1975.

Murdoch, W.W.; Bence, J. General predators and unstable prey populations. In: Kerfoot, W.C.; Sih, A., eds. Predation: direct and indirect impacts on aquatic communities. Hanover, NH: University Press of New England; 1987:17–30.

Newman, R.M. Herbivory and detritivory on freshwater macrophytes by invertebrates: a review. J. North Am. Benth. Soc. 10:89–114; 1991.

Newman, R.M.; Hanscom, Z.; Kerfoot, W.C. The watercress glucosinolate-myrosinase system: a feeding deterrent to caddisflies, snails and amphipods. Oecologia 92:1–7; 1992.

Newman, R.M.; Kerfoot, W.C.; Hanscom, Z.A.C. Watercress allelochemical defends high nitrogen foliage against consumption: effects on freshwater invertebrate herbivores. Ecology 77:2312–2323; 1996.

Nichols, S.A. The interaction between biology and the management of aquatic macrophytes. Aquat. Bot. 41:225–252; 1991.

Ogden, J.C.; Brown, R.A.; Salesky, N. Grazing by the echinoid *Diadema antillarum* Philippi: formation of halos around West Indian patch reefs. Science 182:715–717; 1979.

Otto, C. Adaptations to benthic freshwater herbivory. In: Wetzel, R.G., ed. Periphyton of freshwater ecosystems. Boston: Dr. W. Junk Publ.; 1983:199–205.

Owen-Smith, N.; Cooper, S.M. Palatability of woody plants to browsing ruminants in a South African savanna. Ecology 68:319–331; 1987.

Painter, D.S.; McCabe, K.J. Investigation into the disappearance of Eurasian water-milfoil from the Kawartha Lakes. J. Aquat. Plant Manage. 26:3–12; 1988.

Palmer, R.S. Handbook of North American birds. New Haven, CT: Yale University Press; 1976.

Parmenter, R.R. Effects of food availability and water temperature on the feeding ecology of pond sliders. Copeia 3:503–514; 1980.

Parmenter, R.R.; Avery. 1990. The feeding ecology of the slider turtle. In: Parmenter, R.R.; Avery, H.W., eds. Life history and ecology of the slider turtle. Washington, DC: Smithsonian Institution Press; 1990:257–265.

Pelikan, J.; Svoboda, J.; Kvet, J. Relationship between the populations of muskrats (*Ondatra zibethica*) and the primary production of cattail (*Typha latifolia*). Hydrobiologia 12:177–180; 1971.

Persson, L.; Diehl, S.; Johansson, L.; Andersson, G.; Hamrin, S.F. Shifts in fish communities along the productivity gradient of temperate lakes—patterns and the importance of size-structured interactions. J. Fish. Biol. 38:281–293; 1991.

Polunin, N.V.C.; Klumpp, D.W. Algal food supply and grazer demand in a very productive coral reef zone. J. Exp. Mar. Biol. Ecol. 164:1–15; 1992.

Power, M.E.; Stewart, A.J.; Matthews, W.J. Grazer control of algae in an Ozard Mountain stream: effects of short-term exclusion. Ecology 69:1894–1898; 1988.

Prejs, A. Herbivory by temperate freshwater fishes and its consequences. Environ. Biol. Fish. 10:281–296; 1984.

Prince, J. Limited effects of the sea urchin *Echinometra mathaei* (de Blainville) on the recruitment of benthic algae and macroinvertebrates into intertidal rock platforms at Rottnest Island, Western Australia. J. Exp. Mar. Biol. Ecol. 186:237–258; 1995.

Provenza, F.D. Tracking variable environments: there is more than one kind of memory. J. Chem. Ecol. 21:911–924; 1995.

Randall, J.E. Overgrazing of algae by herbivorous marine fishes. Ecology 42:812; 1961.

Randall, J.E. Grazing effect on seagrasses by herbivorous reef fishes in the West Indies. Ecology 46:255–260; 1965.

Reinecke, K.J.; Kaminski, R.M.; Moorhead, D.J.; Hodges, J.D.; Nassar, J.R. Mississippi Alluvial Valley. In: Smith, L.M.; Pederson, R.L.; Kaminski, R.M., eds. Habitat management for migrating and wintering waterfowl in North America. Lubbock, TX: Texas Tech University Press; 1989.

Rhoades, D.F. Offensive–defensive interactions between insects and plants: their relevance in herbivore population dynamics and ecological theory. Am. Nat. 125:205–238; 1985.

Richardson, M.J.; Whoriskey, F.G.; Roy, L.H. Turbidity generation and biological impacts of an exotic fish *Carassius auratus,* introduced into shallow seasonally anoxic ponds. J. Fish. Biol. 47:576–585; 1995.

Risley, L.S.; Crossley, D.A. Herbivore-caused greenfall in the southern Appalachians. Ecology 69:1118–1127; 1988.

Rockwell, R.; Abraham, K.; Jefferies, R. Tundra under siege. Nat. History 105:20–21; 1996.

Rose, P.M. Midwinter waterfowl counts. In: Pose, P.M., ed. Western Palearctic and South-West Asia waterfowl census 1994. Slimbridge, Gloucester, UK: Waterfowl and Wetland Res. Bur. IWRB; 1994.

Rosenthal, G.A.; Berenbaum, M.R., eds. Herbivores: their interactions with secondary plant metabolites. Vol. II: Evolutionary and ecological processes. New York: Academic Press; 1992.

Russ, G.R. Is the rate of removal by grazers reduced inside territories of tropical damsel-fishes? J. Exp. Mar. Biol. Ecol. 110:1–17; 1987.

Sammarco, P.W. *Diadema* and its relationship to coral spat mortality: grazing, competition, and biological disturbance. J. Exp. Mar. Biol. Ecol. 45:245–272; 1980.

Sammarco, P.W. Effects of grazing by *Diadema antillarum* Philippi (Echinodermata: Echinoidea) on algal diversity and community structure. J. Exp. Mar. Biol. Ecol. 65:83–105; 1982.

Sammarco, P.W.; Levinton, J.S.; Ogden, J.C. Grazing and control of coral reef community structure by Diadema antillarum Philippi (Echinodermata: Echinoidea) : a preliminary study. J. Mar. Res. 32:47–53; 1974.

Santelices, B.; Martinez, E. Effects of filter-feeders and grazers on algal settlement and growth in mussel beds. J. Exp. Mar. Biol. Ecol. 118:281–306; 1988.

Sarnelle, O.; Kratz, K.W.; Cooper, S.D. Effects of an invertebrate grazer on the spatial arrangement of a benthic microhabitat. Oecologia 96:208–218; 1993.

Scheffer, M.; Hosper, S.H.; Meijer, M.-L.; Moss, B.; Jeppesen, E. Alternative equilibria in shallow lakes. Trends Ecol. Evol. 8:275–279; 1993.

Schreiner, E.G.; Krueger, K.A.; Happe, P.J.; Houston, D.B. Understory patch dynamics and ungulate herbivory in old-growth forests of Olympic National Park, Washington. Can. J. For. Res. 26:255–265; 1996.

Schutten, J.; van der Velden, A.; Smit, H. Submerged macrophytes in the recently freshened lake system Volkerak-Zoom (The Netherlands), 1987–1991. Hydrobiologia 275/276: 207–218; 1994.

Scott, M.L.; Haskins, J.L. Effects of grazing by chrysomelid beetles on two wetland herbaceous species. Bull. Torrey Bot. Club 114:13–17; 1987.

Shaffer, G.P.; Sasser, C.E.; Gosselink, J.G.; Rejmanek, M. Vegetation dynamics in the emerging Atchatalaya Delta, Louisiana, USA. J. Ecol. 80:677–687; 1992.

Sheldon, S.P. The effects of herbivorous snails on submerged macrophyte communities in Minnesota lakes. Ecology 68:1920–1931; 1987.

Sheldon, S.P. More on freshwater snail herbivory: a reply to Brönmark. Ecology 71:1215–1216; 1990.

Shireman, J.V. Control of aquatic weeds with exotic fishes. In: Courtenay, W.R.; Stauffer, J.R., eds. Distribution, biology, and management of exotic fishes. Baltimore, MD: Johns Hopkins University Press; 1984: 302–312.

Sih, A. The behavioral response race between predator and prey. Am. Nat. 123:143–150; 1984.

Smith, L.M.; Kadlec, J.A. Fire and herbivory in a Great Salt marsh. Ecology 66:259–265; 1985.

Smith, T.J.; Odum, W.E. The effects of grazing by snow geese on coastal salt marshes. Ecology 62:98–106; 1981.

Søndergaard, M.; Bruun, L.; Lauridsen, T.; Jeppesen, E.; Madsen, T.V. The impact of grazing waterfowl on submerged macrophytes: in situ experiments in a shallow eutrophic lake. Aquat. Bot. 53:73–84; 1996.

Soska, G.J. Ecological relations between invertebrates and submerged macrophytes in the lake littoral. Ekol. Pol. 23:393–415; 1975.

Squires, J.R. Trumpeter swam food habitats, forage processing, activities and habitat use. Ph.D. thesis, Univ. Wyoming, Laramie; 1991.

Steinberg, P.D.; Estes, J.A.; Winter, F.C. Evolutionary consequences of food chain length in kelp forest communities. Proc. Natl. Acad. Sci. U.S.A. 92:8145–8148; 1995.

Steinman, A.D. Effects of grazers on freshwater benthic algae. In: Stevenson, R.J.; Bothwell, M.L.; Lowe, R.L., eds. Algal ecology: freshwater benthic ecosystems. New York: Academic Press; 1996:533–573.

Sterry, P.R.; Thomas, J.D.; Patience, R.L. Behavioural responses of *Biomphalaria glabrata* (Say) to chemical factors from aquatic macrophytes including decaying *Lemna paucicostata* (Hegelm ex Engelm). Freshwat. Biol. 13:465–476; 1983.

Strong, D.R.; Lawton, J.H.; Southwood, R. Insects on plants: community patterns on plants. Cambridge, MA: Harvard University Press; 1984.

Suren, A.M.; Lake, P.S. Edibility of fresh and decomposing macrophytes to three species of freshwater invertebrate herbivores. Hydrobiologia 178:165–178; 1989.

Taylor, J.N.; Courtenay, W.R.; McCann, J.A. Known impacts of exotic fishes in the continental United States. In: Courtenay, W.R.; Stauffer, J.R., eds. Distribution, biology, and management of exotic fishes. Baltimore, MD: Johns Hopkins University Press; 1984:322–373.

Taylor, K.L.; Grace, J.B. The effects of vertebrate herbivory on the plant community structure in the coastal marshes of the Pearl River, Louisiana, USA. Wetlands 15:68–73; 1995.

Taylor, K.L.; Grace, J.B.; Guntenspergen, G.R.; Foote, A.L. The interactive effects of herbivory and fire on an oligohaline marsh, Little Lake, Louisiana, USA. Wetlands 14:82–87; 1994.

Underwood, G.J.C.; Thomas, J.D.; Baker, J.H. An experimental investigation of interactions in snail-macrophyte-epiphyte systems. Oecologia 91:587–595; 1992.

Urban, E. The mining fauna in four macrophyte species in Mikolajskie Lake. Ekol. Pol. 23:417–438; 1975.

Urbanc, B.O.; Blejec, A. Aquatic macrophytes of Lake Bled: changes in species composition, distribution and production. Hydrobiologia 262:189–194; 1993.

Valentine, J.F.; Heck, K.L. The role of sea urchin grazing in regulating subtropical seagrass meadows: evidence from field manipulations in the northern Gulf of Mexico. J. Exp. Mar. Biol. Ecol. 154:215–230; 1991.

Vance, R.R. Effects of grazing by the sea urchin, *Centrostephanus coronatus,* on prey community composition. Ecology 60:537–546; 1979.

van der Velde, G. Initial decomposition of *Nymphoides peltata* (Gmel.) O. Kuntze (Menyanthaceae), as studied by the leaf-marking method. Hydrobiol. Bull. 16:51–60; 1982.

van Donk, E.; Gulati, R.D. Transition of a lake to turbid state six years after biomanipulation: mechanisms and pathways. Wat. Sci. Techn. 32:197–206; 1995.

van Donk, E.; Otte, A. Effects of grazing by fish and waterfowl on the biomass and species composition of submerged macrophytes. Hydrobiologia 340:285–290; 1996..

van Donk, E.; De Deckere, E.; Klein-Breteler, J.G.P.; Meulemans, J.T. Herbivory by waterfowl and fish on macrophytes in a biomanipulated lake: effects on long-term recovery. Verh. Int. Verein. Limnol. 25:2139–2143; 1994.

van Tamelan, P.G. Early successional mechanisms in the rocky intertidal: the role of direct and indirect mechanisms. J. Exp. Mar. Biol. Ecol. 112:39–48; 1987.

Verhoeven, J.T.A. Natural regulation of plant biomass in a *Ruppia*-dominated system. Proc. EWRS 5th Symp. on Aquatic Weeds; 1978:53–61.

Vermaat, J.E. Periphyton removal by freshwater micrograzers. In: van Vierssen, W.; Hootsmans, M.; Vermaat, J.E., eds. Lake Veluwe, a macrophyte-dominated system under eutrophication stress. Boston: Kluwer Academic Publishers; 1994:213–249.

Wallace, J.B.; O'Hop, J. Life on a fast pad: waterlilly beetle impact on water lilies. Ecology 66:1534–1544, 1985.

Wanders, J.B.W. The role of benthic algae in the shallow reef of Curacao (Netherlands Antilles) III: the significance of grazing. Aquat. Bot. 3:357–390; 1977.

Whicker, A.D.; Detling, J.K. Ecological consequences of prairie dog disturbances. Bioscience 38:778–785; 1988.

Whitaker, J.O. The Audubon Society field guide to mammals. New York: Knopf, Inc. 1980.

Willey, C.H.; Halla, B.F. Mute swans of Rhode Island. Wildl. Pamphlet No. 8, Rhode Island Dept. Nat. Res., Div. of Fish and Wildl.; 1972.

Wilsey, B.J.; Chabreck, R.H. Nutritional quality of nutria diets in three Louisiana wetland habitats. Northeast Gulf Sci. 12:67–72; 1991.

Wilson, E.O. The diversity of life. New York: W.W. Norton & Co.; 1992.

Wilson, U.W.; Atkinson, J.B. Black brant winter and spring-staging use at two Washington coastal areas in relation to eelgrass abundance. Condor 97:91–98; 1995.

Witkowski, E.T.F. The defoliation of woody vegetation by large herbivores at Nylsvley. Dissertation, Univ. Witwatersrand, Johannesburg, South Africa; 1980.

Wood, R.; Gelston, W.L. Preliminary report: the mute swans of Michigan's Grand Traverse Bay region. Mich. Dept. Nat. Resour. Rep. 2683; 1972.

Wootton, J.T.; Oemke, M.P. Latitudinal differences in fish community trophic structure, and the role of fish herbivory in a Costa Rican stream. Environ. Biol. Fish. 35:311–319; 1992.

Ydenberg, R.C.; Prins, H.H. Spring grazing and the manipulation of food quality by barnacle geese. J. Appl. Ecol. 18:443–453; 1981.

9. Interactions Between Grazing Birds and Macrophytes

Stuart F. Mitchell and Martin R. Perrow

Introduction

In the past, aquatic birds were largely overlooked by limnologists, receiving scant attention in hydrobiological journals and no more than passing mention in limnology texts. There has recently been rapid growth in interest in their roles in lake ecosystems, with the integration of bird studies into intensive limnological programs, comparative investigations over large groups of lakes, an increase in the number of experimental studies, and increasing contact with water bird biologists (Kerekes and Pollard, 1994; Faragó and Kerekes, 1997). Much of this interest stems from recent scientific focus on the factors that lead to shallow eutrophic lakes being dominated alternatively by phytoplankton or by macrophytes (Scheffer et al., 1993) and management investment in the restoration of eutrophic lakes to a clear macrophyte-dominated state (e.g., National Research Council [USA], 1992; Broads Authority [UK], 1994; National Environmental Research Institute [Denmark], 1994). It is now clear that aquatic bird populations may, at times, be very sensitive to ecological changes in lakes and that they can also play significant roles in producing such changes.

Our objectives are to review methods for quantifying bird–macrophyte interactions, to discuss examples selected to illustrate some of the interactions that have been quantified, and to point to areas in which further study may be particularly fruitful. We emphasize the functional links indicated in Figure 9.1 and do not consider how the interactions may affect such structural properties as plant or bird

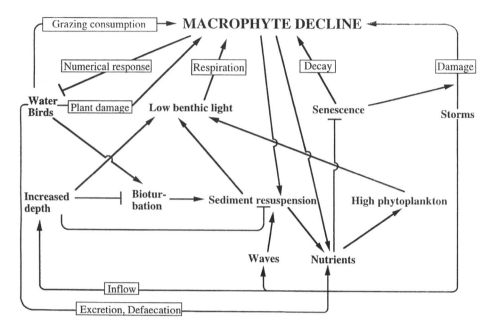

Figure 9.1. Direct effects and indirect physical and chemical effects of water birds on macrophytes in relation to some other factors regulating macrophyte abundance in lakes. Stimulatory effects →; inhibitory effects —|. (Redrawn from Mitchell and Wass, 1996a, with kind permission of Elsevier Science-NL, Sara Burgerhartstraat 25, 1055 KV Amsterdam, The Netherlands.)

community diversity or species richness or the potential role of birds in the dispersal of macrophytes. Although such effects may themselves have functional consequences, as may feeding interactions among bird species, and food chain interactions involving carnivorous birds, little is known of them in lakes. We also neglect the potential effects of the complex food chain interactions that arise from omnivory in birds, although among aquatic birds only some geese and swans appear to be strict herbivores (Owen and Black, 1990; Baldassarre and Bolen, 1994). Even those species are not confined to eating aquatic macrophytes, with geese most typically feeding terrestrially.

Birds directly affect the dynamics of plant biomass by consuming macrophyte tissue. They may also affect the productivity or rates of change in biomass indirectly by grazing selectively, damaging the remaining plants, and cycling nutrients or by a variety of other mechanisms that modify the plants or their habitats (Fig. 9.1). The effect on the dynamics of the plant community (interaction strength) depends on grazing consumption, the net effect of these various indirect positive and negative feedbacks, and the timing of consumption within the plant growth cycle. These interactions, the ways in which population densities of aquatic birds respond to changes in trophic status and macrophyte abundance and the role of birds in determining phytoplankton and macrophyte dominance, provide the focus for our study.

Water Bird Abundance and Trophic Status

Indices of trophic status such as phytoplankton (chlorophyll *a*) and plant nutrient concentrations are available for many lakes, as is information on aquatic bird populations. Few attempts have been made, however, to collate large data sets on aquatic bird population densities and trophic status. Notable exceptions include the studies of Hoyer and Canfield (1994) on shallow subtropical lakes in Florida and Suter (1994) on deep Swiss lakes. Hoyer and Canfield estimated annual average population densities of 50 bird species at 46 lakes of widely varying trophic status. They found significant correlations of total bird numbers and biomasses with the trophic indicators total phosphorus (TP), total nitrogen (TN), and chlorophyll *a* ($r = 0.56$–0.61). Nilsson and Nilsson (1978) and Murphy et al. (1984) have recorded similar correlations of water bird densities to chlorophyll *a* and/or TP concentrations in southern Swedish and Alaskan lakes. Suter (1994) surveyed wintering bird populations over 12 years in 20 generally large (3–580 km^2) alpine lakes, in relation to trophic status, in categories defined by TP concentrations. Abundances per unit lake surface area were significantly related to trophic status only for mallard (*Anas platyrhynchos*) among the eight herbivorous

crocorax carbo). Similar relationships to trophic status have been noted for other piscivorous birds in lakes in Sweden (Nilsson, 1978), Nova Scotia (Kerekes, 1990), and Alaska (Heglund et al., 1994). Effects of lake morphometry have also been noted, with population densities sometimes being more closely related to shore length than to surface area in lakes of similar trophic status (Nilsson, 1978; Suter, 1994; Kerekes, in press).

Empirical modeling offers considerable promise for predicting water bird densities, although different feeding guilds and different species within guilds are likely to show different types of response. The consistent positive reponses of piscivores to trophic status over a wide range of lake depths may be due to fish biomass generally increasing with increasing eutrophication (Hanson and Peters, 1984) but being largely independent of lake depth (Jeppesen et al., 1997).

Herbivorous and omnivorous species cannot, however, be expected to show continuous positive responses to increases in TP, as macrophytes, after increasing progressively with increasing eutrophication (Canfield and Hoyer, 1992), often then decline abruptly owing to shading by phytoplankton or epiphytes (e.g., Scheffer et al., 1993). Thus in Tomahawk Lagoon, New Zealand, the winter black swan (*Cygnus atratus*) population size is related inversely to phytoplankton productivity in the previous summer, owing to the inverse variations of phytoplankton and perennial macrophytes (McKinnon and Mitchell, 1994). There is some evidence that primary productivity of phytoplankton and macrophytes combined may be higher in shallow eutrophic lakes during phytoplankton dominance than when macrophytes predominate (Roijackers, 1985; Mitchell, 1989). It is unclear to what extent this effect, if it is a general one, may flow up through the food chain to regulate populations of omnivorous or carnivorous birds. Primary productivity is, however, likely to be a better predictor of such populations than total phosphorus, from which it is liable to become uncoupled in shallow, highly eutrophic lakes (e.g., Mitchell et al., 1988).

Water Birds and Macrophyte Abundance

Among the first to note an effect of macrophyte abundance on water bird populations were Allison and Newton (1972/3), who observed progressive decline in the populations of mute swan (*Cygnus olor*), coot (*Fulica atra*), and pochard (*Aythya ferina*) from 1950–1971, in Loch Leven, Scotland, in parallel with a decline in macrophytes brought about by eutrophication. Dirksen et al. (1991) have similarly reviewed the wax and wane of Bewick's swan (*Cygnus columbianus bewicki*) populations in The Netherlands in relation to the establishment and decline of freshwater macrophytes after polder construction and subsequent eutrophication. A violent storm that destroyed macrophyte beds in the large (180 km^2) Lake Ellesmere, New Zealand, in 1968 led to a decline in the population of black swans from 40,000–80,000 in the mid-1960s to 4,000 by 1986 (Williams, 1979; McKinnon and Mitchell, 1994). In Lake Christina, Minnesota, populations of various dabbling and diving ducks, american coots (*Fulica americana*), and Canada geese (*Branta canadensis*) all declined throughout a progressive 10-year decline in macrophyte cover and recovered in parallel with the macrophytes in the following 5 years, after a fish removal biomanipulation (Hanson and Butler, 1994). Giles (1994) showed similar effects of fish removal in a biomanipulated gravel pit by comparison with an unmanipulated control pit. Similar large, long-term changes in vegetation and parallel changes in populations of coots, swans, and herbivorous and omnivorous ducks have been reported, among others, by Hargeby et al. (1994), Lauridsen et al. (1994), Schutten et al. (1994), van Donk et al. (1994), and Søndergaard et al. (this volume, Chapter 20). In Hawksbury Lagoon, New Zealand, black swan population densities followed the week-to-week changes in biomass of aquatic vegetation remarkably closely during a 7-month period (Mitchell and Wass, 1996a). In surveys of groups of lakes, brood densities of black duck (*Anas rubripes*) were significantly correlated with macrophyte cover in 32 Nova Scotian lakes (Staicer et al., 1994), and winter black swan populations were significantly correlated with macrophyte biomass in seven New Zealand lakes (McKinnon and Mitchell, 1994).

These results suggest that population densities of water birds are often closely related to the abundance of macrophytes, and further development of general empirical models for the relationships may prove very useful. Lakes provide many examples of simple quantitative relationships linking various functional or taxonomic groups. The phosphorus–chlorophyll relationship (e.g., OECD, 1982), which has proved a powerful tool for lake research and management, is one example, but there are many others (e.g., Hanson and Leggett, 1982; Hanson and Peters, 1984). Aquatic birds are unlikely to prove an exception, although the generality of models for birds will be affected by such seasonal factors as territoriality and migration. For example, coot densities may become related to macrophyte abundance only in autumn, after territories break up (Perrow et al., 1997).

Until now, most attention has been given to demonstrating the existence of such relationships in particular lakes. Some of the studies cited above have been semiquantitative, with only general observations of macrophyte abundance, or are quantified and presented in ways that do not readily allow comparisons with other systems. Other studies have shown no clear relationships between aquatic bird populations and macrophyte biomass (Hoyer and Canfield, 1994) or other macrophyte indices (Hoyer and Canfield, 1994; Lillie and Evrard, 1994), and it is unclear to what extent differences in the results and our present inability to make general predictions about aquatic bird populations from macrophyte data reflect the true complexity of the relationships, or whether they result merely from the paucity of quantitative information expressed in the appropriate units.

What the appropriate units might be is a matter for conjecture. It seems likely, however, that the closest general relationships involving waterfowl will be revealed by studies of plant biomass available as food, rather than shoot density, percentage cover, or other variables such as shore length or the extent of the littoral zone. Macrophytes at depths beyond the feeding reach of a species are not available to the birds as food, so that an appropriate unit for water bird population densities might be

Total number of birds on the lake/(ha of lake at depths within feeding reach)

with the available macrophytes being estimated as g/m^2 for the area within feeding reach. Use of these units resulted in better predictions of black swan population densities than simple numbers and biomasses per unit area (McKinnon and Mitchell, 1994). An alternative unit that may be even more appropriate for tall-growing macrophytes is g/m^2 in the upper part of the water column that is within feeding reach.

There is scattered evidence of discontinuities in the numerical responses of some water bird populations at both high and low macrophyte biomasses. Beekman et al. (1991) observed that migratory Bewick's swans abandoned feeding on *Potamogeton* tubers and left a lake when they had reduced the tuber density to about 7 g DW/m^2 and coots abandoned feeding on *Ruppia* at a similar biomass (Verhoeven, 1980). Black swan numbers on Hawksbury Lagoon show evidence of a similar discontinuity at filamentous algal biomasses of about 2–3 g DW/m^2 (Mitchell and Wass, unpublished data). Black swan populations also tend to decline when macrophytes or filamentous algae form dense emergent patches that occupy a large part of the lake surface (e.g., Wass and Mitchell, this volume, Chapter 18). Use of the algal patches by swans, as indicated by rates of fecal deposition, was only about 25% of their use of the intervening relatively bare patches (Mitchell and Wass, unpublished data). This effect might be due simply to swans finding it difficult to swim through dense plant stands or, alternatively, to differences in food quality.

An increase in macrophytes that results in more food for many water bird species may represent a decrease in available habitat or feeding opportunity for others, and negative relationships between macrophyte cover and bird abundance have been recorded for several bird species associated with Florida lakes (Hoyer and Canfield, 1994).

Estimating Food Consumption

If the role of birds in lake ecosystems is to be understood and quantified, it is essential to know, among other things, how much they eat in nature. Until adequate estimates become available for rates of food consumption and food preferences and how they vary among species, seasonally, with age, with food availability, and with food quality, our understanding of the roles of birds in aquatic ecosystems will be hampered.

In the absence of direct estimates of food consumption, there has been widespread reliance on bioenergetic estimates that take little or no account of the above sources of variation or interspecific variation in the birds' anatomy, digestive physiology, or feeding behavior. Such estimates often have very wide 95% confidence intervals, which may span an order of magnitude (Furness, 1978; Peters, 1983). Bioenergetic modeling essentially involves determining size-based standard metabolic rates (e.g., Zar, 1969; Aschoff and Pohl, 1970) or size-based rates for metabolism of birds confined to small cages (Kendeigh, 1970), which are then corrected for the various types of activity in nature, and for the effects of temperature on metabolism (e.g., Woollhead, 1994). These relationships, in turn, have been used to derive size–energy expenditure relationships in nature (King, 1974). In attempting to estimate food consumption from modeled energy expenditure, the already wide confidence intervals are further extended by the need to account for energy directed to growth, and for differences in the efficiency of assimilation. Published assimilation efficiencies for waterfowl range from about 20–70%, suggesting that this is a substantial source of additional error. There appear to be systematic differences in the food consumption of swan and goose species, even after adjustment for the effects of body size on metabolism, which can be related to differences in their digestive morphology and feeding behavior (Mitchell and Wass, 1995).

Energy expenditure by individual birds in nature may be estimated by determining the rates of elimination of experimentally administered doses of isotopically enriched (double labeled) water (Nagy, 1987). This method, used in conjunction with a knowledge of the assimilability of the food, has strong possibilities for estimating food consumption by waterfowl. The less direct bioenergetic methods offer no more than crude approximations, and are perhaps best avoided in most circumstances.[1]

---

[1]Although crude, such estimates are greatly preferable to scaling for body weight by simple proportions, which can lead to large errors. Metabolism normally varies with body weight to the power of about 0.65–0.75 (e.g., Bertalanffy, 1968).

There is a wide range of alternative methods available. Geese and ducks lend themselves readily to experimental studies of food consumption in a variety of conditions ranging from small cages to near-natural conditions (e.g., Marriott and Forbes, 1970; Mattocks, 1971; Gere and Andrikovics, 1994), and controlled feeding experiments have also been done with captive mute swans (Mathiasson, 1973) and coots (Verhoeven, 1980).

Food consumption may also be calculated by measuring rates of deposition of feces and analyzing the food and feces for cellulose, on the assumption that it is not digested (Ebbinge et al., 1975; Cargill and Jeffries, 1984; Mitchell and Wass, 1995). Any digestion of cellulose will lead to food consumption being underestimated. In food mass balance studies with geese, Buchsbaum et al. (1986) report that up to 28% of cellulose was digested, but Marriot and Forbes (1970) and Sedinger et al. (1989) found no evidence of cellulose digestion, and Mattocks (1971) could detect no cellulase activity in the gut ceca.

Beekman et al. (1991) estimated grazing consumption by Bewick's swans by measuring rates of decline of dormant tubers of *Potamogeton* in naturally grazed areas and ungrazed control areas, and relating the difference to the abundance of the birds. They note, however, that other species may at times accompany the swans and compete for tubers that they have uprooted. When birds do not feed continuously, it may be possible to identify recently ingested material in the gut and thereby estimate short-term rates of consumption (e.g., Gauthier, 1993). Perrow et al. (1997) estimated food consumption by coots from direct observa-

determining the biomass per unit bill length for various macrophyte species and other food items. Most waterfowl species, however, handle food items in ways that preclude this method.

Little is also known of the extent and causes of food selectivity among waterfowl. The food species eaten by water birds often change seasonally (e.g., Black and Rees, 1984; Owen and Black, 1990; Baldassarre and Bolen, 1994; Grant et al., 1994). This may often be no more than a response to changing availability (Perrow et al., 1997), but ontogenetic shifts occur in ducks (Baldassarre and Bolen, 1994) and in coots, in which young are fed predominantly on high-protein insects before they become self-sufficient herbivores or omnivores (Perrow et al., 1997). Female ducks increase both their food intake and the proportion of animals in the diet during laying (Noyes and Jarvis, 1985) and the diet of adult coots shows similar changes during parental care (Perrow et al., 1997). Food preferences might be related to protein content (Rees, 1990), texture, secondary plant metabolites, or other chemical constituents (cf. Lodge et al., this volume, Chapter 8; Cronin, this volume, Chapter 21). McKinnon (1989) found no indication of any selectivity between charophytes and angiosperms among black swans on four lakes, but the results were highly variable, and the discriminatory power of the indices used was low. The relative abundances of different stable isotopes of carbon and nitrogen in the tissues of birds and their potential food species provide promising opportunities for unraveling complex food linkages (Hoyer et al., this volume, Chapter 23).

Little is also known of individual variations in food consumption in nature. It must, however, be a cause for concern when estimates are based on small numbers of individual birds. For example, Marriott and Forbes (1970) found an almost fourfold range of variation in daily food intake by captive Cape Barren geese fed ad libidum. Similarly, functional feeding responses have not been investigated for natural water bird populations. Food consumption varies with food availability in many animals, but such responses might be relatively unimportant in water birds, first because of their strong numerical responses—if food levels become suboptimal they may simply move elsewhere. Second, feeding occupies only a small part of the day in many species (Black and Rees, 1984; Baldassarre and Bolen, 1994), and additional time might be devoted to it when food is in short supply, to maintain a constant daily ration.

Given the difficulties of determining rates of food consumption in nature, authors, having obtained an estimate, have generally been content to regard it as a constant. Energy requirements are, however, known to vary with temperature, and phases of the molt and reproductive cycles, and the extent to which these variations are translated into differing rates of food consumption or accommodated by changes in body fat requires further study.

Water Birds as Consumers

Even at their most abundant, birds appear to be a minor component of the biomass of lakes. Bird biomasses at 46 shallow lakes in subtropical Florida ranged from 0.001 to 0.47 g/m^2 annual average live weight (0.01–4.7 kg/ha) (Hoyer and Canfield, 1994). The mean was only 0.11 g/m^2 (ca 0.04 g DW/m^2) by comparison with the mean wet weight biomass of submerged and floating leaved macrophytes of 3,100 g/m^2.

From cooler regions, Pöysä (1983) reports a figure equivalent to 0.16 g/m^2 live weight for a shallow, eutrophic, Finnish lake. In Hawksbury Lagoon (46° S), the black swan density becomes as high as 25 birds/ha, and the average of 10 birds/ha during long phases of macrophyte dominance corresponds to 1.9 g DW/m^2 (5.6 g/m^2 or 56 kg/ha live weight), which is higher than for any of the Florida lakes cited and again very low in relation to the maximum macrophyte (macroalgal) biomass of 200 g DW/m^2. These figures may be near the upper limit for annual average water bird biomasses (Mitchell and Wass, 1995), although flamingos (Hurlbert and Chang, 1983), some roosting populations, and populations that are artificially fed may be exceptions. Bird biomasses may be seasonally three to four times higher than this on lakes where terrestrially feeding migratory geese gather (from Manny et al., 1994). Bird biomasses also appear to be low in relation to those commonly observed for phytoplankton and benthic fauna (frequently >10 g DW/m^2), and lower than those often recorded for fish (100–400 kg live weight/ha or ca 3–12 g DW/m^2) and zooplankton (0.1–0.5 g DW/m^2) (data from various sources). These figures suggest a rather minor role for birds as consumers in lake ecosystems. As they are among the largest of the lake fauna, the role of birds in biological productivity will be further reduced by inverse size/metabolism relationships but increased by the additional energy demands of endothermy.

Kiørboe (1980), who used bioenergetic estimates of food consumption in Ringkøbing Fjord, Denmark, suggested limits for consumption of macrophytes by birds of between 15–60% of the annual productivity but also cautioned about the use of this method. Woollhead (1994), using similar methods, suggests a figure of 13% for eutrophic Lake Esrom, Denmark. Using more direct methods, Clausen (1994) obtained a figure of 12% (range, 8–21%) for Brent geese (*Branta bernicla*) grazing on *Zostera,* and black swan consumption of macroalgal productivity at Hawksbury Lagoon was 12 and 16% during different years (Mitchell and Wass, 1996a; Wass and Mitchell, this volume, Chapter 18). Although many of the 19 figures cited by Cyr and Pace (1993) are substantially higher than these, those estimates do not represent consumption (Mitchell and Wass, 1996b). Figures for total annual consumption expressed as a fraction of maximum plant biomass also overstate the fraction of net primary productivity consumed, as consumption during biomass increase is a component of the productivity, which should be added to the net biomass increase.

Little information is available on consumption by birds in relation to that by other herbivores. In Lake Zwemlust, The Netherlands, calculated annual consumption of macrophytes by rudd (*Scardinius erythrophthalamus*) varied only from 170 to 360 kg, from 1990 to 1996, whereas consumption by coots varied from 30 to 1,200 kg as the coot population changed in tandem with variations in

their low biomasses and generally small role as consumers suggest that possibilities for birds to have major effects on the dynamics of plant biomass may be limited to temperate waters in periods in which macrophyte productivity is low—either early in the growth phase of seasonal species, when biomass is low, or in autumn, when seasonal plant growth has slowed or ceased. Alternatively, it requires that the net indirect effect of birds on the plants should be negative and large in relation to the direct effect of consumption and/or that the timing of grazing should be important.

Considerations of the Timing of Grazing

Consumption of plants removes not only plant material but also the future productive potential of that material. Consumption of tissue during active growth affects rates of plant biomass increase, but grazing after seasonal growth has ended is inconsequential, at least for that season's productivity (Kiørboe, 1980). For this reason alone, biomass consumption may give misleading indications of interaction strength.

Because of the interdependence of biomass and productivity, the effect of removing growing tissue is nonlinear. If the productivity/biomass quotient (P/B) remains constant, the effect will be exponential, and the effects of earlier consumption will become disproportionately larger as time progresses. This effect

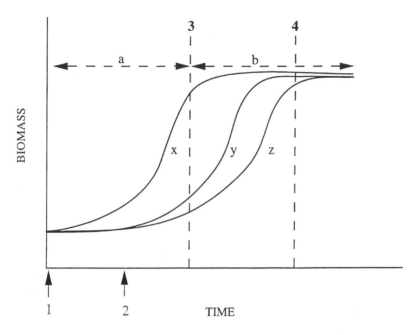

Figure 9.2. Generalized time course of plant biomass increase to equilibrium biomass in herbivore exclusion experiments to illustrate the effects of exponential growth (period a) and density limitation (period b) on the results for experiments with two different starting times (1 and 2) and two alternative finishing times (3 and 4). x, y, ungrazed biomasses; z, grazed biomasses.

("biomass compounding") may cause even a small amount of consumption early in the growth cycle to have a large effect (Fig. 9.2). The ultimate effect of grazing on biomass will be insignificant, however, if the plant community can attain the carrying capacity of the system despite being grazed. In the absence of grazing, it would merely reach it sooner. Approach to the carrying capacity is marked by declining P/B and declining rates of biomass increase to produce the familiar sigmoid seasonal biomass curve (Fig. 9.2).

Thus even in systems with identical grazing rates and plant growth rates, the outcomes may be quite different, ranging from a very large effect on biomass to none at all. The significance of exponential growth is illustrated by the study of Wass and Mitchell (this volume, Chapter 18), in which the biomass of experimentally ungrazed macroalgae became five times higher than that in the natural grazed system, although grazing consumption was only 12% of the net plant productivity. The major effect of grazing arose not from the biomass consumed but from the loss of the future growth potential of that material. That material was only ever represented as potential, and it was never present in the real grazed system to be either eaten (cf. Cyr and Pace, 1993) or dislodged by the herbivores (cf. Cattaneo and Mousseau, 1995). The importance of density limitation is illustrated by the fact that growth of the natural community continued after the experimentally protected plants had become density limited, and the ultimate biomass difference was effectively zero. The early occurrence of density limitation in the ungrazed plants was also an experimental artifact, in the sense that it occurred prematurely, long before any such effect might have been detected in the real grazed system (cf. Fig. 9.2). The effect of grazing on biomass is therefore highly time dependent and highly dependent on the plant growth cycle (Mitchell and Wass, 1996b).

The problems posed by biomass compounding and density limitation can be solved by assessing grazing, not in terms of biomass or ultimate biomass outcomes but as a dynamic process affecting rates of change in biomass, and with proper regard to the growth cycle of the plants. These objectives can be achieved by exponential modeling, with growth, interaction strength, and grazing all being expressed as plant tissue–specific rates.[2] These effects have important implications for the ways in which herbivore exclusion experiments are conducted and interpreted (see below).

At a different level, grazing consumption may show large seasonal variations. With the arrival of migratory flocks at many north temperate waters in autumn, grazing impacts increase dramatically, and water birds may contribute substantially to the decline in plant biomass after seasonal growth has ended (Anderson and Low, 1976; Kiørboe, 1980; Esler, 1989). Conversely, during the breeding season many species, including coots and most swan species, are territorial, which reduces grazing impacts (Perrow et al., 1997).

Indirect Feedback Effects of Birds on Macrophytes

Nutrient Inputs and Cycling

The small role of birds in the biomass of lakes suggests that their body wastes will also play only a minor part in nutrient dynamics, and that appears to be normally true. Marion et al. (1994) report contributions by bird feces to annual loads of 0.4 and 0.7% of TN and 2.4 and 6.6% of TP for two study periods on shallow Lake Grand-Lieu, France. Nutrients in mallard feces are equivalent to less than 1% of the external TP entering Lake Kis-Balaton and 2% of the TN (from Gere and Andrikovics, 1994). Fecal inputs of phosphorus to Hawksbury Lagoon by a dense black swan population (10/ha) were sufficient to generate concentrations of 15–30 mg/m³ on the OECD (1982) loading-concentration model (Mitchell and Wass, 1995) or about 5–10 mg/m³ when only the soluble P is considered. These con-

---

[2]The model may at times be fitted to seasonal biomass data. At other times, the essentially exponential nature of growth may be obscured by changing environmental conditions (including changes in grazing rates), the interplay between aboveground and underground components of the biomass (the latter often neglected), and progressive germination, coupled with high sampling variation (see, e.g., data of Jupp and Spence 1977; Søndergaard et al., 1996). It may still be fitted to individual sampling intervals during continuous growth, however.

centrations, although significant by the standards of many lakes, were small in relation to the annual average TP concentration of 340 mg/m^3. N inputs were similarly small, and weekly fecal loadings of inorganic N and soluble reactive P were less than 1% of the maximum observed weekly increases in concentrations. Other studies, based on bioenergetic estimates or proportionately scaled body weights, tend to support these findings. For example, the annual contribution by bird feces (much of it as an internal load) averaged only 2.4% of the annual external phosphorus load in 14 Florida lakes and 6% of the phosphorus present in the water and macrophytes in a larger sample of 46 lakes (Hoyer and Canfield, 1994).

There are, however, striking exceptions. Large overwintering flocks of terrestrially feeding Canada geese contributed 70% and 27%, respectively, of the annual TP and TN loads to Wintergreen Lake, Michigan, and the lake's very dense phytoplankton populations are attributed to their influence (Manny et al., 1994). The eutrophication and loss of macrophytes in Hickling Broad, England (120 ha), are also attributed to nutrient inputs by flocks of roosting gulls that have reached abundances of 250,000 (Leah et al., 1978; Moss and Leah, 1982). Resident bird populations may be unimportant, but that is certainly not always true for roosting flocks and perhaps also for the high populations that are sometimes sustained by artificial feeding. Furthermore, the inputs from birds that feed elsewhere represent an external load, rather than a recycling of nutrients already present in the system.

These studies, the rather large fractions of total N and P that are present in soluble forms (Mitchell and Wass, 1995), and the apparently rapid mobilization of sedimented faecal nutrients (Moss and Leah, 1982) all support the view that fecal deposition in lakes benefits principally the phytoplankton (and possibly epiphytes), which compete with macrophytes for light, and is therefore potentially harmful to macrophytes in such guanotrophic systems.

Relief of Density Suppression of Growth

Plants may become density limited through a variety of mechanisms such as resource depletion or self-shading (e.g., Lodge, 1991). Grazing at these times may increase tissue-specific growth rates. This effect has been demonstrated for water lilies grazed by invertebrates (Wallace and O'Hop, 1985) and for salt marsh vegetation grazed by geese (Cargill and Jeffries, 1984). It is unlikely, however, that the continual replacement of photosynthetic tissue could be achieved without cost to the underground reserves, which were not considered in either study. In the salt marsh, the underground biomass vastly exceeded the aboveground biomass, and it is unlikely that any significant increase in plant productivity would have been demonstrated. Daily grazing consumption by even large waterfowl populations is only a minute fraction of the macrophyte biomass when plant biomass is at the carrying capacity and density limited (see above), and this effect is unlikely to be significant.

Plant Damage and Selective Grazing

There are various published references to grazing water birds being wasteful feeders that consume only a portion of what they break off or uproot (e.g.,

Berglund et al., 1963; Owen and Black, 1990). Selective grazing on the most actively growing tissues will similarly have a greater effect on the plants than would be predicted from a simple consideration of the amount of tissue consumed. These effects are widely postulated to be important (e.g., Lodge, 1991), but they have not been quantified. Filamentous algae, charophytes, and aquatic angiosperms can be expected to differ in their sensitivities to these effects, owing to differences in their morphological complexity. Thus to an algal filament, each cell is worth as much as any other, but angiosperms must maintain a high level of morphological integrity, and charophytes can be expected to occupy an intermediate position.

Physical Effects

Bioturbation by birds swimming, feeding, or wading in shallow water might affect the light climate of macrophytes directly and also, indirectly, by increasing the rate of nutrient supply from the sediment to phytoplankton (cf. Søndergaard et al., 1992). Mitchell and Wass (1996a) concluded that direct optical effects from black swans in a very shallow, wind-swept 25-ha lake were insignificant in relation to those of sediment resuspension by waves and high phytoplankton concentrations, but they are potentially more important in smaller water bodies where wave resuspension is less frequent or where dense populations of waders gather (Comin and Herrera, 1997).

Effects of Birds on Macrophyte Dynamics: Interaction Strengths

Interaction strengths can be determined from controlled herbivore exclusion experiments. In studies using this method, attempts to overcome the problem of high sampling variability and difficulties associated with the need for destructive sampling to estimate biomass have led to use of a wide variety of exclosure types and numbers, sampling intervals, experimental periods, and macrophyte parameters measured, making comparisons among studies often difficult. Comparison of population density- or species-specific impacts is further inhibited by the paucity of information on bird use of the grazed control areas, and the concern has been largely with (biomass) outcomes, rather than the dynamic processes leading to them, which also presents difficulties of interpretation.

Biomass effects are sensitive to when experiments both begin and end in relation to plant growth cycles. A later start will produce a smaller difference in biomass between the grazed and ungrazed communities after any time interval up to equilibrium (Fig. 9.2) and may also affect the ultimate biomass outcome (i.e., the later the start to herbivore exclusion, the later the carrying capacity will be reached and the greater the probability that seasonal or other factors will intervene to stop growth before it is attained). Similarly, for experiments in which herbivores are added rather than excluded or when transplants are used, the longer the plants are grown before the addition, the less will be the apparent effect after any given time interval. Experiments in which only initial and final biomasses are presented allow neither interaction strengths nor ultimate biomass outcomes to be estimated, owing to the time course of the effect on biomass being nonlinear (Fig. 9.2).

Ultimate biomass outcomes are important, particularly for lake management, although it must be recognized that they reflect not just herbivory but the interactions of herbivory, carrying capacity, growth cycles, and the starting time of experiments. They are likely to be highly lake- and experiment-specific. Published figures indicate a range of effects from near zero (Anderson and Low, 1976; Kiørboe, 1980; Perrow et al., 1997; Wass and Mitchell, this volume, Chapter 18), to almost 90% (Van Wijk, 1988; Lauridsen et al., 1993).

The problems of expressing herbivore impacts adequately in units of biomass are intractable—a static concept cannot be used to express a dynamic effect. Exponential modeling (or the less attractive logistic alternative) solves these problems, and its difficulties are minor by comparison. Further advantages are that the potentially large effects of any small initial difference in biomass between control and experimental plots are removed and that with the units being additive, the effects of grazing on plant growth can be compared directly with those of other factors such as light.

Interaction strength can be expressed as the difference between the instantaneous growth rates (g plant/g plant/day) of the ungrazed (B_{iu}) and grazed (B_{ig}) plants. The relative effect on the plants (relative interaction strength) might be expressed as B_{ig} / B_{iu}. The concept of a per capita interaction strength is no longer appropriate, as the effect becomes related to population density rather than population size. We shall use the term *specific interaction strength,* which can be expressed as the difference between the instantaneous growth rates of the ungrazed and grazed plants per unit herbivore, or

$$(B_{iu} - B_{ig}) / N$$

where N is the population density of the herbivore (no./ha or biomass/ha). To obtain this information requires no more than well-designed herbivore exclusion experiments, adequate monitoring of the birds, and use of the exponential model. Coupled with estimates of grazing consumption, it may also allow the direct effects of herbivory to be isolated from the indirect (Mitchell and Wass, 1996b). With herbivory expressed in these units, it may be possible to make better comparisons among ecosystems and among different plant and herbivore species.

A reanalysis of some exclosure studies in which time course data were provided and which fit closely to a seasonal exponential model indicates that water bird grazing may have substantial dynamic effects on aquatic plant communities, with instantaneous rates of biomass increase in the grazed plants being as low as 29% of those for the ungrazed plants in experiments with transplants in pots (where the relatively high plant biomasses may have attracted increased grazing attention from the birds) or 50% in natural plant communties (Table 9.1).

Table 9.1. Interaction Strengths ($B_{iu} - B_{ig}$) and Relative Interaction Strengths (B_{ig}/B_{iu}) Calculated from Published Figures for Waterfowl Grazing on Submerged Macrophytes during Spring-Summer Growth[a]

Lake	Grazers	B_{iu}	B_{ig}	$(B_{iu} - B_{ig})$	B_{ig}/B_{iu}
Lauwersmeer[b]	Coots, etc.	0.035	0.020	0.015	0.57
Væng[c] a	Coots	0.045	0.020	0.025	0.44
b		0.048	0.014	0.034	0.29
c		0.083	0.064	0.019	0.77
d		0.071	0.059	0.021	0.83
Stigsholm[d]	Coots	0.044	0.027	0.017	0.63
Hawksbury[e]	Swans	—	0.055–0.089	0.007	0.89–0.93
Hawksbury[f]	Swans	0.037–0.055	0.027–0.029	0.008–0.028	0.49–0.78

[a]B_{iu}, B_{ig}, exponential growth rates of ungrazed and grazed plants, respectively (per day).
[b]Data from Verhoeven, 1980.
cData from Lauridsen et al., 1993; transplants in pots; derived from total shoot length per pot, which correlated closely with biomass; a, sheltered sites, mud substrate; b, sheltered sites, sand substrate; c, exposed sites, mud substrate; d, exposed sites, sand substrate.
[d]Data from Søndergaard et al., 1996.
[e]Data from Mitchell and Wass, 1996a; calculated from food consumption estimate.
[f]Wass and Mitchell, this volume, Chapter 18, spring-summer.

Interactions and Threshold Effects: Grazing and Stable States

Aquatic plant production may at different times (e.g., seasonally) lie either above or below the grazing consumption. When plants are close to this threshold, and without the buffer represented by accumulated biomass reserves, even a slight change in grazing pressure by birds may have major effects on the ecosystem. On one hand, it may lead to complete suppression of increase in the plants, on the other to escape from grazing control and ultimate growth to the maximum biomass that the system can sustain. These different outcomes have been observed within the same lake (Wass and Mitchell, this volume, Chapter 18), though such fine balances may not be common or persist for long, as at the typical spring-summer growth rates of macrophytes and filamentous algae, the biomass neccessary for escape from complete suppression by grazing may be only a few grams dry weight per square meter.

Even if grazing birds do not by themselves tip the balance of a lake in favor of phytoplankton dominance, they may do so in conjunction with other factors such as low benthic illuminance. Coots, and even swans, may graze to depths greater than the euphotic depth and have the potential to maintain macrophytes in a permanently light-starved condition. Anything that delays seasonal macrophyte growth or recovery after biomanipulation allows more time for the occurrence of stochastic events that favor phytoplankton dominance, such as resuspension of lake sediments by storms.

There have been few investigations of bird grazing in relation to other factors affecting macrophyte dynamics, but in Lake Væng, Denmark, the effect of coots grazing on macrophyte transplants grown in pots was similar in magnitude to that of exposure to waves and greater than the effect of substrate type (Table 9.1; Lauridsen et al., 1993). In Hawksbury Lagoon, grazing effects were small in relation to those of variations in the benthic light climate produced by fluctuations in phytoplankton and sediment resuspension (Mitchell and Wass, 1996a). Nor are birds the only grazers. The calculated relative consumption by rudd and coots varies widely in Lake Zwemlust, and their combined effect may have contributed to a decline in macrophytes and a change in the species composition (van Donk and Otte, 1996). There is also evidence that some water bird populations may be greatly affected by the abundance of fish that consume the same food species (Giles, 1994). The unraveling of such food web complexities and their relation to stable states offers fascinating problems at a variety of scales (see other chapters, this volume).

The question of whether the effects of heavy grazing by birds on the over-wintering stages of plants in one year carry over to reduce plant productivity in the next year is of considerable interest for the long-term stability of the macrophyte-dominated state (Søndergaard et al., 1996; van Donk and Otte, 1996; Perrow et al., 1997). Beekman et al. (1991) showed that swans abandoned grazing on tubers of *Potamogeton* before they reduced the tuber density sufficiently to affect the next year's plant productivity, and Clausen (1994) obtained similar results for *Zostera* grazed by brent geese. Anderson and Low (1976), however, demonstated that water bird grazing on *Potamogeton* tubers in one year could have substantial, albeit local, effects on biomass development in the following year. Apart from this potential effect, which requires more study, the main role of birds in promoting phytoplankton dominance may lie in complementing other factors that also inhibit macrophyte growth.

Conclusions

Aquatic bird populations are often strongly affected by changes in macrophyte abundance and changes in the trophic status of lakes. It would be very useful to quantify these relationships, produce predictive models, and define the limits of those models. Such a framework may allow outliers to be recognized and permit the influence of both food and other factors to be better evaluated.

The biomasses of birds appear to be normally small in relation to those attained by phytoplankton, macrophytes, benthic invertebrates, fish, and zooplankton. As would therefore be expected, birds typically contribute little to nutrient budgets and consume only a small fraction of the annual macrophyte production. These simple facts belie their sometimes substantial roles in regulating the rates at which macrophyte biomass increases during spring-summer growth and the ultimate biomass attained. Their major influence appears to derive not from what they consume but from the loss of the future growth potential of that material, compounded through the growth period. In angiosperm communities in particular, this negative effect on the plants will be reinforced by plant damage and wastage during feeding. The positive feedback effect of birds due to reduction in self-shading, which may be important for tissue-specific plant growth rates in other systems, is likely to be insignificant. Nutrient recycling, a positive feedback in other systems, becomes negative for macrophytes in lakes but is, in any case, significant only for some lakes where birds that feed elsewhere use the lake for roosting. The net indirect feedback effect of birds on macrophytes is thus almost certain to be negative. Although this effect has not been quantified, a theoretical and practical framework exists that would allow it to be done.

In theory, birds may by themselves induce a loss of macrophytes and cause lakes to switch to phytoplankton dominance. The requirement is simply that the interaction strength should exceed the true plant growth rate. This occurs when plants stop growing in autumn, but at that time there is often a large accumulated biomass, including seeds and other resting stages, to provide protection from extinction. The plants' survival may also be aided by changes in bird feeding behavior at low threshold biomasses. The extent to which heavy grazing on the overwintering stages of macrophytes may influence plant production in the following spring and summer remains unclear and requires study in a range of different lakes. Interaction strengths can also exceed the true growth rates of actively growing plants, at least briefly, but this situation may not be common. Tropical systems with persistently high macrophyte biomasses presumably lack this vulnerability to grazing by birds.

To better understand and quantify these interactions, there is a need to quantify food consumption by water birds in nature and to determine how it varies with size, age, season, species, food availability, and food quality and to relate this information to the biomasses and growth rates of available macrophytes. There is a need to isolate the feedback effects of the birds on the plants, to quantify relative and specific interaction strengths, to compare these features among lakes and among plant and bird species, and to produce, test, and refine predictive models for these effects. There is a need for further studies in which bird herbivory is investigated, not in isolation but as part of integrated studies on other factors also affecting macrophyte dynamics. Multifactor experiments may answer the important question of the extent to which grazing effects are additive with other factors influencing plant growth rates or interactive with them.

To achieve these objectives, limnologists must recognize the unique features of birds among the lake fauna and adapt their methods accordingly. The macrophyte biomass parameters that are relevant to zooplankton or fish may not always be so for birds, and the weekly or two-weekly sampling schedules that we commonly use for other things might be quite inappropriate for animals whose use of different regions of a lake may vary by the hour.

The task of producing quantitative, predictive models for macrophyte–bird interactions should not be underestimated. Apart from the complexities of the relation of food consumption to interaction strengths, food consumption per bird may vary diurnally, seasonally, with age, and with other variables. Birds' use of different regions of a lake may vary on short time scales, and human disturbance of birds may obscure relationships that would otherwise be significant. The plants are frequently patchy. They also often vary regionally within lakes, and they are difficult to sample quantitatively. Not all of them may be accessible, because of the the birds' restricted feeding depths. Even without the massive complexities of omnivory, the birds may complicate matters further by feeding on marginal emergent plants or terrestrial vegetation, as well as submerged plants, by seasonal or local migration, aggregation, territoriality, or other complex behaviors that may influence both their effect on the plant community and their responses to changes in it.

If macrophyte–bird interactions represent a challenge, then they also present opportunities that may not be readily available in other plant–herbivore systems. The dramatic losses of aquatic macrophytes from lakes, relating to eutrophication, and their recovery through human intervention or natural processes can be expected to have equally large effects on herbivorous bird populations. Birds' mobility means that the option of emigration to seek a better habitat is always available, and recruitment is not restricted to a particular brief breeding season as in many other vertebrate herbivores. Time lags and imbalances between populations and their food resources, which hamper simple analysis in many other systems (e.g., Carpenter et al., 1985), may be reduced, and with populations being closer to equilibria, simple relationships are more likely to emerge. The hypothesis that bottom-up effects of macrophytes on aquatic birds are relatively strong should therefore be readily testable, with opportunities to use neighboring biomanipulated and unmanipulated lakes as controls.

The converse hypothesis, that grazing birds have large effects on the structure and dynamics of aquatic plant communities, offers similar opportunities for testing on natural systems. The sudden arrival of large migratory flocks of birds on many northern waters facilitates "before-and-after" studies of their effects, again with opportunities for replication at the ecosystem level by manipulations to inhibit bird use. Stands of aquatic plants, often essentially monospecific, which may be grazed by a bird as the only large herbivore species, offer a simplicity that is rare in plant–large herbivore systems, with a wide range of possibilities for field enclosure/exclosure experiments. The different sensitivities of filamentous algae, charophytes, and angiosperms to different indirect feedback effects of herbivores offer unique opportunities for isolating feedbacks from each other and from the primary effect of consumption. Macrophyte–bird interactions present wide opportunities for formulating and testing hypotheses that may be significant not only for lake ecology but for ecology in general.

Acknowledgments. We are very grateful to the Danish National Environmental Research Institute for their invitation to attend the workshop and for financial assistance, to the Workshop Organising Committee for their splendid efforts, and to other colleagues who attended for much stimulating discussion. We also thank Mark Hoyer, Joe Kerekes, David Lodge, Torben Lauridsen, and Peter Webb for their comments on this chapter. Rob Wass kindly redrew the figures.

References

Allison, A.; Newton, I. Waterfowl at Loch Leven, Kinross. Proc. R. Soc. Edinb. (B), 74 (24):365–379; 1972/73.

Anderson, M.G.; Low, J.B. Use of sago pondweed by waterfowl on the Delta Marsh, Manitoba. J. Wildl. Manage. 40:233–242; 1976.

Aschoff, J.; Pohl, H. Der Ruheumsatz von Vögeln als Funktion der Tageszeit und der Körpergrösse. J. Ornithol. 111:38–47; 1970.

Baldassare, G.A.; Bolen, E.G. Waterfowl ecology and management. New York: Wiley; 1994.

Beekman, J.H.; van Eerden, M.R.; Dirksen, S. Bewick's swans *Cygnus columbianus bewickii utilising the changing resource of Potamogeton pectinatus* during autumn in The Netherlands. Wildfowl (Suppl. 1):238–248; 1991.

Berglund, B.E.; Curry-Lindahl, K.; Luther, H.; Olsson, V.; Rodhe, W.; Sellerberg, G. Ecological studies on the mute swan (*Cygnus olor*) in southeastern Sweden. Acta Vertebratica 2; 1963.

Black, J.M.; Rees, E.C. The structure and behaviour of the whooper swan population wintering at Caerlaverock, Dumphries and Galloway, Scotland; an introductory study. Wildfowl 35:21–36; 1984.

Broads Authority (UK). The Broads plan—no easy answers. Norwich, England: Broads Authority Consultation Document, Broads Authority; 1994.

Buchsbaum, R.; Wilson, J.; Valiela, I. Digestibility of plant constituents by Canada geeses and Atlantic brant. Ecology 67:386–393; 1986.

Canfield, D.E., Jr.; Hoyer, M.V. Aquatic macrophytes and their relation to the limnology of Florida lakes. SP115. Institute Food and Agricultural Sciences, University of Florida; 1992.

Cargill, S.M.; Jefferies, R.L. The effects of grazing by lesser snow geese on the vegetation of a subarctic salt marsh. J. Appl. Ecol. 21:669–686; 1984.

Carpenter, S.R.; Kitchell, J.F.; Hodgson, J.R. Cascading trophic interactions and lake productivity. BioScience 35:634–638; 1985.

Cattaneo, A.; Mousseau, B. Empirical analysis of the removal rate of periphyton by grazers. Oecologia 103:249–254; 1995.

Clausen, P. Waterfowl as primary consumers in shallow water fiord areas. PhD thesis, National Environment Research Institute/University of Aarhus, Kalø, Denmark; 1994.

Comin, F.A.; Herrera, J.A. Enhanced release of nutrients from the sediment to the water column by flamingos' footprints. In: Faragó, S.; Kerekes, J., eds. Workshop Proceedings. Limnology and Waterfowl, Monitoring, Modelling and Management. Working Group on Aquatic Birds, Societas Internationalis Limnologiae and International Waterfowl and Wetlands Research Bureau. Sarród/Sopron, Hungary, 21–23 November 1994. Wetlands Int. Publ. 43; 1997.

Cyr, H.; Pace, M.L. Magnitude and patterns of herbivory in aquatic and terrestrial ecosystems. Nature 361:148–150; 1993.

Dirksen, S.; Beekman, J.H.; Slagboom, T.H. Bewick's swans *Cygnus columbianus bewickii* in The Netherlands: numbers, distribution and food choice during the wintering season. Wildfowl (Suppl. 1):228–237; 1991.

Ebbinge, B.; Canters, K.; Drent, R. Foraging routines and estimated daily food intake in barnacle geese wintering in the northern Netherlands. Waterfowl 26:5–19; 1975.

Esler, D. An assessment of American coot herbivory of Hydrilla. J. Wildl. Manage. 53: 1147–1149; 1989.

Faragó, S.; Kerekes, J., eds. Workshop Proceedings. Limnology and Waterfowl, Monitoring, Modelling and Management. Working Group on Aquatic Birds, Societas Internationalis Limnologiae and International Waterfowl and Wetlands Research Bureau. Sarród/Sopron, Hungary, 21–23 November 1994. Wetlands Int. Publ. 43; 1997.

Furness, R.W. Energy requirements of seabird communities: a bioenergetics model. J. Anim. Ecol. 47:39–53; 1978.

Gauthier, G. Feeding ecology of nesting snow geese. J. Wildl. Manage. 57:216–223;1993.

Gere, G.; Andrikovics, S. Feeding of ducks and their effects on water quality. Hydrobiologia 279/280:157–161; 1994.

Giles, N. Tufted duck (*Atyhya fuligula*) habitat use and brood survival increases after fish removal from gravel pit lakes. Hydrobiologia 279/280:387–392; 1994.

Grant, T.A.; Henson, P.; Cooper, J.A. Feeding ecology of trumpeter swans breeding in south central Alaska. J. Wildl. Manage. 58:774–780; 1994.

Hanson, J.M.; Leggett, W.C. Empirical prediction of fish biomass and yield. Can. J. Fish. Aquat. Sci. 39:257–263; 1982.

Hanson, J.M.; Peters, R.H. Empirical prediction of crustacean biomass and profundal macrobenthos biomass in lakes. Can. J. Fish. Aquat. Sci. 41:439–445; 1984.

Hanson, M.A.; Butler, M.G. Responses to food web manipulation in a shallow waterfowl lake. Hydrobiologia 279/280:457–466; 1994.

Hargeby, A.; Andersson, G.; Blindow, I.; Johansson, S. Trophic web structure in a shallow eutrophic lake during a dominance shift from phytoplankton to submerged macrophytes. Hydrobiologia 279/280:83–90; 1994.

Heglund, P.J.; Jones, J.R.; Frederickson, L.H.; Kaiser, M.S. Use of boreal forested wetlands by Pacific loons (*Gavia pacifica* Lawrence) and horned grebes (*Podiceps auritus* L.): relations with limnological characteristics. Hydrobiologia 279/280:171–183; 1994.

Hoyer, M.V.; Canfield, D.E., Jr. Bird abundance and species richness of Florida lakes; influence of trophic status, lake morphology, and aquatic macrophytes. Hydrobiologia 279/280:107–119; 1994.

Hurlbert, S.H.; Chang, C.C.Y. Ornitholimnology: effects of grazing by the Andean flamingo (*Phoenicoparrus andinus*). Proc. Natl. Acad. Sci. U.S.A. 80:4766–4769; 1983.

Jeppesen, E.; Jensen, J.P.; Søndergaard, M.; Lauridsen, T.L.; Pedersen, L.J.; Jensen, L. Top down regulation in freshwater lakes with special emphasis on the role of fish, submerged macrophytes and water depth. Hydrobiologia 342/343:151–164; 1997.

Jupp, B.P.; Spence, D.H.N. Limitations of macrophytes in a eutrophic lake, Loch Leven. 2. Wave action, sediments and waterfowl grazing. J. Ecol. 65:431–446; 1977.

Kendeigh, S.C. Energy requirements for existence in relation to size of bird. Condor 72:60–65; 1970.

Kerekes, J.J. Possible correlation of summer common loon *(Gavia immer)* population with the trophic state of a water body. Verh. Int. Verein. Limnol. 24:349–353; 1990.

Kerekes, J.J. Problems associated with prediction of aquatic bird biomass from total phosphorus concentration. Verh. Int. Verein. Limnol. 26; (in press).

Kerekes, J.J.; Pollard, J.B., eds. Aquatic birds in the trophic web of lakes. Proceedings of a symposium held in Sackville, New Brunswick, Canada, in August 1991. Hydrobiologia 279/280; 1994.

King, J.R. Seasonal allocation of time and energy resources in birds. In: Paynter, R.A., ed. Avian energetics. Cambridge, MA: Nuttal Ornithol. Club; 1974:4–70.

Kiørboe, T. Distribution and production of submerged macrophytes in Tipper Grund (Ringkøbing Fjord, Denmark), and the impact of waterfowl grazing. J. Appl. Ecol. 17:675–687; 1980.

Lauridsen, T.L.; Jeppesen, E.; Andersen, F.Ÿ. Colonization of submerged macrophytes in shallow fish manipulated Lake Vaeng: impact of sediment composition and waterfowl grazing. Aquat. Bot. 46:1–15, 1993.

Lauridsen, T.L.; Jeppesen, E.; Søndergaard, M. Colonisation and succession of submerged macrophytes in shallow Lake Væng during the first five years following fish manipulation. Hydrobiologia 275/276:233–242; 1994.

Leah, R.T.; Moss, B.; Forrest, D.E. Experiments with large enclosures in a fertile, shallow, brackish lake. Int. Rev. Ges. Hydrobiol. 63:291–310; 1978.

Lillie, R.A.; Evrard, J.O. Influence of macroinvertebrates on waterfowl utilization of wetlands in the Prairie Pothole Region of northwestern Wisconsin. Hydrobiologia 279/280:235–246; 1994.

Lodge, D.M. Herbivory on freshwater macrophytes. Aquat. Bot. 41:195–224; 1991.

McKinnon, S.L.; Mitchell, S.F. Eutrophication and black swan (Cygnus atratus Latham) populations: tests of two simple relationships. Hydrobiologia 279/280:163–170; 1994.

McKinnon, S.L.C. The interrelationship between phytoplankton, submerged macrophytes and black swans (Cygnus atratus) in New Zealand lakes—tests of two models. Unpubl. MSc thesis, University of Otago, New Zealand; 1989.

Manny, B.A.; Johnson, W.C.; Wetzel, R.G. Nutrient additions by waterfowl to lakes and reservoirs: predicting their effects on productivity and water quality. Hydrobiologia 279/280:121–132; 1994.

Marion, L.; Clergeau, P.; Brient, L.; Bertru, G. The importance of avian-contributed nitrogen (N) and phosphorus (P) to Lake Grand-Lieu, France. Hydrobiologia 279/280:133–147; 1994.

Marriott, R.W.; Forbes, D.K. The digestion of lucerne chaff by Cape Barren geese, Cereopsis novaehollandiae Latham. Aust. J. Zool. 18:257–263; 1970.

Mathiasson, S. A moulting population of non-breeding mute swans with special reference to flight feather moult, feeding ecology and habitat selection. Wildfowl 24:43–53; 1973.

Mattocks, J.C. Goose feeding and cellulose digestion. Wildfowl 22:107–113; 1971.

Mitchell, S.F. Primary production in a shallow eutrophic lake dominated alternatively by phytoplankton and by submerged macrophytes. Aquat. Bot. 33:101–110; 1989.

Mitchell, S.F.; Wass, R.T. Food consumption and faecal deposition of plant nutrients by black swans (Cygnus atratus Latham) in a shallow New Zealand lake. Hydrobiologia 306:189–197; 1995.

Mitchell, S.F.; Wass, R.T. Grazing by black swans, physical factors and the growth and loss of aquatic vegetation in a shallow lake. Aquat. Bot. 55:205–215; 1996a.

Mitchell, S.F.; Wass, R.T. Quantifying herbivory—grazing consumption and interaction strength. Oikos 76:573–576; 1996b.

Mitchell, S.F.; Hamilton, D.P.; MacGibbon, W.S.; Nayar, P.K.B.; Reynolds, R.N. Interrelations between phytoplankton, submerged macrophytes, black swans (Cygnus atratus) and zooplankton in a shallow New Zealand lake. Int. Rev. Ges. Hydrobiol. 73:145–170; 1988.

Moss, B.; Leah, R.T. Changes in the ecosystem of a guanotrophic and brackish shallow lake in eastern England; potential problems in its restoration. Int. Rev. Ges. Hydrobiol. 67:625–659; 1982.

Murphy, S.M.; Kessel, B.; Vining, L.J. Waterfowl populations and limnologic characteristics of taiga ponds. J. Wildl. Manage. 48:1156–1163; 1984.

Nagy, K.A. Field metabolic rate and food requirement scaling in mammals and birds. Ecol. Monogr. 57:111–128; 1987.

National Environmental Research Institute (Denmark). Report and activities 1993–94. Roskilde, Denmark: Ministry of the Environment; 1994.

National Research Council (USA). Restoration of Aquatic Ecosystems—Science, Technology and Public Policy. Washington, DC: National Academy Press, USA; 1992.

Nilsson, L. Breeding waterfowl in eutrophicated lakes in south Sweden. Wildfowl 29:101–110; 1978.

Nilsson, S.G.; Nilsson, I.N. Breeding bird community densities and species richness in lakes. Oikos 31:214–221; 1978.

Noyes, J.H.; Jarvis, R.L. Diet and nutrition of breeding female redhead and Canvasback ducks in Nevada. J. Wildl. Manage. 49:203–211; 1985.

Organisation for Economic Cooperation and Development (OECD). Eutrophication of waters—monitoring, assessment and control. Paris: OECD; 1982.

Owen, M.; Black, J.M. Waterfowl biology. Glasgow: Blackie; 1990.

Perrow, M.R.; Schutten, J.; Howes, J.R.; Holzer, T.; Madgwick, F.J. Interactions between coot (Fulica atra) and submerged macrophytes: the role of birds in the restoration process. Hydrobiologia 342/343:241–256; 1997.

Peters, R.H. The ecological implications of body size. Cambridge Studies in Ecology, 2. Cambridge: Cambridge University Press; 1983.

Pöysä, H. Resource utilization pattern and guild structure in a waterfowl community. Oikos 40:295–307; 1983.

Rees, E.C. Bewick's swans: their feeding ecology and coexistence with other grazing Anatidae. J. Appl. Ecol. 27:939–951; 1990.

Roijackers, R.M.M. Phytoplankton studies in a nymphaeid-dominated system: with special reference to the effects of the presence of nymphaeids on the functioning and structure of the phytoplankton communities/ Rudolphus Martinus Maria Roijackers.—(S.I. : s.n.) (Meppel : Krips Repro).—Ill. Thesis Nijmegen. With ref.-With summary in Dutch; 1985.

Scheffer, M.; Hosper, S.H.; Meijer, M-L.; Moss, B.; Jeppesen, E. Alternative equilibria in shallow lakes. Trends Ecol. Evol. 8:275–279; 1993.

Schutten, J.; van der Velden, J.A.; Smit, H. Submerged macrophytes in the recently freshened lake system Volkerak-Zoom (The Netherlands), 1987–1991. Hydrobiologia 275/276:207–218; 1994.

Sedinger, J.S.; White, J.G.; Manns, F.E.; Burris, F.A.; Kedowski, R.A. Apparent metabolising of alfalfa components by herbivorous yearling pack brant. J. Wildl. Manage. 53:726–734; 1989.

Staicer, C.A.; Freedman, B.; Srivastava, D.; Dowd, N.; Kilgar, J.; Hayden, J.; Payne, F.; Pollock, T. Use of lakes by black duck broods in relation to biological, chemical, and physical features. Hydrobiologia 279/280:185–199; 1994.

Suter, W. Overwintering waterfowl on Swiss lakes: how are abundance and species richness influenced by trophic status and lake morphology? Hydrobiologia 279/280:1–14; 1994.

Søndergaard, M.; Kristensen, P.; Jeppesen, E. Phosphorus release from suspended sediment in the shallow and wind-exposed Lake Arresø, Denmark. Hydrobiologia 228:91–99; 1992.

Søndergaard, M.; Bruun, L.; Lauridsen, T.; Jeppesen, E.; Madsen, T.V. The impact of grazing waterfowl on submerged macrophytes: in situ experiments in a shallow lake. Aquat. Bot. 53:73–84; 1996.

van Donk, E.; Otte, A. Effects of grazing by fish and waterfowl on the biomass and species composition of submerged macrophytes. Hydrobiologia 340:285–290; 1996.

van Donk, E.; De Deckere, E.; Klein Breteler, J.G.P.; Meulemans, J.T. Herbivory by waterfowl and fish on macrophytes in a biomanipulated lake: effects on long term recovery. Verh. Int. Verein. Limnol. 25:2139–2143; 1994.

van Wijk, R.J. Ecological studies on *Potamogeton pectinatus* L. 1. General characteristics, biomass production and life cycles under field conditions. Aquat. Bot. 31:211–258; 1988.

Verhoeven, J.T.A. The ecology of *Ruppia*-dominated communities in western Europe. III. Aspects of production, consumption and decomposition. Aquat. Bot. 8:209–253; 1980.

von Bertalanffy, L. General system theory. New York: Braziller; 1968.

Wallace, J.B.; O'Hop, J. Life on a fast pad: waterlily leaf beetle impact on waterlilies. Ecology 66:1534–1544; 1985.

Williams, M.J. The status and management of black swans *Cygnus atratus* Latham at Lake Ellesmere since the "Wahine" storm, April 1968. N.Z. J. Ecol. 2:34–41; 1979.

Woollhead, J. Birds in the trophic web of Lake Esrom, Denmark. Hydrobiologia 279/280:29–38; 1994.

Zar, J.H. The use of the allometric model for avian standard metabolism–body weight relationships. Comp. Biochem. Physiol. 29:227–234; 1969.

10. Effects of Submerged Aquatic Macrophytes on Nutrient Dynamics, Sedimentation, and Resuspension

John W. Barko and William F. James

Introduction

Accelerated eutrophication due to excessive nutrient (particularly P) loadings has led to great interest in the role of submerged macrophytes in the nutritional economy of freshwater aquatic systems. Submerged macrophytes are unique among rooted aquatic vegetation because they link the sediment with overlying water. This linkage is responsible for great complexities in nutrition and has important implications for nutrient cycling. Despite increased attention to vegetated shallow water systems within the past 20 years, no consensus exists on whether submerged macrophytes function as sources or sinks for particular nutrients. As a result, it has been necessary to evaluate quantitatively nutrient source–sink relationships, involving both soluble and particulate nutrient fractions.

In this chapter, we address sediment nutrient interactions with submerged macrophyte growth and metabolism, macrophyte influences on littoral-pelagic nutrient dynamics, and macrophyte influences on sedimentation/sediment re-suspension. Our objective is to integrate pertinent information, primarily from our own studies, but with attention to related and complementary work of others. Our intention is not to provide an exhaustive review of the literature in these areas.

We focus primarily on shallow freshwater systems (lakes, reservoirs, and rivers) in northern temperate regions in which rooted submerged macrophytes constitute structurally and metabolically important components of the underwater landscape. Furthermore, our focus is limited to physical and chemical relationships. We address the role of macrophytes on nutrient budgets with limited attention to potential influences on phytoplankton. The role of macrophytes in affecting phytoplankton is considered more comprehensively in Søndergaard and Moss (this volume, Chapter 6).

Sediment Nutrient Interactions with Macrophyte Growth

For many years, controversy has existed regarding the role of roots versus shoots and sediment versus open water in the nutrition of submerged aquatic macrophytes (reviewed by Sculthorpe, 1967; Carignan and Kalff, 1980; Denny, 1980; Smart and Barko, 1985; Agami and Waisel, 1986; Barko et al., 1986, Barko et al., 1991). Quantification of the relative contribution of sediment and water to nutrient uptake by submerged macrophytes has been critical to an improved understanding of littoral nutrient cycling and littoral-pelagic nutrient exchanges. Based on a variety of information sources, a generalized synthesis of sources of nutrient uptake by rooted submerged macrophytes was provided in Barko et al. (1991), as summarized below.

Phosphorus and nitrogen have been studied most extensively, and for these nutrients, sediment is the primary source for uptake. Sediment appears to be the principal site for uptake of iron, manganese, micronutrients, and trace metals as well (Barko and Smart, 1986; Jackson et al., 1994a,b). These latter elements tend to coprecipitate and are usually present in extremely low concentrations in oxygenated surface waters. Dissolution products of relatively abundant salts are taken up principally from the open water. Among these elements, potassium and calcium are potentially most important in affecting submerged macrophyte growth. Potassium can be obtained from the sediment but is taken up mainly from the open water (Barko 1982; Huebert and Gorham, 1983; Barko et al., 1988). Under some conditions, potassium can be exchanged by submerged macrophyte roots for ammonium ions in sediment (Barko et al. 1988). Calcium is a component of the carbonate system and plays an important role in dissolved inorganic carbon uptake during photosynthesis (Lowenhaupt, 1956; Smart and Barko, 1986).

Given the ecological significance of N and P in aquatic systems and the importance of sediment in supplying N and P to submerged macrophytes, it is important to evaluate the effects of macrophyte growth on the availability of these nutrients. Results indicate that rooted submerged macrophytes, even with relatively diminutive root systems, are capable of significantly depleting pools of N and P in sediments (Prentki, 1979; Trisal and Kaul, 1983; Carignan, 1985; Barko et al., 1988; Chen and Barko, 1988). The magnitude of nutrient reductions in sediment due to uptake by macrophytes can be impressive. For example, Barko et al. (1988) reported greater than 90% and greater than 30% reductions in concentrations of exchangeable N and acid-extractable P, respectively, from sediment on which *Hydrilla verticillata* was grown over two consecutive 6-week periods.

From fertilization experiments involving sediments depleted of nutrients due to uptake by macrophytes, subsequent growth has been shown to be limited prin-

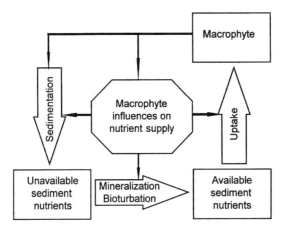

Figure 10.1. Conceptual diagram of macrophyte influences on nutrient supply as an interactive function of sedimentation and sediment processing.

cipally by the availability of sediment N (Anderson and Kalff, 1986; Barko et al., 1988, 1991; Rogers et al., 1995). However, this generalization may not apply to oligotrophic hardwater systems such as Lawrence Lake, Michigan (Moeller et al., this volume, Chapter 22) or to oligotrophic systems with sandy sediments, such as Lake Hampen, Denmark (Christiansen et al., 1985), where P appears to be more important in limiting macrophyte growth.

As emphasized in Barko et al. (1991) and here (see below), sedimentation in concert with mixing and mineralization (Fig. 10.1) provides a potentially important source of nutrient renewal in macrophyte beds, as nutrient losses due to macrophyte uptake must be balanced by inputs to maintain the vigor of continued macrophyte growth. The growth of *Vallisneria americana* on intrinsically infertile sediments (i.e., coarse sand) in Lake Onalaska, Wisconsin, appears to depend greatly on supply of N through sedimentation (Rogers et al., 1995, and Rogers, unpublished data). Notably, a near-complete collapse of this and other submerged macrophyte species from Lake Onalaska occurred following the drought years of 1988 and 1989, when sedimentation was likely minimal. Submerged macrophytes have since recovered in the lake during recent years with normal flow. Greater net sedimentation with less erosion in gently sloped rather than sharply sloped littoral regions may partially account for the relationship established between littoral slope and the biomass of submerged macrophyte communities (Duarte and Kalff, 1986).

The vigor of submerged macrophyte beds is likely maintained by nominal inputs of sediment providing a nutritional subsidy (Fig. 10.1). However, excessive inputs of sediment can result in macrophyte declines due to burial or unfavorable underwater irradiance. Because aquatic systems subject to high rates of sediment loading are frequently turbid, with associated constraints on photosynthetic activity, macrophytes can be expected to grow best under conditions of intermediate

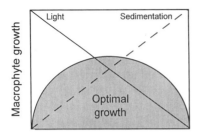

Figure 10.2. Conceptual diagram of interacting roles of underwater light and sedimentation in affecting macrophyte growth.

sediment loadings (Fig. 10.2). It is under these conditions that macrophytes are most likely to have the greatest influence on nutrient dynamics.

Given the demonstrated capacity of submerged macrophytes to mobilize nutrients from sediments directly via root uptake followed by subsequent release during seasonal senescence and decomposition, vegetation of the littoral zone needs to be viewed as a potential direct source of nutrients to the water column (Barko and Smart, 1980; Carpenter, 1980; Landers, 1982; Smith and Adams, 1986). High productivity and biomass turnover of macrophytes in fertile systems contribute to high rates of nutrient mobilization from sediments, particularly with rapidly growing species such as *Myriophyllum spicatum* (Smith and Adams, 1986). These processes considered collectively, in combination with effects of macrophyte metabolic activity (see below), can have significant effects on lacustrine nutrient budgets. However, in less fertile systems, in which nutritionally more conservative macrophyte species tend to dominate, effects on nutrient budgets are probably less pronounced (Barko et al., 1991). Likewise, in large deep lakes, where macrophytes are less abundant relative to lake volume, effects on nutrient budgets are probably negligible (Gasith and Hoyer, this volume, Chapter 29).

Sediment Nutrient Interactions with Macrophyte Metabolism

Increased attention in recent years to the role of submerged aquatic macrophytes in the nutritional economy (particularly P) of lacustrine systems reflects the unparalleled importance of P in the eutrophication process (Schindler, 1974, 1977). In addition to P mobilization directly, as discussed above, submerged macrophytes can also mobilize sediment P indirectly via metabolic activities that alter pH and redox conditions in the surrounding water (Andersen, 1975; Drake and Heaney, 1987; James and Barko, 1991b; James, et al., 1996).

Until recently, littoral sediments have been regarded as a net sink for P because surface sediment layers in littoral regions are usually oxidized. Classic "iron-phosphorus" theories (Mortimer, 1941) indicate that iron oxide-hydroxides (Fe^{+3}) contained in the surface microzone of littoral sediments should have a high binding affinity for P (Lijklema, 1977), greatly reducing the potential for P flux into the water column. In addition, the sorption capacity for P is high, whereas P release is

Figure 10.3. Daily variations in pH (A), dissolved oxygen (mean, minimum, and maximum value, (B), and estimated rates of P release from sediments (C) during the summer 1994 for the inlet region of Lake Delavan, Wisconsin. (Modified with permission from James et al., 1996.)

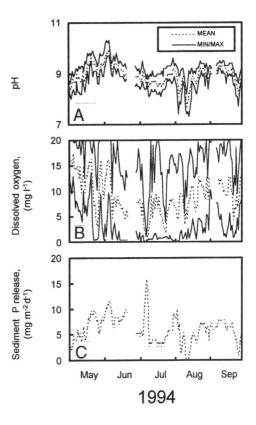

low for sediments with high Fe:P ratios (i.e., >15 by weight), as long as the microzone remains oxidized (Boström et al., 1982; Jensen et al., 1992).

Phosphorus release from littoral sediments can be enhanced, however, at high pH, even under oxic conditions, by replacement of PO_4 with OH^- on iron or aluminum oxides according to the following general equilibrium equation:

$$2FeOH + H_2PO_4^- \rightleftarrows Fe_2OH \approx PO_3^- + H_2O + OH^-$$

Increases in pH from about 8.0 to about 9.0 can result in at least a doubling in the rate of P release from oxic littoral sediments (Boers, 1991). Photosynthesis and respiration in submerged macrophytes can result in dramatic diel variations in pH and oxygen in the water column. For example, diel pH fluctuations between 7–10 were measured in densely vegetated *Hydrilla* beds in the Potomac River, Washington, DC (Carter et al., 1988). In such enriched systems, substantially elevated pH conditions can result in significant increases in the rate of P release from sediments (Boström et al., 1982; James and Barko, 1991b; James et al., 1996).

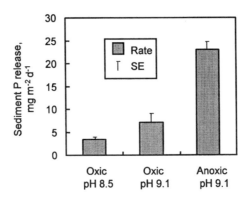

Figure 10.4. Variations in mean ($\pm$ 1 SE) rates of P release from littoral sediments as a function of pH and dissolved oxygen (oxic vs. anoxic) for sediments collected in the inlet region of Lake Delavan, Wisconsin. Rates of P release doubled under oxic conditions as pH increased by 0.6 units. Under anoxic conditions at high pH, rates of P release increased threefold, compared with the rate measured under oxic conditions at the same pH. (Modified with permission from James et al., 1996.)

From studies we conducted in a littoral inlet region of Lake Delavan, Wisconsin, it is apparent that the submerged macrophyte community can promote marked seasonal and daily fluctuations in pH and oxygen (James et al., 1996). In this inlet, pH became greater than 10.0 in May and early June and fluctuated near 9.0 for the remainder of the summer (Fig. 10.3A). During the night, several periods of anoxia occurred near the sediment surface between late June and late July, providing a mechanism for anoxic P release from the sediments (Fig. 10.3B). Using ranges in the rates of P release from sediments measured in the laboratory as a function of pH and redox conditions (Fig. 10.4), James et al. (1996) estimated pronounced variations in the rate of internal loading of P to Lake Delavan (Fig. 10.3C). Oxic P release from sediments under conditions of elevated pH was the dominant means of sediment P flux in the inlet region, accounting for more than 97% of the sediment internal load over the summer compared with only 3% contributed as a result of P release under anoxic conditions.

Macrophyte influences that enhance P mobilization from sediment can account for a significant portion of the total mass of P loading internally in aquatic systems. In the inlet of Lake Delavan, for instance, submerged macrophytes mobilized about 600 kg P (about 6 mg/m^2/day) from the littoral sediments during the summer indirectly by altering pH and dissolved oxygen conditions. An additional 600 kg P was mobilized from the sediments by submerged macrophytes directly via root uptake (James et al., 1995). Together, P release from sediments plus macrophyte tissue P release was equivalent to twice the external P load contributed to Lake Delavan from the watershed (James et al., 1996).

Macrophyte Influences on Littoral-Pelagic Phosphorus Dynamics

Phosphorus released from littoral sediments can result in considerable P accumulation in the water column within macrophyte beds (James and Barko, 1991b). Much of this P is in a soluble form that can be transported directly into the upper mixed layer of adjacent pelagic regions. Linkages between macrophyte-mediated P mobilization in the littoral zone and nutrient dynamics in the pelagic zone are established by horizontal water exchanges between the two zones. Mechanisms that potentially govern horizontal P transport are (1) wind-driven patterns that lead to general water circulation (Weiler, 1978), upwelling/downwelling phenomena, and differential deepening (Imberger and Patterson, 1990) and (2) convectively driven patterns (Monismith et al., 1990; James and Barko, 1991a,b; James et al., 1994). Prentki et al. (1979) suggested that water circulation patterns created by wind shears could transport P derived from the littoral zone into the pelagic zone of Lake Wingra, Wisconsin, at fluxes of 0.5–5.0 mg/m^2/day. Measurements further suggest that water exchanges can be facilitated also by wind shears that create upwelling/downwelling (i.e., seiche activity) and differential deepening of the surface mixed layer (Imberger and Parker, 1985). Convective circulation appears to be the dominant means of driving horizontal exchanges between littoral and pelagic zones in small, wind-sheltered lakes and embayments because it can occur in the complete absence of wind (Stefan et al., 1989; James and Barko, 1991a,b; James et al., 1994). Even under calm conditions, the residence time of littoral regions can be on the order of 1 day or less as a result of convective exchanges (Stefan et al., 1989).

Convective circulation is driven primarily by horizontal temperature (and density) gradients that develop between shallow littoral regions and deeper pelagic regions as a result of differential heating and cooling along a depth gradient (Monismith et al., 1990; James and Barko, 1991a,b; James et al., 1994). On a daily basis, shallow regions of aquatic systems typically heat and cool more rapidly than deeper regions, resulting in the development of unstable horizontal water temperature gradients (Fig. 10.5). During differential cooling, water in shallow regions cools more rapidly than in deeper regions, resulting in the horizontal movement of shallow water and associated nutrients as an underflow below warmer pelagic water (Fig. 10.6). During differential heating, the opposite pattern occurs. Shallow regions heat more rapidly than deeper regions, resulting in horizontal movement of shallow water and associated nutrients as a surface flow over cooler water located in the pelagic zone (Fig. 10.6). The presence of submerged macrophytes in shallow littoral regions can contribute substantially to the development of strong thermal gradients during the day in both the vertical and lateral planes, as foliage near the surface converts solar irradiance to heat, further promoting differential heating, with associated potential transport of soluble nutrients due to convection (Fig. 10.7).

Dye studies conducted at Eau Galle Reservoir, Wisconsin, serve as a good example of linkages between macrophyte-mediated processes that mobilize sedi-

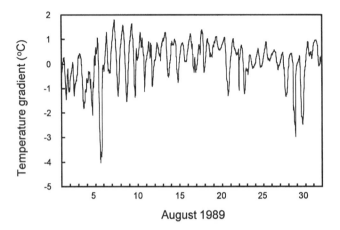

Figure 10.5. Diel variations in horizontal temperature gradients between the surface waters of the littoral and pelagic zone of Eau Galle Reservoir, Wisconsin, in August 1989. A positive temperature gradient indicates that surface water temperatures in the littoral zone are greater than temperatures in the pelagic zone. A negative temperature gradient indicates that surface water temperatures in the littoral zone are less than temperatures in the pelagic zone.

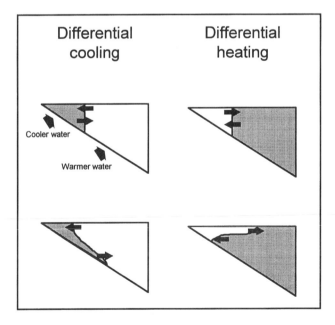

Figure 10.6. Conceptual diagram of convective water exchanges driven by differential cooling and differential heating.

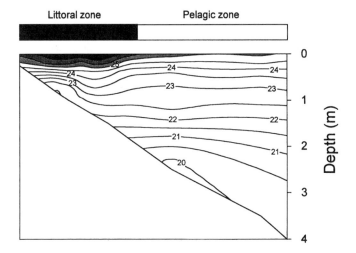

Figure 10.7. Longitudinal and vertical variations in water temperature in the littoral zone during a period of differential heating. Submerged macrophytes promote the development of a thin intensive surface mixed layer in the littoral zone through attenuation of solar radiation that results in the establishment of unstable horizontal water temperature gradients in the surface waters. These gradients give rise to horizontal water exchanges that transport dissolved constituents.

ment P in the littoral zone and exchanges with the pelagic zone (James and Barko, 1991b). Differential cooling at night usually results in the development of an underflow that moves along the littoral slope (Fig. 10.8), transporting P accumulated in the water column as a result of P release from oxic littoral sediments at high pH (i.e., pH about 9.1–9.4). Intrusion of P-rich littoral water into the pelagic

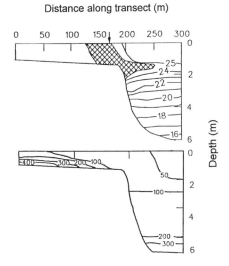

Figure 10.8. Upper panel: Contours of water temperature (°C) in the littoral and pelagic zone and the position of a dye cloud approximately 12 hours after dye injection as a result of convective water exchanges during a period of differential cooling. The arrow indicates the initial dye injection point. Lower panel: Contours of total phosphorus concentration (μg/L) in the littoral and pelagic zones. (Modified with permission from James and Barko, 1991b.)

zone occurs as an interflow, confined usually to the top of the metalimnion, but it can also be mixed into the epilimnion via wind-generated turbulence. Thus, littoral contributions to the P economy of phytoplankton communities in surface waters of adjacent pelagic regions can be important. Internal loadings from the littoral zone of Eau Galle Reservoir into the adjacent upper mixed layer of the pelagic zone appear to affect not only the reservoir nutrient budget (James and Barko, 1993) but also vertically migrating phytoplankton communities (Taylor et al., 1988; James et al., 1992). Implications of convective circulation as a mechanism of water exchange between littoral and pelagic zones are ecologically far-reaching, because all kinds of dissolved constituents can be moved with water.

Macrophyte Influences on Sedimentation

In addition to influences on nutrient dynamics in the water column of aquatic systems, submerged macrophytes play an important role also in mediating sedimentation dynamics in littoral regions and in shallow lakes. The presence or absence of submerged macrophytes in shallow regions of lakes appears to be very important in explaining different sediment accretion and composition patterns along depth gradients. For instance, in lakes with no vegetated littoral zone, sediment accretion is generally greatest in deeper areas, with rates declining in shallower water to a minimum near the shoreline, due to erosional mechanisms that result in sediment resuspension and focusing (Davis, 1968, 1973; Davis and Brubaker, 1973; Likens and Davis, 1975; Håkanson, 1977; Evans and Rigler, 1980, 1983; Davis et al., 1984; Hilton, 1985; Hilton et al., 1986). However, in lakes with vegetated littoral regions anomalies frequently occur with respect to this generalized depositional pattern. Sediment accretion in littoral regions is generally greater than expected (Moeller and Wetzel, 1988; Anderson, 1990), and patterns of sediment composition are different than expected, due to macrophyte influences on sedimentation dynamics (Petticrew and Kalff, 1991, 1992, and see below).

In Eau Galle Reservoir, sediment accretion and composition in nonvegetated regions vary near-linearly along a depth gradient (Fig. 10.9; James and Barko, 1990) in a manner consistent with classic theories for depositional zones (i.e., zones of erosion, transport, and accumulation) developed by Håkanson (1977). Using variations in sediment moisture content (Håkanson, 1977), three depositional zones can be identified in nonvegetated regions of Eau Galle Reservoir: (1) an accumulation zone located between 6–9 m (moisture content >75%), (2) a transport zone located between 3.5–6 m (moisture content, 50–75%), and (3) an erosional zone located at nonvegetated depths between 2.5–3.5 m (moisture content <50%). However, patterns of sediment accretion and composition in the vegetated littoral zone at depths less than 2.5 m differ greatly from those that would be expected on the basis of depth alone in this reservoir. High moisture content, low sediment density, high organic matter content, and high nutrient content in the littoral sediments parallel relatively high rates of sediment accretion compared with

Figure 10.9. Variations in sediment accretion rates (upper panel) and moisture content of the surface sediment (lower panel) in Eau Galle Reservoir. (Modified with permission from James and Barko, 1990.)

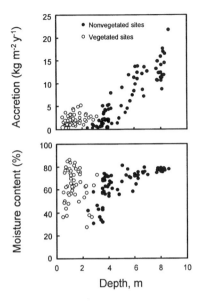

conditions in the erosional zone. These patterns collectively indicate that submerged macrophytes are influencing sedimentation dynamics in shallow regions by reducing sediment erosion and/or promoting sediment accretion.

Aquatic macrophytes can reduce sediment resuspension and erosion and promote accretion by reducing and/or redirecting turbulent water currents (Fonseca et al., 1982; Gregg and Rose, 1982; Madsen and Warncke, 1983; Eckman et al., 1989). They can also serve as effective sediment traps via interception of suspended sediment (Patterson and Brown, 1979; Wetzel, 1979; Carpenter, 1981). Moeller and Wetzel (1988) have suggested that sedimentation of algae and periphyton from macrophyte leaf surfaces may provide an important link for transfer of nutrients absorbed from the water (by algae) to the sediment. Similarly, it has been reported that under conditions of nutrient enrichment, decomposing filamentous algae can provide major inputs of N and P to littoral sediments (Howard-Williams, 1981). Finally, the role of aquatic macrophytes in promoting accretion and retention of sediment and nutrients in an otherwise erosional environment provides a positive feedback mechanism for increasing the sediment surface area colonizable by macrophytes (Carpenter, 1981).

Macrophyte Influences on Sediment Resuspension

By inhibiting sediment resuspension and erosion, submerged aquatic macrophytes can play an important role in regulating water quality in shallow lakes and impoundments. These systems, in the absence of aquatic macrophytes, are often dominated by sediment resuspension induced by wind and/or benthic fishes,

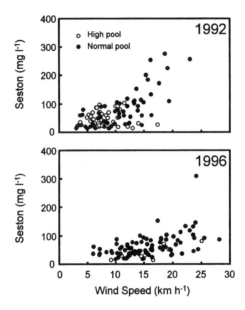

Figure 10.10. Relationships between wind velocity and seston concentrations in Marsh Lake (USA) during 1992 when aquatic macrophytes were absent (upper panel) and 1996 when the lake was densely vegetated with *Potamogeton* sp. (lower panel). The presence of submerged macrophytes changed dramatically the critical wind threshold required to resuspend sediment in this lake at nominal pool elevations. We normalized the data with respect to pool elevation to account for differences in sediment resuspension as a function of water depth at our sampling location. Seston concentrations during periods when pool elevation exceeded nominal levels, due to periods of storm inflow, are denoted by an open circle. Seston concentrations during periods of nominal pool elevation are denoted by a solid circle. (Modified with permission from James and Barko, 1994.)

leading to sediment-related water quality problems such as enhanced nutrient cycling, reduced water clarity, and high phytoplankton biomass (Dillon et al., 1990; Maceina and Soballe, 1990; Hellström, 1991; Søndergaard et al., 1992). The establishment or occurrence of macrophytes in these systems tends to be associated with a clearwater state and low phytoplankton biomass (Hosper, 1989; Dieter, 1990; Scheffer, 1990), due to macrophyte-mediated reduction in sediment resuspension and other macrophyte-mediated factors that inhibit phytoplankton growth, such as provision of refugia for zooplankton, release of allelopathic substances, and nutrient scavenging (Scheffer et al., 1993). Thus, the establishment and maintenance of aquatic macrophyte communities is critical for water quality in these systems (Hosper and Jagtman, 1990; Hanson and Butler, 1994).

Marsh Lake, Minnesota, an impoundment on the Minnesota River, provides an example of the role submerged macrophytes can play in reducing sediment resuspension in a shallow lake (James and Barko, 1994, and unpublished data). Resuspension and export of sediment downstream were examined in this lake in

1992, when submerged macrophyte biomass was absent, and in 1996, when the entire lake was densely vegetated with *Potamogeton pectinatus*. Based on a theoretical wave model (Carper and Bachmann, 1984), nearly the entire sediment surface area (81–100%) of Marsh Lake can be disturbed by wave activity at nominal pool elevations at wind velocities as low as 11–15 km/hr blowing from any direction. During the year when macrophytes were absent from the system (1992), observed critical thresholds of wind velocity required to resuspend sediment were about 12 km/hr at nominal pool elevations, as seston concentrations in the water column increased substantially above this wind velocity (Fig. 10.10). Sediment export from Marsh Lake to downstream Lac Qui Parle Reservoir was also observed when wind velocities exceeded this critical threshold, which had a detrimental impact by exacerbating accretion rates and accelerating reduction in water storage capacity in this reservoir. An almost complete absence of sediment resuspension above nominal seston levels in this lake (Fig. 10.10) was associated with the presence of dense submerged macrophyte beds in 1996, indicating clearly that macrophytes are beneficial in reducing the occurrence of sediment resuspension by dampening wave activity. Seston concentrations remained relatively low at nominal pool elevations, compared to the year 1992, even at wind velocities exceeding 20 km/hr, which is well above the critical velocity predicted by the wave model to cause sediment resuspension. In addition, discharge of sediment from the lake was much less in 1996, compared with 1992.

Concluding Remarks

Nitrogen is a key element in the growth of rooted aquatic macrophytes. Thus, attention to this particular element needs to be elevated to the same level as for P. The role of submerged macrophytes in the N economy of aquatic systems also needs to be more thoroughly investigated. A variety of physical, chemical, and biological processes (e.g., sedimentation, mineralization, and particulate matter processing by benthic invertebrates) that potentially contribute to sediment N availability need to be evaluated within the context of macrophyte nutrition.

Rooted submerged macrophytes play an important role in nutrient cycling in aquatic systems by mediating fluxes of nutrients from sediments into the water column. Aquatic macrophytes mobilize nutrients directly from sediments through root uptake and senescence. They mobilize nutrients indirectly from sediments by causing marked fluctuations in pH and oxygen through metabolic activities, which enhance the rate of P release from sediments. In particular, high pH values (about 9–10) associated with macrophyte photosynthesis can result in ligand exchange with P adsorbed to iron oxide-hydroxides on sediment particles, thus enhancing the rate of P release from sediments. During the night, productive macrophyte beds can also deplete the water column of dissolved oxygen through respiratory activities, thereby promoting P release from sediments under anoxic conditions. These processes can result in substantial accumulation of P in the littoral water column. Water circulation induced by diel heating and cooling of surface water or other means in aquatic plant beds facilitates nutrient exchanges with the adjacent open water of aquatic systems. These processes can result in enhanced phytoplankton production and deterioration in water quality and must be considered in the nutrient budgets for aquatic systems.

In shallow high-energy environments, potential negative effects on water quality by macrophytes (i.e., enhanced nutrient cycling) may be overshadowed by the ability of macrophytes to moderate current and wave energies, thereby producing a positive water quality effect by reducing sediment resuspension, turbidity, and concentrations of suspended particulate materials. Results of studies reported here for Marsh Lake suggest that the maintenance of stands of submerged aquatic macrophytes may be an effective management tool for limiting wind-driven sediment and associated nutrient resuspension and discharge in shallow water systems. Thus, macrophyte growth in some lakes (particularly shallow wind-swept basins) should perhaps be encouraged rather than discouraged, despite their often negative effects on water use.

It is apparent that submerged aquatic macrophytes, through a variety of mechanisms, can have important influences on sedimentation dynamics and can mediate nutrient fluxes in aquatic systems. Studies of nutrient cycling, hydraulic transport, and sedimentation dynamics in macrophyte beds are of great value in providing information on rates and volumes of nutrients and sediment being exchanged with the open water of aquatic systems. Interactions between macrophytes and phytoplankton in aquatic systems need to be examined more fully through consideration of littoral-pelagic hydraulic interactions. Hydrological factors and watershed activities that influence seasonal dynamics and magnitudes of sediment transport in aquatic systems need to be evaluated within the context of their effects on submerged macrophyte growth. Finally, the effects (favorable versus unfavorable) of aquatic plants on water quality conditions in aquatic systems need to be considered within the context of basin morphometry, hydrology, and local climate.

Acknowledgments. We gratefully acknowledge the Danish Natural Science Research Council for sponsoring the workshop in Silkeborg, Denmark (June 1996) entitled "The Role of Submerged Macrophytes in Structuring the Biological Community and Bio-geochemical Dynamics in Lakes," at which this work was presented. We also thank Dr. Erik Jeppesen for his assistance and advice during manuscript preparation and Dr. John D. Madsen and two anonymous referees for reviewing the manuscript. Funding for this research was provided by U.S. Army Engineer Waterways Experiment Station through the Aquatic Plant Control Research Program, the Water Operations Technical Support Program, and the Water Quality Research Program. Additional funding was provided by the U.S. Army Engineer District, St. Paul, and the Wisconsin Department of Natural Resources. Permission to publish this information was granted by the Chief of Engineers.

References

Agami, M.; Waisel, Y. The ecophysiology of roots of submerged vascular plants. Physiol. Veg. 24:607–624; 1986.

Andersen, J.M. Influence of pH on release of phosphorus from lake sediments. Arch. Hydrobiol. 76:411–419; 1975.

Anderson, M.R.; Kalff, J. Nutrient limitation of *Myriophyllum spicatum* growth in situ. Freshwat. Biol. 16:735–743; 1986.

Anderson, N.J. Spatial pattern of recent sediment and diatom accumulation in a small monomictic, eutrophic lake. J. Paleolimnol. 3:143–168; 1990.

Barko, J.W. Influence of potassium source (sediment vs. open water) and sediment composition on the growth and nutrition of a submersed freshwater macrophyte (*Hydrilla verticillata* (L.f.) Royle). Aquat. Bot. 12:157–172; 1982.

Barko, J.W.; Smart, R.M. Mobilization of sediment phosphorus by submersed freshwater macrophytes. Freshwat. Biol. 10:229–238; 1980.

Barko, J.W.; Smart, R.M. Sediment-related mechanisms of growth limitation in submersed macrophytes. Ecology 67:1328–1340; 1986.

Barko, J.W.; Adams, M.S.; Clesceri, N.L. Environmental factors and their consideration in the management of submersed aquatic vegetation: a review. J. Aquat. Plant Manage. 24:1–10; 1986.

Barko, J.W.; Smart, R.M.; McFarland, D.G.; Chen, R.L. Interrelationships between the growth of *Hydrilla verticillata* (L.f.) Royle and sediment nutrient availability. Aquat. Bot. 32:205–216; 1988.

Barko, J.W.; Gunnison, D.; Carpenter, S.R. Sediment interactions with submersed macrophyte growth and community dynamics. Aquat. Bot. 41:41–65; 1991.

Boers, P.C.M. The Influence of pH on phosphate release from lake sediments. Wat. Res. 25:309–311; 1991.

Boström, B.; Jansson, M.; Forsberg, C. Phosphorus release from lake sediments. Arch. Hydrobiol. Beih. Ergebn. Limnol. 18:5–59; 1982.

Carignan, R. Nutrient dynamics in a littoral sediment colonized by the submersed macrophyte *Myriophyllum spicatum*. Can. J. Fish. Aquat. Sci. 42:1303–1311; 1985.

Carignan, R.; Kalff, J. Phosphorus sources for aquatic weeds: water or sediment. Science 207:987–989; 1980.

Carpenter, S.R. Enrichment of Lake Wingra, Wisconsin, by submersed macrophyte decay. Ecology 61:1145–1155; 1980.

Carpenter, S.R. Submersed vegetation: an internal factor in lake ecosystem succession. Am. Nat. 118:372–383; 1981.

Carper, G.L.; Bachmann, R.W. Wind resuspension of sediments in a prairie lake. Can. J. Fish. Aquat. Sci. 41:1763–1767; 1984.

Carter, V.; Barko, J.W.; Godshalk, G.L.; Rybicki, N.B. Effects of submersed macrophytes on water quality in the Tidal Potomac River, Maryland. J. Freshwat. Ecol. 4:493–501; 1988.

Chen, R.L.; Barko, J.W. Effects of freshwater macrophytes on sediment chemistry. J. Freshwat. Ecol. 4:279–289; 1988.

Christiansen, R.; Skøvmand Friis, N.J.; Søndergaard, M. Leaf production and nitrogen and phosphorus tissue content of *Littorella uniflora* (L.) Aschers in relation to nitrogen and phosphorus enrichment of the sediment in oligotrophic Lake Hampen, Denmark. Aquat. Bot. 23:1–11; 1985.

Davis, M.B. Pollen grains in lake sediments: redeposition caused by seasonal water circulation. Science 162:796–799; 1968.

Davis, M.B. Redeposition of pollen grains in lake sediment. Limnol. Oceanogr. 18:44–52; 1973.

Davis, M.B.; Brubaker, L.B. Differential sedimentation of pollen grains in lakes. Limnol. Oceanogr. 18:635–646; 1973.

Davis, M.B.; Moeller, R.E.; Ford, J. Sediment focusing and pollen influx. In: Haworth, E.Y.; Lund, J.W., eds. Lake sediment and environmental history. Leicester, England: Leicester University; 1984.

Denny, P. Solute movement in submerged angiosperms. Biol. Rev. 55:65–92; 1980.

Dieter, C.D. The importance of emergent vegetation in reducing sediment resuspension in wetlands. J. Freshwat. Ecol. 5:467–473; 1990.

Dillon, P.J.; Evans, R.D.; Molot, L.A. Retention and resuspension of phosphorus, nitrogen, and iron in a central Ontario lake. Can. J. Fish. Aquat. Sci. 47:1269–1274; 1990.

Drake, J.C.; Heaney, S.I. Occurrence of phosphorus and its potential remobilization in the littoral sediments of a productive English lake. Freshwat. Biol., 17:513–523; 1987.

Duarte, C.M.; Kalff, J. Littoral slope as a predictor of maximum biomass of submerged macrophyte communities. Limnol. Oceanogr. 31:1072–1080; 1986.

Eckman, J.E.; Duggins, D.O.; Sewell, A.T. Ecology of understory kelp beds. I. Effects of kelps on flow and particle transport near the bottom. J. Exp. Mar. Biol. Ecol. 129:173–187; 1989.

Evans, R.D.; Rigler, F.H. Measurement of whole lake sediment accumulation and phosphorus retention using lead-210 dating. Can. J. Fish. Aquat. Sci. 37:817–822; 1980.

Evans, R.D.; Rigler, R.H. A test of lead-210 dating for measurement of whole lake soft sediment accumulation. Can. J. Fish. Aquat. Sci. 40:506–515; 1983.

Fonseca, M.S.; Fisher, J.S.; Zieman, J.C.; Thayer, G.W. Influence of the sea grass, *Zostera marina* L., on current flow. Est. Coast. Shelf Sci. 15:351–364; 1982.

Gregg, W.W.; Rose, F.L. The effects of aquatic macrophytes on the stream microenvironment. Aquat. Bot. 14:309–324; 1982.

Håkanson, L. The influence of wind, fetch and water depth on the distribution of sediments in Lake Vänern, Sweden. Can. J. Earth Sci. 14:397–412; 1977.

Hanson, M.A.; Butler, M.G. Responses of plankton, turbidity, and macrophytes to biomanipulation in a shallow prairie lake. Can. J. Fish. Aquat. Sci. 51:1180–1188; 1994.

Hellström, T. The effect of resuspension on algal production in a shallow lake. Hydrobiologia 213:183–190; 1991.

Hilton, J. A conceptual framework for predicting the occurrence of sediment focusing and sediment redistribution in small lakes. Limnol. Oceanogr. 30:1131–1143; 1985.

Hilton, J.; Lishman, J.P.; Allen P.V. The dominant processes of sediment distribution and focusing in a small, eutrophic, monomictic lake. Limnol. Oceanogr. 31:125–133; 1986.

Hosper, S.H. Biomanipulation, new perspectives for restoration of shallow, eutrophic lakes in The Netherlands. Hydrobiol. Bull. 23:5–10; 1989.

Hosper, S.H.; Jagtman, E. Biomanipulation additional to nutrient control for restoration of shallow lakes in The Netherlands. Hydrobiologia 200/201:523–534; 1990.

Howard-Williams, C. Studies on the ability of a *Potamogeton pectinatus* community to remove dissolved nitrogen and phosphorus compounds from lake water. J. Appl. Ecol. 18:619–637; 1981.

Huebert, D.B.; Gorham, P.R. Biphasic mineral nutrition of the submersed aquatic macrophyte *Potamogeton pectinatus* L. Aquat. Bot. 16:269–284; 1983.

Imberger, J.; Parker, G. Mixed layer dynamics in a lake exposed to a spatially variable wind field. Limnol. Oceanogr. 30:473–488; 1985.

Imberger, J.; Patterson, J.C. Physical limnology. Adv. Appl. Mech. 27:303–475; 1990.

Jackson, L.J.; Rowen, D.J.; Cornett, R.J.; Kalff, J. *Myriophyllum spicatum* pumps essential and nonessential trace elements from sediment to epiphytes. Can. J. Fish. Aquat. Sci. 51:1769–1773; 1994a.

Jackson, L.J.; Rasmussen, J.B.; Kalff, J. A mass-balance analysis of trace metals in two weedbeds. Wat. Air Soil Pollut. 75:107–119; 1994b.

James, W.F.; Barko, J.W. Macrophyte influences on the zonation of sediment accretion and composition in a north-temperate reservoir. Arch. Hydrobiol. 120:129–142; 1990.

James, W.F.; Barko, J.W. Estimation of phosphorus exchange between littoral and pelagic zones during nighttime convective circulation. Limnol. Oceanogr. 36:179–187; 1991a.

James, W.F.; Barko, J.W. Littoral-pelagic phosphorus dynamics during nighttime convective circulation. Limnol. Oceanogr. 36:949–960; 1991b.

James, W.F.; Barko, J.W. Analysis of summer phosphorus fluxes within the pelagic zone of Eau Galle Reservoir, Wisconsin. Lake Reserv. Manage. 8:61–71; 1993.

James, W.F.; Barko, J.W. Macrophyte influences on sediment resuspension and export in a shallow impoundment. Lake Reserv. Manage. 10:95–102; 1994.

James, W.F.; Taylor, W.D.; Barko, J.W. Production and vertical migration of *Ceratium hirundinella* in relation to phosphorus availability in Eau Galle Reservoir, Wisconsin. Can. J. Fish Aquat. Sci. 49:694–700; 1992.

James, W.F.; Barko, J.W.; Eakin, H.L. Convective water exchange during differential heating and cooling: implications for dissolved constituent transport. Hydrobiologia. 294:167–176; 1994.

James, W.F.; Smith, C.S.; Barko, J.W.; Field, S.J. Direct and indirect influences of aquatic macrophyte communities on phosphorus mobilization from littoral sediments of an inlet region in Lake Delavan, Wisconsin. Technical Report W-95-2. U.S. Army Engineer Waterways Experiment Station, Vicksburg, MS; 1995.

James, W.F.; Barko, J.W.; Field, S.J. Phosphorus mobilization from littoral sediments of an inlet region in Lake Delavan, Wisconsin. Arch. Hydrobiol. 138:245–257; 1996.

Jensen, H.S.; Kristensen, P.; Jeppesen, E.; Skytte, A. Iron:phosphorus ratio in surface sediment as an indicator of phosphate release from aerobic sediment in shallow lakes. Hydrobiologia 235:731–743; 1992.

Landers, D.H. Effects of naturally senescing aquatic macrophytes on nutrient chemistry and chlorophyll *a* of surrounding waters. Limnol. Oceanogr. 27:428–439; 1982.

Lijklema, L. Interaction of orthophosphate with iron (III) and aluminum hydroxides. Environ. Sci. Tech. 5:537–541; 1977.

Likens, G.E.; Davis, M.B. Post-glacial history of Mirror Lake and its watershed in New Hampshire USA: an initial report. Verh. Int. Verein. Theor. Angew. Limnol. 19:982–993; 1975.

Lowenhaupt, B. The transport of calcium and other cations in submerged aquatic plants. Biol. Rev. 31:371–395; 1956.

Maceina, M.J.; Soballe, D.M. Wind-related limnological variation in Lake Okeechobee, FL. Lake Reserv. Manage. 6:93–100; 1990.

Madsen, T.V.; Warncke, E. Velocities of currents around and within submerged aquatic vegetation. Arch. Hydrobiol. 97:389–394; 1983.

Moeller, R.E.; Wetzel, R.G. Littoral vs profundal components of sediment accumulation: contrasting roles as phosphorus sinks. Verh. Int. Verein. Theor. Angew. Limnol. 23:386–393; 1988.

Monismith, S.; Imberger, J.; Morison, M. Convective motions in the sidearm of a small reservoir. Limnol. Oceanogr. 35:1676–1702; 1990.

Mortimer, C.H. The exchange of dissolved substances between mud and water in lakes. J. Ecol. 29:280–329; 1941.

Patterson, K.J.; Brown, J.M.A. Growth and elemental composition of *Lagarosiphon major* in response to water and substrate nutrients. Prog. Water Techn. 2:231–246; 1979.

Petticrew, E.L.; Kalff, J. Predictions of surficial sediment composition in the littoral zone of lakes. Limnol. Oceanogr. 36:384–392; 1991.

Petticrew, E.L.; Kalff, J. Water flow and clay retention in submerged macrophyte beds. Can. J. Fish. Aquat. Sci. 49:2483–2489; 1992.

Prentki, R.T. Depletion of phosphorus from sediment colonized by *Myriophyllum spicatum* L. In: Breck, J.E.; Prentki, R.T.; Loucks, O.L., eds. Aquatic plants, lake management, and ecosystem consequences of lake harvesting. Madison, WI: Institute for Environmental Studies, University of Wisconsin; 1979:161–176.

Prentki, R.T.; Adams, M.S.; Carpenter, S.R.; Gasith, A.; Smith, S.C.; Weiler, P.R. The role of submersed weedbeds in internal loading and interception of allochthonous materials in Lake Wingra, Wisconsin, USA. Arch. Hydrobiol. Suppl. 57:221–250; 1979.

Rogers, S.J.; McFarland, D.G.; Barko, J.W. Evaluation of the growth of *Vallisneria americana* Michx. in relation to sediment nutrient availability. Lake Reserv. Manage. 11:57–66; 1995.

Scheffer, M. Multiplicity of stable states in freshwater systems. Hydrobiologia 200/201: 475–486; 1990.

Scheffer, M.; Hosper, S.H.; Meir, M-L.; Moss, B.; Jeppesen, E. Alternative equilibria in shallow lakes. Trends Ecol. Evol. 8:275–279; 1993.

Schindler, D.W. Eutrophication and recovery in experimental lakes: implications for lake management. Science 184:897–898; 1974.

Schindler, D.W. Evolution of phosphorus limitation in lakes. Science 195:260–262; 1977.

Sculthrope, C.D. The biology of aquatic vascular plants. London: Edward Arnold; 1967.

Smart, R.M.; Barko, J.W. Laboratory culture of submersed freshwater macrophytes on natural sediments. Aquat. Bot. 21:251–263; 1985.

Smart, R.M.; Barko, J.W. Effects of water chemistry on aquatic plants: growth and photosynthesis of *Myriophyllum spicatum* L. Technical Report A-86–2, Environmental Laboratory, U.S. Army Engineer Waterways Experiment Station, Vicksburg, MS; 1986.

Smith, C.S.; Adams, M.S. Phosphorus transfer from sediments by *Myriophyllum spicatum.* Limnol. Oceanogr. 31:1312–1321; 1986.

Søndergaard, M.; Kristensen, P.; Jeppesen, E. Phosphorus release from resuspended sediment in the shallow and wind-exposed Lake Arresø, Denmark. Hydrobiologia 228:91–99; 1992.

Stefan, H.G.; Horsch, G.M.; Barko, J.W. A model for the estimation of convective exchange in the littoral region of a shallow lake during cooling. Hydrobiologia 174:225–234; 1989.

Taylor, W.D.; Barko, J.W.; James, W.F. Contrasting diel patterns of vertical migration in the dinoflagellate *Ceratium hirundinella* in relation to phosphorus supply in a north temperate reservoir. Can. J. Fish. Aquat. Sci. 45:1093–1098; 1988.

Trisal, C.L.; Kaul, S. Sediment composition, mud-water interchanges and the role of macrophytes in Dal Lake, Kashmir. Int. Rev. Ges. Hydrobiol. 68:671–682; 1983.

Weiler, P.R. Littoral-pelagic exchange in Lake Wingra, Wisconsin, as determined by a circulation model. Madison, WI: University of Wisconsin, Madison, Inst. Environ. Stud. Rep. 100; 1978.

Wetzel, R.G. The role of the littoral zone and detritus in lake metabolism. Arch. Hydrobiol. Beih. Ergebn. Limnol. 13:145–161; 1979.

2. Case Studies

11. Macrophyte Structure and Growth of Bluegill (*Lepomis macrochirus*): Design of a Multilake Experiment

Stephen R. Carpenter, Mark Olson, Paul Cunningham,
Sarig Gafny, Nathan Nibbelink, Tom Pellett, Christine Storlie,
Anett Trebitz, and Karen Wilson

Introduction

Experimental manipulations of whole ecosystems can be a powerful test of ecological understanding. In particular, ecosystem-scale manipulations can evaluate basic ecological ideas in ways that complement comparative studies, models, and smaller-scale experiments (Carpenter et al., 1995a). From an applied perspective, ecosystem experiments can also give unique insights into what works at a scale directly relevant to managers (Kitchell, 1992). When management actions are coupled with scientific studies of the response of the ecosystem, learning may lead to improved management practices (Gunderson et al., 1995). Here we present early results of an experiment to test the idea that nuisance macrophytes can be managed to enhance fish growth.

In North American lakes, dense beds of an exotic macrophyte, Eurasian water milfoil (*Myriophyllum spicatum*) are often associated with stunted populations of bluegill (*Lepomis macrochirus*) and their predators such as largemouth bass (*Micropterus salmoides*) (Heck and Crowder, 1991). One explanation for this association is that dense vegetation inhibits foraging by piscivores, allowing bluegill to reach high densities that deplete benthic invertebrates and thereby reducing growth rates of individual bluegill (Crowder and Cooper, 1982; Trebitz et al., 1997). Intermediate macrophyte densities may allow predators to control bluegill densities while also providing habitat that supports good growth by surviving bluegill (Cooper and Crowder, 1979; Wiley et al., 1984; Trebitz and Nibbelink,

1996; Trebitz et al., 1997). Lake managers are interested in both controlling milfoil and improving growth of bluegill. We proposed that both goals could be met by harvesting channels through milfoil beds to make bluegill more accessible to piscivores. This hypothesis is being tested with a multilake experiment.

Our experiment uses a set of lakes in which bluegill growth rates were measured before and after manipulation. The manipulation involved harvesting channels through the macrophyte beds in some of the lakes, while leaving the other lakes as unmanipulated controls. This experiment was scrutinized closely because it entailed a substantial commitment of staff and resources and conspicuous interventions in public waters. Design of the experiment was a complex process that required us to balance scientific goals with the realities of funding and the expectations of managers and the public. Therefore, we conducted a series of planning exercises to clarify the scientific goals and discuss them with interested managers and the public. These exercises consisted of ecosystem models to help plan the program, statistical power analysis to determine the number of lakes to be manipulated, and comparative studies to select the experimental lakes from a large number of candidates. Here we summarize the design process, the macrophyte manipulation, and responses of bluegill growth in the first year after manipulation.

The Role of Modeling

We used models to reach a common understanding of the problem and our approach, to plan the macrophyte harvesting, and to synthesize results.

Early in the project, a series of meetings of managers and scientists developed a common understanding of macrophyte–fish interactions. Our goal was to develop a conceptual framework that would help us decide what to measure and how to manipulate the lakes. The resulting model used a daily time step to simulate dynamics of zooplankton, four classes of benthic and plant-associated invertebrates, and size-structured populations of bluegill and largemouth bass for 10 years (Trebitz et al., 1997). Extensive analyses of this model suggested that harvesting a moderate amount of milfoil in narrow channels could increase growth rates of bluegill and population densities of largemouth bass (Trebitz et al., 1997). A valuable result of the modeling exercise was discussion among our diverse team of the key state variables, the time scales of ecosystem response, and a general strategy for the experiment.

To plan the macrophyte harvests, we needed a simpler model that could be explained to diverse stakeholders and tied closely to field data. Our approach combined principles of fish bioenergetics with the geometry of the littoral zone (Trebitz and Nibbelink, 1996). This model indicated that the width of cut channels and the percentage of vegetation removed interacted to determine bluegill growth response (Fig. 11.1). Bluegill growth is predicted to be maximized by cutting channels 2 or 3 m wide (near the minimum width possible with available equipment) that collectively remove about one-third of the area of the milfoil bed. These calculations are contingent on assumptions about the relative availability of

Figure 11.1. Predicted feeding rate of 2-year-old bluegill in Fish Lake, Wisconsin, as a function of width of harvested channels and percentage of the macrophyte bed that is removed. Feeding rates are expressed as proportion of the maximum (Hewett and Johnson, 1992). The heavy line marks the average feeding rate in Fish Lake before manipulation (0.45). Effect size refers to the relative availability of invertebrate prey in the edge habitat (which extends 2 m from open water) versus the milfoil bed interior (Trebitz and Nibbelink, 1996).

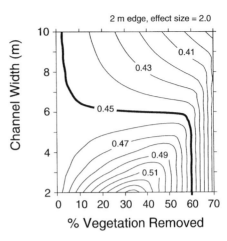

macroinvertebrates in edge habitats versus the interior of the milfoil bed. This variable is difficult (perhaps impossible) to measure in the field. Trebitz and Nibbelink (1996) used many simulations over a wide range of conditions to show that harvesting recommendations were not sensitive to assumptions about availability of invertebrate prey. Over a wide range of assumptions, optimal bluegill growth occurred when relatively narrow channels were cut to remove 20–40% of the area of the milfoil beds. Although these differences in growth rate may appear small, over the lifetime of a fish they lead to large differences in size at age (Trebitz and Nibbelink, 1996).

Several models are being used to synthesize results of the study. A well-known fish bioenergetics model (Hewett and Johnson, 1992) will be used with growth and diet data to test the hypothesis that consumption of bluegill by bass increased after manipulation. An individual-based model for bluegill has been developed to assess the implications of changes in habitat usage by bluegill after manipulation (Nibbelink, 1996).

Power to Detect Effects

Statistical power, the probability of detecting a given effect, is an important consideration for fisheries management experiments (McAllister and Peterman, 1992). Power is defined as the probability of detecting a specified effect by rejecting a null hypothesis of no effect at a specified significance level, given the error variance and degrees of freedom (Cohen, 1988). Although Bayesian statistics can evaluate our experiment more logically, usefully, and completely than frequentist statistics (Howson and Urbach, 1989), and power is a frequentist concept, we viewed power analysis as a useful exercise. Power analysis provided us with "rules of thumb" for choosing sample sizes and deciding which variates to include in the study. Also, power analysis provided insurance against possible criticisms of the experiment from a frequentist perspective.

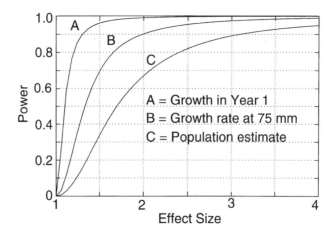

Figure 11.2. Power (the probability of detecting the effect at the 5% significance level) versus effect size (postmanipulation value divided by the premanipulation value). These power curves assume five manipulated lakes and five control lakes. Power curves are shown for bluegill growth (mm) during year 1 (A), growth rate (g/yr) for bluegills 75 mm long at the start of the year (B), and population density (bluegills/ha) (C).

Power analysis showed that an experiment with about 10 lakes (five treatment, five control) was adequate to detect effects deemed important by lake managers (Carpenter et al., 1995b). Even smaller effects could be detected with larger sample sizes, but resolution of such subtle effects was not thought to be worth the cost of additional replicates. More complex experiments, including a gradient of harvest intensities and designs to test various interactions, were considered but rejected as too costly at this stage of the research (Carpenter et al., 1995b).

Power to detect fish growth responses was generally greater than power to detect changes in fish population size (Fig. 11.2), because errors in measuring growth are smaller than errors in estimating population size. Population data are far more costly to collect than growth data. Adding a lake to the study for population estimates cost more than 50× more person-hours than adding a lake to the study for growth rate estimates (Carpenter et al., 1995b). This important insight prompted us to study growth and size structure in a larger number of lakes and to abandon population size as a response variate.

Lake Selection .

The process of selecting lakes for the experiment was based on a sequence of independent criteria. To receive initial consideration, the fish community had to be dominated by largemouth bass and bluegill. In addition to being two of the most important sport fish in Wisconsin, our predicted responses to macrophyte harvesting were based on expected changes in the interaction between these two species.

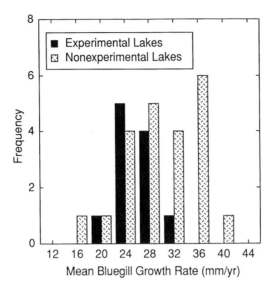

Figure 11.3. Frequency distribution of mean bluegill growth rates from 33 of the lakes surveyed in 1993. Mean growth rates were estimated from total length versus age regressions (mean sample size ± 1 standard deviation was 60.5 ± 24.3, and mean r^2 was 0.90 ± 0.04). Of the lakes surveyed, 11 were chosen for the experiment.

We sought relatively small lakes (<150 ha) to facilitate sampling and manipulation. We also chose lakes that lacked surface water connections to other lakes to eliminate the possibility that our results could be confounded by interlake fish movements.

Drawing on the expertise of regional fish biologists and lake managers, we generated a pool of more than 400 lakes (from a total of 15,000 lakes in Wisconsin). To this set of lakes, we applied several scientific, political, and logistical criteria. Because we planned a manipulation of littoral zone macrophytes, we searched for lakes that had extensive littoral zones dominated by Eurasian milfoil. We also eliminated shallow lakes that possessed a high risk of winterkill. Winterkill events can dramatically change fish communities (Tonn and Paszkowski, 1986; Hall and Ehlinger, 1989) and could potentially confound our experimental manipulation. For continued eligibility, a lake had to have public access and adequate boat launch facilities. This criterion reflects agency goals of improving public fisheries. Furthermore, we avoided lakes where our experiment might conflict with other research and/or stocking projects.

Using these criteria, the set of 400 lakes was narrowed to 37 candidates. These lakes were sampled in 1993 to better characterize the fish and macrophyte communities and to quantify patterns of bluegill growth. Bluegill were collected by electroshocking, and scales were collected to back-calculate length-at-age using the Fraser-Lee method (Tesch, 1968). Mean annual growth rates were estimated by determining the slope of total length versus age regressions for each lake. These estimates were used to make our final selections. To maximize our ability to detect growth responses, we chose our experimental lakes from a narrow range of pretreatment bluegill growth rates (Fig. 11.3). We also selected lakes with slow-growing bluegill to increase the potential for increases in growth after manipulation (Fig. 11.3). After the lakes were selected, five were chosen to be treatment lakes and six served as unmanipulated controls. The allocation of lakes to treatment and control groups was not strictly random but appeared to be unbiased. We had decided a priori that Fish Lake would be harvested, because it was near Madison and could be studied more intensively than the other lakes. Despite this nonrandom decision, the treatment and control groups were not distinguishable in any way that seemed capable of biasing the outcome of the experiment.

Manipulation

The experiment was initiated in August 1994 when four of the five treatment lakes were manipulated (Gibbs Lake was manipulated in 1995). The removal of macrophytes from the littoral zone of each lake was accomplished with an aquatic plant harvester equipped with two cutting bars. The front bar cut a 3-m swath to a maximum depth of 2.5 m. When using this bar, cut macrophytes were collected immediately and transported to shore for removal. The other cutting bar was specially designed to cut to a depth of 5 m. This deep bar cut a narrower swath of 2 m, and macrophytes that floated to the surface were collected during a second pass. Choice of cutting bars depended on the depth of macrophyte growth in a lake and the slope of the littoral zone.

Manipulations of the treatment lakes are summarized in Table 11.1. In each lake, macrophytes were removed in a series of channels. Depending on lake size and type of cutting bar, the number of channels ranged from 108 to 285. Mean channel length also varied from 44 to 123 m, due to lake morphometry and depth

Table 11.1. Summary of Macrophyte Manipulations in the Five Treatment Lakes

Lake	Area (ha)	% of area in littoral zone	Dates of manipulation	Number of channels	Channel width (m)	Mean channel length (m)	Created edge (m)	% of littoral cleared
Fish	102.0	54.5	8/08/94–8/16/94	285	2	123.0	70,104	18.0
Gibbs	26.0	65.2	8/14/95–8/22/95	158	3	60.0	18,960	16.8
Heidemann	10.5	67.2	8/30/94–9/01/94	172	2	43.7	15,030	21.3
Silver	28.3	77.7	8/19/94–8/23/94	127	3	93.2	33,514	15.2
Tuma	7.7	76.9	8/25/94–8/29/94	108	2	46.5	10,044	17.0

of macrophyte growth. As a consequence of variation in channel length and number, the absolute amount of edge varied widely among lakes (Table 11.1). However, the percentage of vegetation removed from littoral zones was similar among lakes (mean = 17.7%, coefficient of variation = 12.8%) and close to our target of 20%.

First Year Responses

To evaluate bluegill growth responses in the first year after our manipulation, we electroshocked each lake in autumn 1995. Scales were collected from a size range of bluegill to back-calculate a fish's size at the start of the growing season, which was then subtracted from the size-at-capture to estimate growth in 1995. Mean growth rates were calculated for age-1 through age-5 bluegill in each lake. Macrophyte densities were measured by the method of Deppe and Lathrop (1992). Results reported here exclude Gibbs Lake, which was not manipulated until 1995 and did not have as much time as the other lakes to respond to manipulation.

Probability distributions of growth differences between cut and control lakes were calculated for the first growing season after manipulation (Fig. 11.4). These

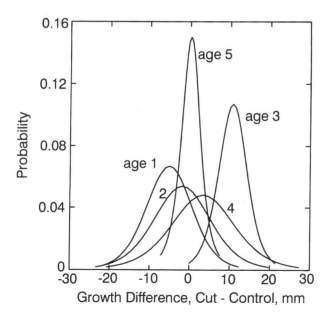

Figure 11.4. Probability distributions of the growth difference (mm) between cut and control lakes for bluegill aged 1, 2, 3, 4, or 5 years. Growth increments were calculated for January–September 1995, a period that includes the first growing season after the experimental macrophyte harvests. Distributions shown here are based on six reference lakes and four manipulated lakes (Fish, Heidemann, Silver, and Tuma).

distributions are Bayesian posterior distributions based on a noninformative prior distribution (Box and Tiao, 1973). Unlike conventional t-tests, which yield the probability of being wrong if the null hypothesis is rejected, these distributions can be interpreted directly as the probability of a given effect (Box and Tiao, 1973). For a specified growth difference, the height of the curve is the probability of that growth difference. The area under the curve to the right of the growth difference is the probability of a higher growth difference, and the area to the left is the probability of a lower growth difference. For example, the probability that age-3 fish grew more in the harvested lakes is nearly 1 (almost the entire area under the curve is to the right of zero). By contrast, the probability that age-5 fish grew more in the harvested lakes is about 0.5 (about half the area below the curve is right of zero, and half is left). If the management goal is a 5-mm growth increment, we can see that the probability of meeting the goal is very low for bluegill aged 1 and 5, about one-third for bluegill aged 2, about 0.5 for bluegill aged 4, and more than 90% for bluegill aged 3. These distributions can also be used to compare effect sizes other than zero. The probability of an age-3 fish growing 10 mm is more than five times greater than the probability of an age-2 fish growing 10 mm (at 10 mm on the X-axis, the Y value for age-3 fish is much higher than the Y value for age-2 fish).

These early results suggest that experimental harvesting had strong positive effects on growth of age-3 bluegill, somewhat positive and variable effects on age-4 bluegill, no effect on growth of age-5 bluegill, mildly negative effects on growth of age-2 bluegill, and more strongly negative effects on growth of age-1 bluegill. The variable growth responses observed in our preliminary analysis may reflect differing patterns of habitat use among age classes. Bluegill often undergo a habitat shift in their ontogeny, in response to size-specific changes in predation risk (Mittelbach, 1981; Werner and Hall, 1988). Small or young bluegill typically use the littoral zone as a refuge from predatory largemouth bass. However, because vulnerability to bass predation decreases with size, larger or older bluegill are able to forage for zooplankton in the energetically profitable pelagic zone (Werner and Hall, 1988). Consequently, intermediate age classes would be expected to respond most strongly to experimental manipulation of edge habitat. The fact that only age-3 bluegill responded positively may indicate that this age class was the only one to benefit from macrophyte harvesting. Over the course of the growing season, these bluegill increased in size from 90- to 110-mm total length. Within this size range, individuals often use the vegetation as a refuge between foraging bouts in open water (Werner and Hall, 1988). Increasing the amount of edge may have significantly increased the available foraging habitat for this age class but not changed (or even decreased) the amount of habitat for younger age classes.

Alternatively, age-1 and age-2 bluegill may not have responded because of a natural decrease in macrophyte density in all lakes. From 1993 to 1995, we observed a significant decrease in overall macrophyte density in all our experimental lakes. Analysis of variance indicated a significant year effect ($F_{1,16} =$ 5.74, $P <.05$), a nonsignificant manipulation effect ($F_{1,16} = 2.00$, $P >.05$), and a marginal interaction ($F_{1,16} = 4.27$, $P \approx .05$). This decrease may have created de facto channels at a small scale, which benefited age-1 and age-2 bluegill equally in both treatment and control lakes. However, age-3 bluegill may have been too large to benefit from the macrophyte decline but of appropriate size to benefit from the manipulation. Future analyses of bluegill growth will directly contrast pre- and postmanipulation growth rates and will enable us to separate the effect of our manipulation from natural variation in macrophyte densities.

Conclusions

The planning exercises had a substantial impact on the field experiment. Modeling studies suggested that fish growth was a key response and that substantial changes in edge habitat and predator–prey interactions could be achieved by removing only about 20% of the macrophyte cover. The power analysis reinforced the idea that it was better to focus on fish growth and include more lakes in the study than to make costly population estimates in fewer lakes. The lake selection process revealed the natural variability that could be seen in the experiment and helped us select a relatively homogeneous set of lakes that could be expected to respond to macrophyte harvest. All these exercises helped us explain the experiment to managers, the public, and scientific referees.

After all this planning, the experiment is underway and we are beginning to see results. As is often the case with large-scale experiments, the results are more complex than we expected (Kitchell, 1992; Carpenter et al., 1995a). Future ecosystem responses and more detailed analyses will improve our understanding of the responses to the manipulations. Because the experimental design has relatively high power, we expect to determine whether macrophyte harvesting does or does not improve bluegill growth.

Acknowledgments. We are grateful for insightful comments by Larry Crowder, Sebastian Diehl, Avital Gasith, and Lennart Persson. The Dane County Department of Public Works built the macrophyte harvester and conducted the harvests. This research was funded by the Federal Aid to Sport Fish Restoration Act through the Wisconsin Department of Natural Resources, Project F-95-P, and the NTL-LTER site.

References

Box, G.E.P.; Tiao, G.C. Bayesian inference in statistical analysis. New York: Wiley; 1973.
Carpenter, S.R.; Chisholm, S.W.; Krebs, C.J.; Schindler, D.W.; Wright, R.F. Ecosystem experiments. Science 269:324–327; 1995a.
Carpenter, S.R.; Cunningham, P.; Gafny, S.; Muñoz-del-Rio, A.; Nibbelink, N.; Olson, M.; Pellett, T.; Storlie, C.; Trebitz, A. Responses of bluegill to habitat manipulations: power to detect effects. North Am. J. Fish. Manage. 15:519–527; 1995b.
Cohen, J. Statistical power analysis for the behavioral sciences. Hillside, NJ: Erlbaum; 1988.

Cooper, W.E.; Crowder, L.B. Patterns of predation in simple and complex environments. In: Stroud, R.H.; Clepper, H., eds. Predator–prey systems in fisheries management. Washington, DC: Sport Fishing Institute; 1979:257–267.

Crowder, L.B.; Cooper, W.E. Habitat structural complexity and the interaction between bluegills and their prey. Ecology 63:1802–1813; 1982.

Deppe, E.R.; Lathrop, R.C. A comparison of two rake sampling techniques for sampling aquatic macrophytes. Research Management Findings 32. Madison, WI: Department of Natural Resources; 1992.

Gunderson, L.H.; Holling, C.S.; Light, S.S., eds. Barriers and bridges to the renewal of ecosystems and institutions. New York: Columbia University Press; 1995.

Hall, D.J.; Ehlinger, T.J. Perturbation, planktivory and pelagic community structure: the consequences of winterkill in a small lake. Can. J. Fish. Aquat. Sci. 46:2203–2209; 1989.

Heck, K.L.; Crowder, L.B. Habitat structure and predator-prey interactions in vegetated aquatic systems. In: Ball, S.S.; McCoy, E.D.; Mushinsky, H.R., eds. Habitat structure: the physical arrangement of objects in space. London: Chapman & Hall; 1991:281–299.

Hewett, S.W.; Johnson, B.L. A generalized bioenergetics model of fish growth for microcomputers. Sea Grant Institute Publication 92–250. Madison, WI: University of Wisconsin; 1992.

Howson, C.; Urbach, P. Scientific reasoning: the Bayesian approach. LaSalle, IL: Open Court Press; 1989.

Kitchell, J.F., ed. Food web management: a case study of Lake Mendota. New York: Springer-Verlag; 1992.

McAllister, M.K.; Peterman, R.M. Experimental design in the management of fisheries: a review. North Am. J. Fish. Manage. 12:1–18; 1992.

Mittelbach, G.G. Foraging efficiency and body size: a study of optimal diet and habitat use by bluegills. Ecology 62:1370–1386; 1981.

Nibbelink, N. Explaining growth and size structure of bluegill: inverse analysis of an individual-based model. M.S. thesis, University of Wisconsin, Madison; 1996.

Tesch, F.W. Age and growth. In: Ricker, W.E., ed. Methods for assessment of fish production in fresh waters. Oxford, England: Blackwell Scientific; 1968:93–123.

Tonn, W.M.; Paszkowski, C.A. Size-limited predation, winterkill, and the organization of *Umbra-Perca* fish assemblages. Can. J. Fish. Aquat. Sci. 43:194–202; 1986.

Trebitz, A.; Carpenter, S.; Cunningham, P.; Johnson, B.; Lillie, R.; Marshall, D.; Martin, T.; Narf, R.; Pellett, T.; Stewart, S.; Storlie, C.; Unmuth, J. A model of bluegill-largemouth bass interactions in relation to aquatic vegetation and its management. Ecol. Model. 94:139–156; 1997.

Trebitz, A.S.; Nibbelink, N. Effect of pattern of vegetation removal on growth of bluegill: a simple model. Can. J. Fish. Aquat. Sci. 53:1844–1851; 1996.

Werner, E.E.; Hall, D.J. Ontogenetic habitat shifts in bluegill: the foraging rate-predation risk trade-off. Ecology 69:1540–1548; 1988.

Wiley, M.J.; Gordon, R.W.; Waite, S.W.; Powless, T. The relationship between aquatic macrophytes and sport fish production in Illinois ponds: a simple model. North Am. J. Fish. Manage. 4:111–119; 1984.

12. Vertical Distribution of In-Benthos in Relation to Fish and Floating-Leaved Macrophyte Populations

Ryszard Kornijów and Brian Moss

Introduction

Several papers have described the vertical distribution of in-benthos in fresh waters (e.g., Kajak and Dusoge, 1971; Becket et al., 1992; van de Bund and Groenendijk, 1994), but the reasons for such distributions of invertebrates as midge larvae or tubificid worms in deep bottom sediments are often obscure. One reason may be reduced abundance in the surface sediment due to fish predation (Hershey, 1985; Lammens et al., 1987; Kornijów, 1997). However, a behavioral response by the in-benthic invertebrates to avoid fish (i.e., migration to deeper sediments) may play an important role as well but has not been tested.

Predation by fish on in-benthos is affected by the density of vegetation (Diehl, 1988, 1992). It is possible that vegetation density also modifies the effects of fish on the vertical distribution of in-benthos. In this chapter we experimentally tested the hypothesis that in-benthos can adjust its vertical distribution in response to foraging pressure by fish. We carried out the study at various densities of vegetation (nymphaeids) to test whether the effect of fish on the vertical distribution of prey can be modified by macrophytes, providing in-benthic invertebrates with a refuge.

Methods

The experiment was carried out in 18 2-m² enclosures, made of curtain netting (mesh size 0.5 mm), in a dense bed of *Nuphar lutea* (L.) Sm. at a water depth of about 75 cm, in the small and shallow lake Little Mere in northwest England. (For more details of the lake and the design of the enclosures, see Kornijów, 1997.) Three densities of the floating leaves were created in enclosures to give 10%, 50%, and 90% coverage of the water surface, by cutting superfluous leaves at the bases of their petioles. On June 16, 1993, six perch (mean length, 14.8 cm; SD = 0.8 cm), collected by seining in Little Mere, were added randomly to each of three enclosures at each manipulated plant density. The experiment ended on August 15. The fish were then caught, and their stomach contents were preserved with 4% formaldehyde solution and examined under a dissecting microscope.

Benthic macroinvertebrates were sampled by means of a perspex tube (area, 15 cm²; length, 120 cm), on July 12–13 and on August 10–11. From each enclosure, six cores were taken and sliced into three depth-fractions (0–2 cm, 2–5 cm, and 5–10 cm) by using the method of van de Bund and Groenendijk (1994). The material collected was passed through a 400-μm sieve to concentrate the invertebrates.

Results

The main invertebrates burrowing in the bottom sediments were Tubificidae and larvae of *Chironomus* f.l *plumosus* L., together constituting on average more than 90% of the total in-benthos density and biomass. *Chironomus* larvae were the main chironomid prey, constituting 42% of all the midge larvae found in the guts of perch. Tubificidae mainly consisted of *Euilyodrilus hammoniensis* Mich. and *Limnodrillus hoffmeisteri* Clap. (ca. 30 and 70%, respectively). They were not encountered in the diet of perch.

There was a highly significant relation between animal density and depth in the sediment (one-way ANOVA, effect of the sediment layer on the density of Tubificidae, $df = 2$, $F = 10.89$, $P = .0001$; effect of the sediment layer on the density of *Chironomus* larvae, $df = 2$, $F = 7.12$, $P = .0016$), with higher densities of animals with depth (Figs. 12.1 and 12.2). The vertical distribution patterns of different size classes of *Chironomus* differed markedly, with bigger larvae being more common at greater depth (Kornijów, 1997).

In July, there were no interactive effects of either plant density or perch on the density of in-benthic animals in any sediment layer (Figs. 12.1 and 12.2; Table 12.1). In August, the densities of Tubificidae in the separate layers (0–2, 2–5, 5–10 cm) still did not depend on the density of vegetation. They were similar in fish enclosures to those in the controls in all the layers, except for the 5–10-cm layer where densities were positively affected by fish (Fig. 12.1; Table 12.1). Densities of *Chironomus* larvae were affected by plant density in all three sediment layers, being generally lowest at highest density of vegetation (Fig. 12.2;

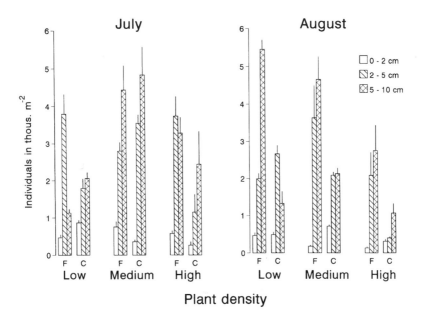

Figure 12.1. Vertical distribution of Tubificidae in the bottom sediments, divided into three layers (0–2, 2–5, and 5–10 cm) in enclosures with perch (F) and in the controls (C) at low, medium, and high densities of *Nuphar* leaves in July and August. Error bars represent standard error (*n* = 3).

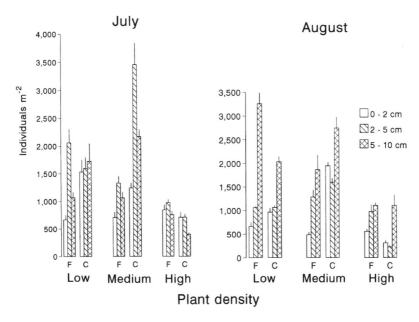

Figure 12.2. Vertical distribution of *Chironomus* larvae in the bottom sediments, divided into three layers (0–2, 2–5, and 5–10 cm) in enclosures with perch (F) and in the controls (C) at low, medium, and high densities of *Nuphar* leaves in July and August. Error bars represent standard error (*n* = 3).

Table 12.1. Two-Way Repeated-Measures ANOVAs of the Effects of *Nuphar lutea* Density and Feeding by Perch on the Density of *Chironomus* Larvae and Tubificidae in Different Sediment Layers (0–2, 2–5, 5–10 cm)

		F ratio *Chironomus*			F ratio Tubificidae		
		0–2	2–5	5–10	0–2	2–5	5–10
Source of variation	df			(sediment depth, cm)			
July							
Plant density	2	0.371^{ns}	3.606^{ns}	0.208^{ns}	0.708^{ns}	0.213^{ns}	1.823^{ns}
Perch	1	1.877^{ns}	0.001^{ns}	1.672^{ns}	0.456^{ns}	1.075^{ns}	0.002^{ns}
Plant density × perch	2	0.089^{ns}	0.282^{ns}	1.509^{ns}	0.264^{ns}	0.840^{ns}	0.079^{ns}
August							
Plant density	2	15.550^{***}	5.442^{**}	6.766^{**}	0.798^{ns}	1.117^{ns}	0.942^{ns}
Perch	1	9.044^{**}	0.462^{ns}	0.095^{ns}	1.803^{ns}	0.084^{ns}	7.104^{*}
Plant density × perch	2	15.900^{***}	2.271^{ns}	2.733^{ns}	0.690^{ns}	0.648^{ns}	0.445^{ns}

$*P < .05$; $** P < .01$; $*** P < .001$; ns, not significant.

Table 12.1). Perch had a significant effect on *Chironomus* larvae in the surface layer at low and medium densities of *Nuphar* but no effect (or even a positive effect) at high density (Fig. 12.2; Table 12.1).

Discussion

The overall vertical distribution of in-benthos found in this study corresponded to that found by other authors (Kajak and Dusoge, 1971; Becket et al., 1992; van de Bund and Groenendijk, 1994). The direct influence of vegetation on Tubificidae was negligible. Densities of Tubificidae were similar in the surficial sediments in the presence and absence of perch but, in the deeper layers, were about twice as high in the presence of fish than in the controls. This suggests a behavioral response to fish predation.

The negative effect of the increased density of *Nuphar* leaves on the densities of *Chironomus* larvae became significant in the second month of the experiment, which suggests that at least 1 month was needed by the larvae to adjust their density to the changed conditions. The larvae can be classified functionally as collector-gatherers or collector-suspension feeders. Their basic food consists of both planktonic algae and those living on the surface of the sediments. Vegetation itself did not influence chlorophyll$_a$ concentrations in the water (R. Kornijów, B. Moss, and J. Measey, unpublished data). However, chlorophyll concentrations were inversely related to plant density in the presence of perch. This could have reduced the food supply to the benthos.

The size structure of the *Chironomus* larvae eaten by perch, which is considered an epi-benthic predator (Mattila, 1992), was most similar to that in the top

2-cm sediment layer (Kornijów, 1997). The abundance of chironomids in the upper sediment layer was negatively affected by perch at low and medium *Nuphar* densities. This is most easily explained as a direct effect of consumption. Larvae formerly at the surface did not seem to have moved deeper to avoid fish predation. Otherwise, larval densities in the deeper sediment layers should have been higher in the presence of fish than in the controls. We cannot exclude the possibility, however, that some vertical migration occurred, but that migration into deeper layers was compensated by some predation also at greater depth. There is evidence that perch were able to penetrate quite deeply, for the effect of their foraging on density of some larval classes was found to be significant even in the deepest, 5–10 cm, sediment layer (Kornijów, 1997).

The lack of effect of perch on the larvae at high vegetation density supports the view that dense vegetation can act as a refuge (Diehl, 1988, 1992; Hershey, 1985). An equally plausible explanation may be that at high vegetation density, fish could exploit additional food resources, epiphytic invertebrates, abundantly living on the plant surface (R. Kornijów, B. Moss, and J. Measey, unpublished data). The impact of perch on the in-benthic community could then be softened.

This chapter reveals that the behavioral responses of *Chironomus* larvae and Tubificidae to perch predation differ. *Chironomus* larvae did not seem to migrate to deeper sediments, as was found for Tubificidae. In addition, the paper supports the hypothesis that a high density of vegetation may modify the influence of perch on the vertical distribution of in-benthic prey.

Finally, it should be stressed that the results presented here, and especially those concerning the influence of perch on the vertical distribution of in-benthos, should be generalized with caution and only to those fish whose feeding behavior is similar to that of perch. It is very likely that the response of benthos might differ from that presented in this chapter if typical in-benthic feeders such as tench or bream had been used in the experiment.

Acknowledgments. The study was supported by the British Council. The authors thank John Measey and Sabri Kilinc for their help in field and laboratory work.

References

Becket, D.C.; Aartila, T.P.; Miller, A.C. Contrasts in the density of benthic invertebrates between macrophyte beds and open littoral patches in Eau Galle Lake, Wisconsin. Am. Midl. Nat. 127:77–90; 1992.

Diehl, S. Foraging efficiency of three freshwater fishes: effects of structural complexity and light. Oikos 53:207–214; 1988.

Diehl, S. Fish predation and benthic community structure: the role of omnivory and habitat complexity. Ecology 73:1646–1661; 1992.

Hershey, A.E. Effects of predatory sculpin on the chironomid communities in an arctic lake. Ecology 66:1131–1138; 1985.

Kajak, Z.; Dusoge, K. The regularities of vertical distribution of benthos in bottom sediments of three Masurian lakes. Ekol. Pol. 19:485–499; 1971.

Kornijów, R. The impact of predation by perch on the size-structure of *Chironomus* larvae: the vertical distribution of the prey, and habitat complexity. Hydrobiol. 342/343:207–213; 1997.

Lammens, E.H.R.R.; Geursen, J.; McGillavry, P.J. Diet shifts, feeding efficiency and coexistence of bream (*Abramis brama*), roach (*Rutilus rutilus*) and white bream (*Blicca björkna*) in eutrophicated lakes. Proceedings of the Fifth Congress of European Ichthyology, 1985, Stockholm:153–162; 1987.

Mattila, J. Can fish regulate benthic communities on shallow soft-bottoms in the Baltic Sea? The role of perch, ruffe and roach. PhD dissertation, Abo Akademi University, Turku, Finland; 1992.

van de Bund, W.J.; Groenendijk, D. Seasonal dynamics and burrowing of littoral chironomid larvae in relation to competition and predation. Arch. Hydrobiol. 132:213–225; 1994.

13.   Horizontal Migration of Zooplankton: Predator-Mediated Use of Macrophyte Habitat

Torben L. Lauridsen, Erik Jeppesen, Martin Søndergaard, and David M. Lodge

Introduction

Aquatic macrophytes have multiple roles in ecosystem function (Carpenter and Lodge, 1986) and in mediating predator–prey interactions involving fish and macroinvertebrates (Crowder and Cooper, 1982; Savino and Stein, 1982). In recent years, authors have suggested that macrophytes also provide a spatial refuge from fish predation for *Daphnia* during daytime (Timms and Moss, 1984; Davies, 1985), thereby contributing to a lower phytoplankton level in shallow lakes (Scheffer et al., 1993). However, migration into macrophyte beds contradicts earlier results suggesting that macrophytes are repellent to daphnids (Hasler and Jones, 1949; Pennak, 1966). The evidence for the use of macrophytes as a refuge and for the role of fish in diel horizontal migration (DHM) by zooplankton is sparse (Davies, 1985; Vuille, 1991).

In this chapter, we describe diel horizontal distribution of *Daphnia* between structured littoral zones and open water in two lakes with fish and one without. In addition, we report results of a laboratory experiment testing the response of *Daphnia magna* to macrophytes in the presence and absence of fish and fish odor.

Study Areas and Methods

Two of the lakes (Lake Ring and Lake Væng) are situated in central Jutland, Denmark (see Lauridsen and Buenk, 1996). Both are shallow and eutrophic and

Table 13.1. Morphometric Data and Fish CPUE[c]

	L. Væng	L. Ring	C. Long L.[b]
Area, ha	15	22	2.1
Mean depth, m	1.2	2.9	2
Max depth, m	1.8	5	4
Retention time, days	15–21	450	Seepage lake
Secchi depth, m	1.4	2.9	3.6
Total P, mg P/L	0.07	0.35	0.02
CPUE[c]	41	10	No fish

[a]Secchi depth and total-P concentrations are average summer values (1 May to 30 September).
[b]Data are courtesy of S.R. Carpenter.
[c]Catch per unit effort, using multiple mesh size gill nets from Lake Væng and Lake Ring.

contain fish (Table 13.1). The third lake, Central Long Lake, is a shallow, meso-trophic, fishless lake at the University of Notre Dame Environmental Research Center, Gogebic County (Michigan). Morphometric and other characteristics are given in Table 13.1.

In Lake Væng and Central Long Lake, zooplankton were sampled once every day (1 PM–3PM) and night (1 AM–3AM) at five stations in a 72-hour period. At each station, sampling was undertaken in a structured environment (0.5–1 m from the shore) and in open water (5–10 m apart from the shore) without structure. Structure consisted of tree roots (primarily *Alnus glutinosa* L.) and emergent macrophytes (*Phragmites australis* L. and *Carex rostrata* L.) in Lake Væng and shrub roots (primarily *Chamaedaphne*) and water lilies (*Nuphar*) in Central Long Lake. Using a core sampler, we took depth-integrated samples from the surface to 3–5 cm above the bottom. We sampled with a core sampler. Differences in day-night densities for each 24-hour period were recorded by means of the Mann-Whitney u-test. For further details, see Lauridsen and Buenk (1996).

In Lake Ring, zooplankton were sampled at 24 stations along a 45-m-long transect, running 15 m from and parallel to the shore. This transect passed through three macrophyte beds and four macrophyte-free areas. Fifteen of the stations were located in the macrophyte beds or at the edges, and the remaining nine stations were situated in the macrophyte-free zones. Sampling was undertaken once every day and night as above during a 60-hour period. In the macrophytes and the open zones, we tested whether day and night densities were different by using the Mann-Whitney u-test. For further details, see Lauridsen and Buenk (1996).

To test the behavioral response of *Daphnia magna* Straus to combinations of the presence of macrophytes, green sunfish (*Lepomis cyanellus*), and sunfish odor, we did experiments in half-filled 38-L tanks. The central oval zone of the tanks was without macrophytes in all treatments. *Daphnia,* fish, or fish odor could be added to this open zone. If fish were added, they were kept in a cage that *Daphnia* were too large to enter. In treatments with macrophytes, the peripheral part of each tank was filled with *Myriophyllum excalbescens* L. In all treatments, the number of daphnids in the central open zone was counted. We tested for differences between treatments using a Tukey's test. For further details, see Lauridsen and Lodge (1996).

Results and Discussion

In the structured littoral in Lake Væng, two of three 24-hour periods showed significantly higher *Daphnia hyalina/galeata* densities during day than night. In the open water, we found the reverse pattern, with significantly higher densities at night (Fig. 13.1). At all times, *Daphnia* densities in the structure were higher than in open water. These data suggest *Daphnia hyalina/galeata* night-time migration toward open water from a narrow structure-filled nearshore zone. Considered as a whole, the narrow structure-filled zone constitutes about 0.5% of the total lake volume. Using the maximum day-to-night differences in density, migration from this zone may result in an increase of *Daphnia* density in open water of maximum 4 individuals/L. Migration from open water from submerged macrophyte beds is, however, undertaken (Lauridsen et al., 1996), implying that limited macrophyte coverage combined with the structured littoral zone may result in a significant increase in *Daphnia* density in open water.

Results were similar in the macrophyte beds of Lake Ring: *Daphnia magna* densities were significantly higher during day than night (Fig. 13.2A). In open water, however, we did not find a reverse pattern: *Daphnia* density was lowest at night (significantly in one of three 24-hour periods) (Fig. 13.2B). By testing *Daphnia* densities in the macrophytes versus in open water, we found significantly higher densities in open water at night. Sampling in the macrophyte environment

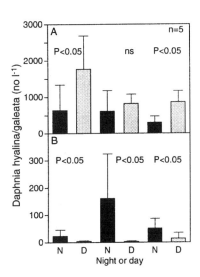

Figure 13.1. Night- and daytime mean densities (*n* = 5) of *Daphnia hyalina/galeata* in Lake Væng. (A) in the structure (tree roots and emergent macrophytes) and (B) in open water (5 m apart from the structure). *P* values and 95% confidence limits given; ns, no significance. (Redrawn from Lauridsen and Buenk, 1996.)

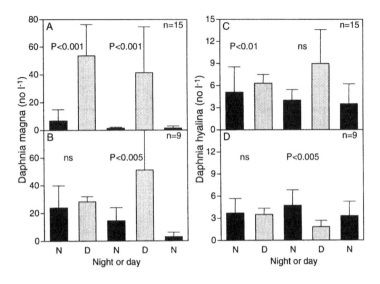

Figure 13.2. Night- and daytime mean densities of *Daphnia magna* (A and B) and
Daphnia hyalina (C and D) in Lake Ring. A and C: in the structure (submerged macrophytes), $n = 15$. B and D: in open water, $n = 9$. *P* values and 95% confidence limits given;
ns, no significance. (Redrawn from Lauridsen and Buenk, 1996.)

about 3–5 m from the shore revealed similar results, with higher densities during
day than night (Lauridsen and Buenk, 1996), demonstrating that daphnids did not
migrate toward the shore at night. It seems likely that they migrated further out
into open water, although we do not have the data to evaluate this possibility. *D.
hyalina/galeata* data from Lake Ring showed the same tendency as for *D. magna*,
with highest density in the macrophytes at daytime (Fig. 13.2C), although it was
only significant in one of the two 24-hour periods.

The large change in density was particularly found around and at the edges of
the macrophyte beds. From day to night the edge mean density of *D. magna* in
Lake Ring was reduced from 64 individuals/L to 3 individuals/L, whereas in the
middle of the beds and open zones, there were no significant changes (Table 13.2).
The results demonstrate the importance of the edge as a refuge for the migrating
Daphnia species.

The weaker day–night differences for *D. hyalina* in Lake Ring relative to those
of *D. magna* are consistent with observations of Walls et al. (1990) that large
cladocerans use DHM to a greater extent. We do not have conclusive evidence for
a fish-induced DHM in the field, but we attribute the DHM of the daphnids to the
presence of planktivorous fish as there was no evidence of oxygen depletion
(Lauridsen and Buenk, 1996), which also may induce migration (Frodge et al.,
1990). Fish-induced DHM may also explain the differentiated migration of species
of *Daphnia* as planktivorous fish select for large-sized prey (Eggers, 1982; Dodson, 1988).

Table 13.2. Daytime and Nighttime Mean Density of *Daphnia magna* at Various Stations Inside (mac) and Outside (open) the Macrophyte Edge in Lake Ring[a]

Distance from edge (m)		Day			Night		
	n	Mean (*n/L*)	95% cl	*n*	Mean (*n/L*)	95% cl	*P*
4 (mac)	6	15.2	15.9	6	11.8	20.4	ns
1 (mac)	12	47.3	26.8	12	1.5	0.7	0.039
Edge	12	63.8	39.8	12	3.0	1.9	0.006
1 (open)	12	45.6	24.2	12	19.7	12.3	0.003
3 (open)	6	28.2	8.74	6	18.5	14.4	ns

[a]ANOVA was used to test for a time effect. As no time (block) effect was found (block$_{day}$ MS = 574, F1,46 = 2.08, *P* = ns; block$_{night}$ MS = 11.02, F1,46 = 0.004, *P* = ns), data are mean values for two nights and two days, respectively. 95% confidence limits and *P* values for day–night differences are given (Student *t*-test).

In previous studies, macrophytes were repellent to zooplankton (Hasler and Jones, 1949; Pennak, 1966). Consequently, zooplankton have to choose between fish and macrophytes. In the experiments testing behavioral response of *D. magna* to the presence of macrophytes, fish, or fish odor, 15% of the *Daphnia* were found in the central zone in control tanks without macrophytes. With macrophytes present, 80% were located in the central unvegetated zone, and with fish or fish odor present, 35% and 45%, respectively, of the daphnids were located in the unvegetated zone (Fig. 13.3). Only the fish and the fish odor treatments were not significantly different. These results demonstrate that avoidance of fish can increase occupancy of *Myriophyllum excalbescens* by *D. magna* and may explain why *Daphnia* apparently use macrophyte habitats in Lake Væng and Lake Ring as a refuge despite their repellent impact. The results also suggest that the response is predominantly chemically mediated, which is consistent with De Meester (1993).

If DHM is fish-induced, we would expect that *Daphnia* would avoid macrophyte night and day in fishless lakes, as they avoided macrophytes in our laboratory experiments. Consistent with this expectation, densities of *Daphnia* in the fishless Central Long Lake were higher in open water than in structure in four of

Figure 13.3. Percentage of *Daphnia magna* in the central open water zone in four different treatments: 1, without macrophytes; 2, with macrophytes; 3, with fish and macrophytes; and 4, with fish odor and macrophytes. Tukey's test results are indicated by ns between treatments that do not differ significantly. (Redrawn from Lauridsen and Lodge, 1996.)

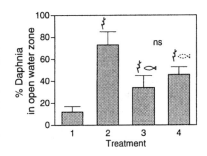

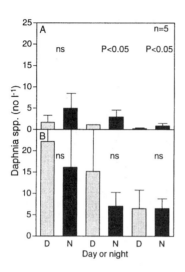

Figure 13.4. Night- and daytime mean densities ($n = 5$) of *Daphnia* spp. in Central Long Lake (A) in the structure (tree roots and emergent macrophytes) and (B) in open water (5–10 m apart from the structure). *P* values and 95% confidence limits given; ns, no significance.

the six samplings. However, at the nearshore stations, we found significantly higher densities of *Daphnia* spp. at night in two of the three 24-hour periods (Fig. 13.4). This pattern suggests that DHM in Central Long Lake was the reverse of what was found in Lake Væng and Lake Ring. The pattern also suggests that *Daphnia* in Central Long Lake either were avoiding something in the open water during night or something in the structured zone during day. We do not know what this might be, but invertebrate predators (e.g., *Chaoborus*) might be a possibility. More data on the diel horizontal patterns of *Daphnia* and of invertebrate predation activity are needed to evaluate that hypothesis. Whatever the exact explanation is, the opposite pattern of horizontal migration in Central Long Lake (relative to the two lakes with fish) is consistent with the hypothesis that *Daphnia* use littoral structure as a refuge from fish predation. However, more shallow lakes with and without fish need to be sampled to assess the generality of the patterns that we have discussed in this paper.

Acknowledgments. We thank Brian Moss and a second anonymous reviewer for valuable comments and suggestions to an earlier version and Kathe Møgelvang and Juana Jacobsen for drawings. The study was financed in part by the Centre for Freshwater Environmental Research and the Danish Natural Science Research Council (grant 9501315).

References

Carpenter, S.R.; Lodge, D.M. Effects of submersed macrophytes on ecosystem processes. Aquat. Bot. 26:341–370; 1986.

Crowder, L.B.; Cooper, W.E. Habitat structural complexity and the interaction between bluegills and their prey. Ecology 63:1802–1813; 1982.

Davies, J. Evidence for a diurnal horizontal migration in *Daphnia hyalina lacustris* Sars. Hydrobiologia 120:103–105; 1985.

De Meester, L. Genotype, fish-mediated chemicals, and planktonic behaviour in *Daphnia magna*. Ecology 74:1467–1474; 1993.

Dodson, S. The ecological role of chemical stimuli for the zooplankton: predator avoidance behavior in *Daphnia*. Limnol. Oceanogr. 33:1431–1439; 1988.

Eggers, D.M. Planktivore preference by prey size. Ecology 63:381–390; 1982.

Frodge, J.D.; Thomas, G.L.; Pauley, G.B. Effects of canopy floating and submergent aquatic macrophytes on the water quality of two shallow Pacific Northwest lakes. Aquat. Bot. 38:231–248; 1990.

Hasler, A.D.; Jones, E. Demonstration of the antagonistic action of large aquatic plants on algae and rotifers. Ecology 30:359–364; 1949.

Lauridsen, T.L.; Buenk, I. Diel changes in the horizontal distribution of zooplankton in two shallow eutrophic lakes. Arch. Hydrobiol. 137:161–176; 1996.

Lauridsen, T.L.; Lodge, D.M. Avoidance of *Daphnia magna* by fish and macrophytes: chemical cues and predator-mediated use of macrophyte habitat. Limnol. Oceanogr. 41:794–798; 1996.

Lauridsen, T.L.; Pedersen, L.J.; Jeppesen, E.; Søndergaard, M. The importance of macrophyte bed size for cladoceran composition and horizontal migration in a shallow lake. J. Plankton Res. 18:2283–2294; 1996.

Pennak, R.W. Stucture of zooplankton populations in the littoral macrophyte zone of some Colorado lakes. Trans. Am. Microsc. Soc. 85:329–349; 1966.

Savino, J.F.; Stein, R.A. Predator–prey interaction between largemouth bass and bluegills as influenced by simulated, submersed vegetation. Trans. Am. Fish. Soc. 111:255–266; 1982.

Scheffer, M.; Hosper, S.H.; Meijer, M.-L.; Moss, B.; Jeppesen, E. Alternative equilibria in shallow lakes. Trends Ecol. Evol. 8:275–279; 1993.

Timms, R.M.; Moss, B. Prevention of growth of potentially dense phytoplankton populations by zooplankton grazing, in the presence of zooplanktivorous fish, in a shallow wetland ecosystem. Limnol. Oceanogr. 29:472–486; 1984.

Vuille, Th. Abundance, standing crop and production of microcrustacean populations (*Cladocera, Copepoda*) in the littoral zone of Lake Biel, Switzerland. Arch. Hydrobiol. 123:165–185; 1991.

Walls, M.; Rajasilta, M.; Sarvala, J.; Salo, J. Diel changes in horizontal microdistribution of littoral cladocera. Limnologica 20:253–258; 1990.

14. Changing Perspectives on Food Web Interactions in Lake Littoral Zones

Larry B. Crowder, Elizabeth W. McCollum, and
Thomas H. Martin

Introduction

New information has modified how we view food web interactions in lake littoral habitats as well as the linkages from the littoral to pelagic habitats. First, productivity of epiphytes in lakes is extremely high, often exceeding production by submerged macrophytes (Wetzel, 1990; Wetzel and Søndergaard, this volume, Chapter 7). Second, a portion of this epiphyte productivity is harvested by littoral herbivores, including snails and grazing insects (Brönmark and Vermaat, this volume, Chapter 3; Jones et al., this volume, Chapter 4) and ultimately transported by fish out of the littoral habitat to the pelagic food web (Schindler et al., 1996; Vanni, 1996). Finally, food web cascades documented in the pelagia of lakes (Carpenter and Kitchell, 1993) may also occur commonly within the littoral habitat. Food web interactions within the littoral habitat, the importance of linkages among littoral habitat patches, and links from the littoral to the pelagic are still poorly understood, but progress in the past 15 years has been rapid.

Species interactions (e.g., competition, predation) cannot be fully evaluated outside a food web context because their outcomes can be modified by other members of the web as well as variable environmental factors (Crowder et al., 1988; Winemiller and Polis, 1996). Over the past 30 years, aquatic ecologists repeatedly have examined predation and competition by using experimental manipulations, but the results often included unanticipated effects (Sih et al., 1985; Kerfoot and Sih, 1987). Most of these surprises relate to having not considered all the important direct and indirect effects in the food web. A food web is "a network of consumer–resource interactions among a group of organisms, populations or aggregate trophic units" (Winemiller and Polis, 1996). Interactions in food webs are often complex, resulting from multiple pathways linking organisms and abiotic resources. These interactions tend to be complex because they involve indirect effects and time lags as well as spatial and temporal variability in per capita interaction strength (Carpenter, 1988; Paine, 1992). Identifying and measuring the strength of interactions via controlled experiments is a difficult but necessary approach to understanding food web dynamics.

The interaction of predation and resource limitation as factors determining the structure and function of aquatic communities is not well understood. Traditionally, ecologists have considered the relative importance of resource-based (bottom-up) and predator-based (top-down) forces in the structure and function of food webs (Hunter and Price, 1992; Power, 1992; Strong, 1992). But the interaction of these processes is more interesting and informative than some reasonable simplistic assessment of their relative importance (Carpenter et al., 1987; Vanni, 1987, 1996). One of the first efforts to explain complex interactions between predation and resource limitation in food webs is the "trophic cascade" hypothesis (Paine, 1980; Carpenter et al., 1985, 1987; Carpenter and Kitchell, 1993). Although the trophic cascade has been demonstrated for relatively simple food chains in the pelagia of lakes, the predictions sometimes fail in more complicated webs (Strong, 1992). Many factors can contribute to this failure, including omnivory, temporal, and spatial variability; physical refuges; size structure; ontogenetic shifts among predators and prey; and chemical defenses of plants to herbivory. Many of these factors could be important in the submerged macrophyte habitat of lakes. Do trophic cascades occur in the littoral habitat of lakes?

Macrophytes Provide Habitat Structure

Early on, we viewed macrophytes as adding physical complexity to lake littoral zones and coastal marine systems (Crowder and Cooper, 1982; Heck and Crowder, 1991). Not only are macrophytes common but they are also a very productive habitat for invertebrates; animal abundances in vegetated habitats are frequently several orders of magnitude higher than in nearby unvegetated areas. The mechanisms most often proposed to explain the extraordinary richness of the fauna on submerged macrophytes are (1) food supplies are greater in vegetated areas than elsewhere (as is borne out by recent information of epiphyte productivity) and (2) survival rates of potential prey items are greater in vegetation than elsewhere due to the refuge provided by dense vegetation. Predator efficiency declines with increased structural complexity (Heck and Crowder, 1991), but prey density and diversity tend to increase with increased structure—this leads to the hypothesis that benthically feeding fishes might eat more and grow better at intermediate vegetation densities (Crowder and Cooper, 1982). Macrophyte density mediates predator–prey interactions between bluegill sunfish (*Lepomis macrochirus*) and their invertebrate prey (Crowder and Cooper, 1982). Sparse macrophytes harbored few prey, and dense macrophytes reduced encounter rates and pursuit and capture probabilities. Intermediate densities of macrophytes provided the highest feeding and growth rates to these littoral fishes (Crowder and Cooper, 1982). But our experiment also provided several surprising results; due to decreases in the abundance of invertebrate predators (e.g., odonates) in the presence of fish, some smaller invertebrate predators increased. Many groups that are typically prey of bluegills declined, but others did not. It was clear even then that indirect effects and behavioral responses of prey complicated the interpretation of our results.

Food Web Interactions in the Macrophyte Habitat

Food web interactions in vegetated littoral zones are complex. Fish can suppress the abundance of large invertebrate predators or grazers potentially leading to indirect (and sometimes cascading) effects (Heck and Crowder, 1991). Until recently, few people have focused on the whole submerged macrophyte food web. Carpenter and Lodge (1986) proposed some hypotheses on the effects of submerged macrophytes on ecosystem processes in freshwater. Heck and Crowder (1991) extended and modified these hypotheses to deal directly with the differences between webs dominated by mesograzers and those dominated by macrograzers. A typical submerged macrophyte ecosystem is sketched in Figure 14.1. Obviously, particular systems will diverge from this figure, and many species shift from one role to another through ontogeny. Large predators such as fish, birds, and some mammals consume small fish and invertebrate predators that eat small herbivores (mesograzers). Large predators may also consume large herbivores (macrograzers). In many systems, the herbivores include both mesograzers (primarily oriented toward epiphytic algae) and macrograzers (primarily oriented toward macrophytes). Epiphytes live on the surface of macrophytes and may derive some nutrients directly from them. But epiphytes and macrophytes can also compete for nutrients (particularly those in the water column) or for light.

In a system containing only mesograzers, the hypothetical effect of decreased abundance of large predators (e.g., due to harvesting) would be that small predators would increase, leading to reductions in mesograzers and increases in epiphyte biomass (Figure 14.2). If the epiphyte load becomes too high, light limitation could lead to losses of submerged vegetation. This parallels the pelagic trophic cascade (Carpenter et al., 1985) and is an alternative explanation to nutrient enrichment for increase in epiphyte biomass and declines in submerged macrophytes. This system may tend to stabilize because as macrophytes are reduced, small predators associated with the vegetation are more exposed to predation by large predators. This should allow some recovery of the mesograzers and some reductions in epiphyte biomass, benefiting the macrophytes (Heck and Crowder, 1991).

In systems dominated by macrograzers, the hypothetical effects of reductions in large predators would be increases in macrograzers, which could lead directly

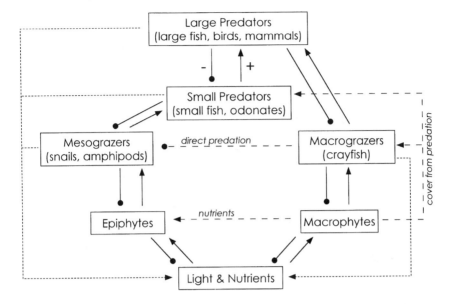

Figure 14.1. Generalized food web for submerged macrophyte systems. Boxes contain the major types of interactors. Lines ending in arrows indicate positive effects to the upper trophic level of energy flow up the web. Lines ending in closed circles indicate negative effects on the lower trophic level due to the level above. These may include the potential for predators to "control" prey densities. Dashed lines indicate potential interactions such a leaking of nutrients from macrophytes to epiphytes or indirect effects of cover on predator–prey interactions. Dotted lines indicate possible routes of nutrient recycling from consumers.

to losses of macrophytes along with their associated epiphytes and fauna (Fig. 14.2). Reduced refuge might also lead to decreased survival of small fish, thus leading to further reductions in the abundance of large fish through poor recruitment. This likely explains why systems dominated by macrograzers that can grow large enough to be relatively immune from predation will have two alternative states: if predators are abundant, macrograzers will be low and macrophytes abundant. Alternatively, if predators are low, macrograzers are abundant and macrophytes are sparse (Carpenter and Lodge, 1986). Many macrograzers are omnivorous and may have other food web effects, particularly on the abundance of mesograzers (in systems for which both grazer types are common) (Lodge et al., 1994).

Another set of linkages involve consumer-mediated nutrient recycling (and transport). Both predators and grazers influence prey directly via consumption, but they also release nutrients that can affect the production and potentially the composition of the algal community (Fig. 14.1). Predators and grazers can also alter the ratios of nutrients available (Sterner, 1990; Sterner et al., 1996), benefiting some producers over others. The idea of consumer-mediated nutrient recycling has received some attention in the pelagic food web of lakes (Vanni, 1996) but basically has not been explored in vegetated systems.

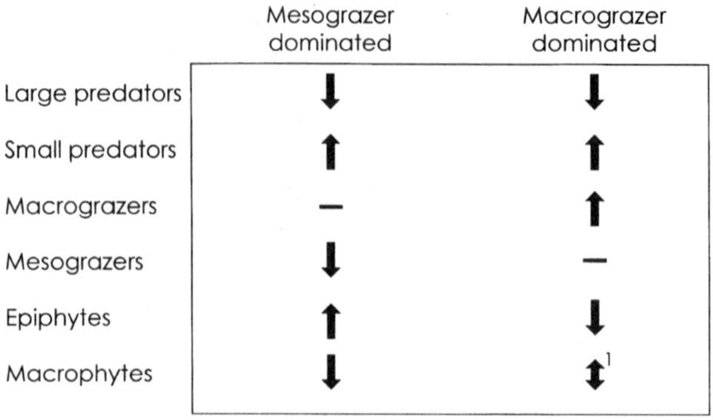

	Mesograzer dominated	Macrograzer dominated
Large predators	↓	↓
Small predators	↑	↑
Macrograzers	−	↑
Mesograzers	↓	−
Epiphytes	↑	↓
Macrophytes	↓	↕[1]

Figure 14.2. Theoretical food web responses to reductions in the abundance of large predators. The left column reflects the response in mesograzer-dominated webs and the right had column reflects macrograzer-dominated webs. [1]Response is dependent on the magnitude and sign of feedback from macrograzers.

Evidence for a Littoral Trophic Cascade

We conducted a 16-month field experiment to examine direct and indirect effects of fish on their littoral prey assemblage in Bay's Mountain Lake, Tennessee. The experiment was conducted in 24-m^2 mesh enclosures in which we manipulated the presence and absence of large redear sunfish (*Lepomis microlophus* >150 mm standard length [SL]) and small sunfish (*L. macrochirus* and *L. microlophus* <50 mm SL). We found that both large redear sunfish and small sunfish suppressed the recruitment of snails to experimental enclosures, but snails (primarily *Helisoma anceps, Physella heterostropha,* and *Gyraulus parvus*) increased significantly during the first 2 months of the experiment in the fish-free controls. Five months into the experiment, the difference in snail biomass between the enclosures with and without fish was 10-fold. By the end of the experiment, enclosures without fish had significantly lower periphyton percentage cover (similar trends were detected in biovolume) and significantly higher macrophyte biomass (Fig 14.3; Martin et al., 1992). Brönmark et al. (1992) also found that fish enhanced periphyton biomass by removing snail grazers.

Direct and Indirect Effects of Fish and Snails on Periphyton

Although Martin et al. (1992) interpreted this littoral cascade as due to indirect effects of fish on periphyton, mediated through snail grazers, periphyton clearly could be enhanced directly by nutrient release from fish. To separate these effects and determine their relative importance, we manipulated the presence and absence of redear sunfish and snails, *P. heterostropha,* in a replicated factorial tank experiment (McCollum et al., 1998). Experimental tanks (72 L) were placed outdoors in

Figure 14.3. Littoral trophic cascade in Bay's Mountain Lake, Tennessee. Fish (both large and small *Lepomis* sp.) apparently suppressed the recuitment of snails to experimental enclosures. Without fish, snail biomass increased, periphyton cover and biomass declined, and macrophyte biomass increased.

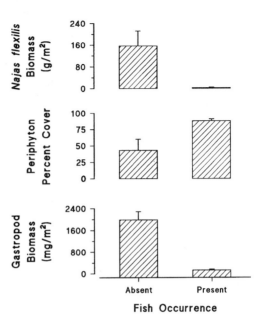

two water baths, which acted as thermal buffers. Each aquarium was divided by a plastic tank divider (perforated with 1-mm holes on 5-mm centers) so that one side of the aquarium held about one-third of the aquarium volume and the other side held the remaining two-thirds (Fig. 14.4).

Artificial plants were used as substrata for the periphyton (Fig. 14.4). The plants were made from bamboo skewers (the main stem) and black polypropylene

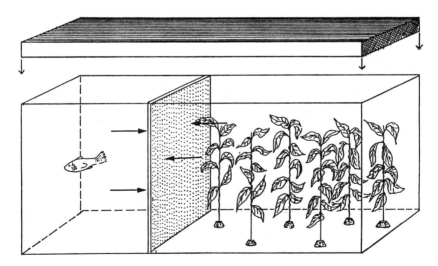

Figure 14.4. Aquarium set up used in experiment to separate the effects of fish feeding and nutrient recycling on periphyton.

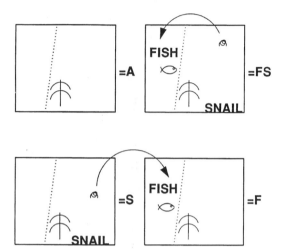

Figure 14.5. Schematic of experimental manipulations in McCollum et al. (1988). The four treatments in the factorial were algae alone (A) (i.e., no animals), snails only (S) (in the large side of the aquarium, were removed at a specified "predation rate" and fed to fish in "fish only" treatments), fish only (F) (in the small side of the aquarium), and fish + snails (FS) (snails removed from large side of the aquarium and fed to fish on the small side of the same aquarium).

ribbon cut into leaves 7.6 cm long (70 cm²); each "plant" supported 12 leaves. The ribbons were gently abraded with emery board to create a textured surface, allowing for a more diverse periphyton community than would have colonized smooth ribbon (Muntenau and Malay, 1981; Pringle, 1990). The artificial plants were constructed to mimic submerged *Polygonum densiflorum,* a common macrophyte in North Carolina littoral zones.

We manipulated the presence of redear sunfish and snails in a 2 × 2 factorial experiment (Fig. 14.5); all treatments contained a periphtyon community. The four treatments in the factorial were algae alone (A) (i.e., no animals), snails only (S) (in the large side of the aquarium, were removed at a specified "predation rate" and fed to fish in "fish only" treatments), fish only (F) (in the small side of the aquarium), and fish + snails (FS) (snails removed from large side of the aquarium and fed to fish on the small side of the same aquarium). The experiment was blocked by water bath; treatments were replicated twice within each block, resulting in four replicates per treatment. An algal inoculum was created by collecting macrophytes from local lakes and rinsing the associated periphyton into deionized water. This slurry was homogenized gently in a loosely fitting tissue grinder to create a homogeneous mixture immediately before addition to the tanks. Periphyton were allowed to colonize and grow for 2 weeks before the initiation of treatments.

In treatments with fish, one redear sunfish was placed in the smaller section of the aquarium; in treatments with snails, 45 snails (360 snails/m²) were added to the larger section. Thus, when fish and snails were together, they were separated by a plastic divider. To simulate fish predation, we removed three snails from each snail tank at 3-day intervals and fed them to the fish in the fish treatments (36 snails removed over the course of the experiment).

During the experiment, we determined nutrient concentrations and periphyton cell number, species composition, cell size, and biovolume every 10 days. Snail behavior (position in the water column) was recorded on five dates, and snail number and sizes were recorded at the end of the 5-week experiment.

Complete results of this experiment are reported elsewhere (McCollum et al., 1988). Briefly, however, concentrations of phosphorus and nitrogen in the water were significantly higher with fish, but this had little direct effect on total periphyton cell number or biovolume. Snails decreased periphyton cell number; they also increased average cell size of periphyton and of green algae in the absence of fish. Snails decreased diatom and blue-green biovolume. Fish also decreased diatom biovolume by decreasing the average cell size of diatoms. Snails increased the percentage of the algal community comprised of gelatinous colonies. Fish suppressed snail grazing independent of snail mortality and also inhibited snail reproduction. Our previous research (Martin et al., 1992) suggested that fish have a positive indirect effect on algae by removing grazers (an interaction chain) (Wooton, 1993). Fish apparently enhance this effect by inhibiting snail reproduction and by suppressing grazing as well (interaction modifications) (Wooton, 1993). Fish can also have an important but less obvious direct effect on algae, via nutrient recycling, possibly by altering competitive outcomes among taxa and growth forms (McCollum et al., 1998).

Concluding Remarks

It is clear that both pelagic and littoral habitats can support trophic cascades, but it is also clear that the interactions in these systems are complex and that interaction chains can be modified in a wide variety of ways. Ultimately, however, we will have to make the additional step of linking the littoral and pelagic food webs in lakes (Lodge et al., 1988). Animal movements (e.g., onshore, offshore migrations of fish, vertical migrations of zooplankters) link the littoral and pelagic habitats both in terms of prey consumption and translocation of nutrients (Carpenter et al., 1992; Schindler et al., 1996; Vanni, 1996). Epiphyte productivity is high and supports substantial secondary productivity, leading to a productive littoral food web. Consumers that feed in the littoral habitat can transform and translocate nutrients, allowing substantial linkages from the littoral to the pelagic.

Acknowledgments. Financial support was provided by NSF BSR-8709108 (L.B.C.) and an NSF graduate fellowship (E.W.M.). Christer Brönmark and Steve Carpenter provided insightful comments on a prevous draft of the manuscript.

References

Brönmark, C.; Klowsiewski, S.P.; Stein, R.A. Indirect effects of predation on a freshwater benthic food chain. Ecology 73:1662–1674; 1992.

Carpenter, S.R., ed. Complex interactions in lake communities. New York: Springer-Verlag; 1988.

Carpenter, S.R.; Kitchell, J.F. The trophic cascade in lakes. Cambridge: Cambridge University Press, 1993.

Carpenter, S.R.; Lodge, D.M. Effect of submersed macrophytes on ecosystem processes. Aquat. Bot. 26:341–370; 1986.

Carpenter, S.R.; Kitchell, J.F.; Hodgson, J.T. Cascading trophic interactions and lake productivity. BioScience 35:634–639; 1985.

Carpenter, S.R.; Kitchell, J.F.; Hodgson, J.T.; Cochran, P.A.; Elser, J.J.; Elser, M.M.; Lodge, D.M.; Kretchmer, D.; He, X.; von Ende, C.N. Regulation of lake primary productivity by food web structure. Ecology 68:1863–1876; 1987.

Carpenter, S.R.; Kraft, C.E.; Wright, R.A.; He, X.; Soranno, P.A.; Hodgson, J.R. Resilience and resistance of a lake phosphorus cycle before and after food web manipulation. Am. Nat. 140:781–798; 1992.

Crowder, L.B.; Cooper, W.E. Habitat structural complexity and the interaction between bluegills and their prey. Ecology 63:1802–1813; 1982.

Crowder, L.B.; Drenner, R.W.; Kerfoot, W.C.; McQueen, D.J.; Mills, E.L.; Sommer, U.; Spencer, C.N.; Vanni, M.J. Food web interactions in lakes. In: Carpenter, S.R., ed. Complex interactions in lake communities. New York: Springer-Verlag, 1988:141–160.

Heck, K.L., Jr.; Crowder, L.B. Habitat structural complexity and predator–prey interactions in vegetated aquatic systems. In: Bell, S.S.; McCoy, E.D.; Mushinsky, H.R., eds. Habitat structure: the physical arrangement of objects in space. London: Chapman & Hall, 1991:281–299.

Hunter, M.D.; Price, P.W. Playing chutes and ladders: heterogeneity and the relative roles of bottom-up and top-down forces in natural communities. Ecology 73:724–732; 1992.

Kerfoot, C.; Sih, A., eds. Predation: direct and indirect impacts on aquatic communities. Hanover, NH: University Press of New England; 1987.

Lodge, D.M.; Barko, J.W.; Strayer, D.; Melack, J.M.; Mittelbach, G.G.; Howarth, R.W.; Menge, B.; Titus, J.E. Spatial heterogeneity and habitat interactions in lake communities. In: Carpenter, S.R., ed. Complex interaction in lake communities. New York: Springer-Verlag; 1988:181–208.

Lodge, D.M.; Kershner, M.W.; Aloi, J.E.; Covich, A.P. Effects of an omnivorous crayfish (*Orconectes rusticus*) on a freshwater littoral food web. Ecology 75:1265–1281; 1994.

Martin, T.H.; Crowder, L.B.; Dumas, C.F.; Burkholder, J.M. Indirect effects of fish on macrophytes in Bays Mountain Lake: evidence for a littoral trophic cascade. Oecologia (Berlin) 89:476–481; 1992.

McCollum, E.W.; Crowder, L.B.; McCollum, S.A. Complex interactions of fish, snails and littoral zone periphyton. Ecology 79; 1998.

Muntenau, N.; Malay, E.J. The effect of current on the distribution of diatoms settling on submerged glass slides. Hydrobiologia 78:278–282; 1981.

Paine, R.T. Food webs: linkage, interaction strength, and community infrastructure. J. Anim. Ecol. 49:667–685; 1980.

Paine, R.T. Food web analysis through field measurements of per capita interaction strength. Nature 355:73–75; 1992.

Power, M.E. Top-down and bottom-up forces in food webs: do plants have primacy? Ecology 73:733–746; 1992.

Pringle, C. Nutrient spatial heterogeneity: effects on community structure, physiognomy, and diversity of stream algae. Ecology 71:905–920; 1990.

Schindler, D.E.; Carpenter, S.R.; Cottingham, K.L.; He, X.; Hodgson, J.R.; Kitchell, J.F.; Soranno, P.A. Food web structure and littoral zone coupling to pelagic trophic cascades.

In: Polis, G.A.; Winemiller, K.O., eds. Food webs: integration of pattern and dynamics. New York: Chapman & Hall; 1996:96–108.

Sih, A.; Crowley, P.; McPeek, M.; Petranka, J.; Strohmeier, K. Predation, competition, and prey communities: a review of field experiments. Annu. Rev. Ecol. Syst. 16:269–305; 1985.

Sterner, R.W. The ratio of nitrogen to phosphorus resupplied by herbivores: zooplankton and the algal competitive arena. Am. Nat. 136:209–229; 1990.

Sterner, R.W.; Elser, J.J.; Chrzanowski, T.H.; Schampel, J.H.; George, N.B. Biogeochemistry and trophic ecology: a new food web diagram. In: Polis, G.A.; Winemiller, K.O., eds. Food webs: integration of pattern and dynamics. New York: Chapman & Hall; 1996:72–80.

Strong, D.R. Are trophic cascades all wet? Differentiation and donor control in speciose ecosystems. Ecology 73:747–754; 1992.

Vanni, M.J. Effects of nutrients and zooplankton size on the structure of a phytoplankton community. Ecology 68:624–635; 1987.

Vanni, M.J. Nutrient transport and recycling by consumers in lake food webs: implications for algal communities. In: Polis, G.A.; Winemiller, K.O., eds. Food webs: integration of pattern and dynamics. New York: Chapman & Hall; 1996:81–95.

Wetzel, R.G. Land–water interfaces: metabolic and limnological indicators. Verh. Int. Verein. Theor. Limnol. 24:6–24; 1990.

Winemiller, K.O.; Polis, G.A. Food webs: what do they tell us about the world? In: Polis, G.A.; Winemiller, K.O., eds. Food webs: integration of pattern and dynamics. New York: Chapman & Hall; 1996:1–22.

Wooton, J.T. Indirect effects and habitat use in an intertidal community: interaction chains and interaction modifications. Am. Nat. 141:71–89; 1993.

15. Bacterioplankton and Carbon Turnover in a Dense Macrophyte Canopy

Morten Søndergaard, Jon Theil-Nielsen, Kirsten Christoffersen,
Louise Schlüter, Erik Jeppesen, and Martin Søndergaard

Introduction

Studies on cascading trophic interactions in lakes have shown that planktonic food web changes may take place to the level of protozoans (reviewed by Carpenter and Kitchell, 1993; Riemann and Christoffersen, 1993). It is more unclear if and how cascading might influence bacterioplankton (Jeppesen et al., 1992; Christoffersen et al., 1993; Pace, 1993). From studies in oligo-mesotrophic temperate lakes, Pace (1993) concluded "that bacteria responded to changes in phytoplankton and increases in nutrients, but not to changes in zooplankton." More generally, it was suggested that "trophic cascades do not have immediately obvious consequences for microbial processes in lakes" (Kitchell and Carpenter, 1993). In accordance, Jeppesen et al. (1992) found that a trophic cascade with high grazing by cladocerans and a four- to sixfold reduction in phytoplankton biomass only slightly altered bacterioplankton production in two fish-manipulated shallow and eutrophic Danish lakes.

Submerged macrophytes aid food web changes by their function as refuge for zooplankton (Timms and Moss, 1984) and are a potential substrate source for bacteria. Jeppesen et al. (1992) suggested that a high biomass of submerged macrophytes could explain deviations at the microbial level from their tentative model of pelagic trophic interactions. Likewise, Pace (1993) recognized organic carbon from allochthonous and littoral sources as a possible explanation of unclear responses by the bacterioplankton to trophic cascades. Thus, one main reason for the apparent uncoupling of bacterioplankton production from the pelagic food web in some shallow lakes could be macrophytes.

There is a lack of quantitative information on the effects of macrophytes and littoral zone production on pelagic microbial communities in lakes. Coveney and Wetzel (1995) found that bacterioplankton production in oligotrophic Lake Lawrence could only be sustained and maintained by an input of organic substrate from littoral production. Furthermore, low planktonic P:R ratios (<1) in oligotrophic and mesotrophic lakes, but not in eutrophic lakes, suggest littoral and allochthonous organic carbon to be of quantitative importance to planktonic metabolism (del Giorgio and Peters, 1994). These results support the increasing recognition of nonphytoplanktonic organic sources in lakes and decoupling of direct metabolic links between phyto- and bacterioplankton (Findlay et al., 1992; Hessen, 1992; Kairesalo et al., 1992; Tranvik, 1992; Wetzel, 1992).

The purpose of the present study was to compare bacterioplankton production and planktonic microbial communities in enclosures with and without submerged macrophytes. The use of enclosures excluded an input of organic matter from other sources than the enclosed communities. Carbon flow scenarios with respect to bacterioplankton carbon demand and planktonic substrate sources were constructed.

Materials and Methods

Large circular enclosures (5-m diameter) were placed at a depth of 60 to 70 cm in the eutrophic and shallow Lake Stigsholm, Denmark. In three enclosure areas, the submerged macrophyte *Potamogeton pectinatus* reached a relative plant colonized volume of about 47% and almost 100% area cover. Three other enclosure areas were kept free from macrophytes. Two days before the experiment (from July 26 to August 3, 1994) the enclosure areas were closed with a curtain made of heavy plastic. All fish were removed by electrofishing and trapping. Planktivorous fish (0^+ perch) at natural lake densities ($4/m^2$) were added to the enclosures on August 1.

Whole water column samples were taken randomly at 25–30 points within each enclosure every second day at noon and on two occasions at midnight. All subsamples were pooled, and processing took place either immediately at the lake shore or within 1 hour. Vertical profiles were measured during one day/night cycle in two enclosures.

Chlorophyll *a* and primary production (the ^{14}C method) were measured according to Schlüter et al. (1997). Particulate organic carbon (POC) was measured in triplicate on GF/F filters (Søndergaard and Middelboe, 1993) and dissolved organic carbon (DOC) was measured in the filtrates with a Shimadzu TOC-5000 (Søndergaard et al., 1995).

Zooplankton abundance was measured with standard microscopy technique. Mesozooplankton clearance rates were measured in short-term uptake experiments by using radiotracer labeled natural assemblages of phytoplankton and bacteria (Jeppesen et al., unpublished data).

Samples for bacterial abundance were fixed with glutaraldehyde (1.5% final conc.), stained with DAPI (Porter and Feig, 1980), and counted by epifluorescence microscopy (at least 500 cells). Individual bacteria cells (at least 25 cells) were sized on enlarged micrographs, and the biovolume was calculated as rods with hemispheres. Two approaches were used to convert biovolume to biomass. In scenario I, we used a constant of 100 fg $C/\mu m^3$ (Fagerbakke et al., 1996; Theil-Nielsen and Søndergaard, unpublished data), and in scenario II a scaling factor according to size was included: $C = 120 \times V^{0.72}$, (C is carbon/cell and V is cell volume; Norland, 1993, based on Simon and Azam, 1989). Both of the chosen conversion factors are in current use (Carlson and Ducklow, 1996; Simek et al., 1996), although they cannot be considered global standards.

Bacterial cell production was measured with the thymidine method (Fuhrman and Azam, 1980). ^{3}H-thymidine was added to a final and saturating concentration of 20 nM, incubated for 30 minutes, and stopped with ice-cold TCA (5% final conc.). Cell production was calculated with the empirical factor 2×10^{18} cells/mol (Smits and Riemann, 1988).

Results and Discussion

For this 10-day experiment in midsummer, the concentration of chlorophyll a was about fivefold and phytoplankton primary production eightfold higher in -M than in +M (Fig. 15.1). The abundance of filter-feeding cladocerans and copepods was about 50- and 4-fold higher in +M than in -M (Fig. 15.2). Mesozooplankton clearance rate for phytoplankton was 3 L/L/day in +M and only about 0.03 L/L/day in –M (Jeppesen et al., unpublished data). As 90% and 60% of the chlorophyll were in particles less than 20 μm in both enclosure types, most of the phytoplankton were presumably available for filter-feeding grazers. The difference in chlorophyll a between +M and –M might be explained by zooplankton grazing, although effects from macrophyte shading and high sedimentation rates cannot be excluded.

The high mesozooplankton abundance in +M affected the abundance of the protozooplankton. The densities of ciliates and flagellates in –M were about 200 and 2,100 cells/ml, respectively, which is 70- and 5-fold higher than in +M (Fig. 15.2). Changes in zooplankton abundance and composition from July 25 to August 1 were negligible, but a decrease in mesozooplankton abundance was observed to take place after the introduction of planktivorous fish on August 1 (Jeppesen et al., unpublished data).

There was a three- to fourfold higher POC concentration in –M than in +M, whereas no measurable difference in DOC concentrations was present at the start of the experiment (Fig. 15.3). However, after about 6 days the concentration in +M increased, whereas DOC decreased in –M after a peak value on July 29 (Fig. 15.3). The DOC concentrations in +M and –M were significantly different except for the two first samples (P <.001, paired t-test).

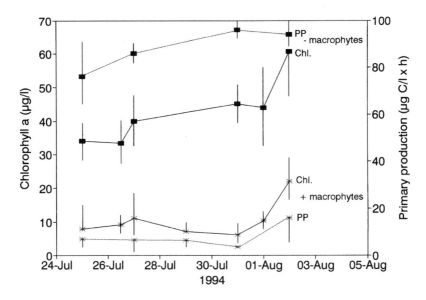

Figure 15.1. Chlorophyll *a* (Chl) and primary production (PP) in enclosures with and without submerged macrophytes. Means ± SD, *n* = 3 (three independent enclosures).

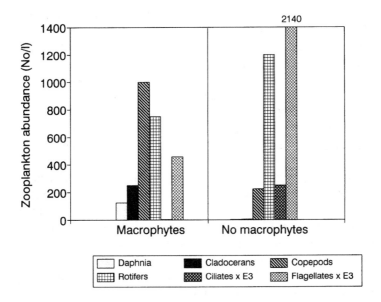

Figure 15.2. Average meso- and protozooplankton abundance in enclosures with and without submerged macrophytes. Averages for three enclosures and the 9-day experimental period. Data on ciliates are provided by Klaus Jürgens.

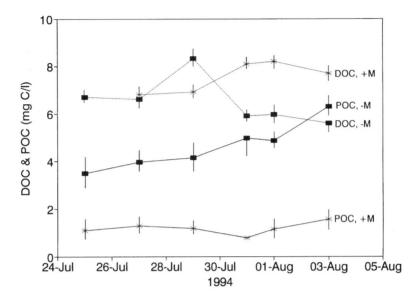

Figure 15.3. Particulate (POC) and dissolved (DOC) organic carbon in enclosures with and without submerged macrophytes. Means ± SD, $n = 9$ (triplicates in three enclosures).

The samples collected in a vertical profile at midday on July 31 and just before sunrise on August 1 showed a tendency toward higher DOC concentrations just above the sediment in both types of enclosures (Fig. 15.4). The profiles indicate a DOC net efflux from the sediments. Furthermore, the DOC concentrations at night were about 0.5–1 mg C/L higher than at midday.

The abundance of bacteria in –M was three- to fourfold higher than in +M (Fig. 15.5A). The bacteria were initially small in all enclosures with an average biovolume of 0.05–0.06 μm^3. After July 31, the average cell volume in +M increased to about 0.17 μm^3, whereas the cells in –M decreased to 0.039 μm^3 (but see Jürgens and Jeppesen, this volume, Chapter 16). The mesozooplankton clearance rate for bacterioplankton in +M was 1.5 L/L/day (Jeppesen et al., unpublished data). In –M, the mesozooplankton clearance rate for bacterioplankton was 100-fold lower. As bacterial abundance during the experimental period did not increase in +M and had a decreasing trend in –M, bacterial loss must at least have equaled bacterial growth rates. The high mesozooplankton clearance rate of bacterioplankton in +M offers a plausible explanation for the loss in +M. The grazing in –M was most probably dominated by flagellates, small ciliates, and rotifers. Other factors such as virus and sedimentation of particle-associated bacteria also contribute to the loss of bacterioplankton.

Despite the large difference in bacterial abundance between +M and –M, the bacterial cell production did not differ markedly—although significantly—except for July 27 (Fig. 15.5B). During the experimental period, the production decreased

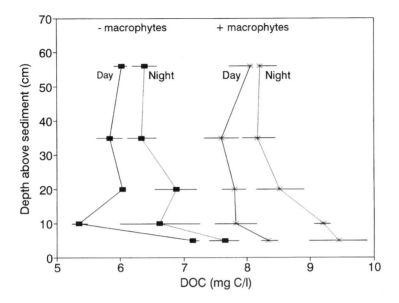

Figure 15.4. Vertical profiles of DOC in enclosures with and without submerged macrophytes. Samples taken on July 31 (day) and August 1 before sunrise (night). Means ± SD, *n* = 3. Absent bars indicate SD within markers.

from about 8 to 2×10^5 cells/ml/h in both enclosure types (Fig. 15.5B). Initially, the bacterial community in +M had a growth rate of 2.4/day, decreasing to about 0.5/day. In –M the growth rate was about threefold lower and varied between 0.8 and 0.2/day. The reason(s) for the decrease in bacterial production during the experiment is unknown, but it might be due to the exclusion of external substrate sources. The generally higher specific growth rate in +M is in accordance with the results presented in Jürgens and Jeppesen (this volume, Chapter 16), where it is shown that removal of zooplankton resulted in a substantial increase of bacteria. The high grazing resulted in a higher specific growth rate, as can be predicted from a logistic growth model, in which growth rates are high when the population biomass is below the carrying capacity and will deminish as the carrying capacity is approached (Wright, 1988). The cell production integrated for the entire experimental period was 91 ± 3 and 100 ± 14 × 10⁶ cells/ml for –M and +M, respectively (means ± SD, *n* = 3 enclosures of each type).

Bacterial Carbon Demand

The conversion of bacterial biovolume to carbon biomass is not trivial (see the Materials and Methods section) and has major consequences for the values used to assess the quantitative importance of bacteria. In –M, the bacterioplankton bio-

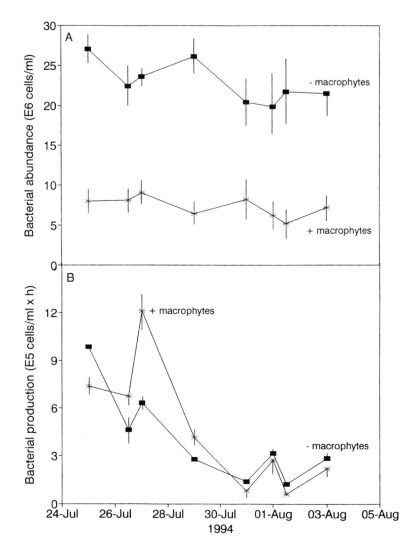

Figure 15.5. Bacterial abundance (A) and cell production (B) in enclosures with and without submerged macrophytes. Means ± SD, $n = 9$ (triplicates in three enclosures). Absent bars indicate SD within markers.

mass decreased from 160 to 80 µg C/L and from 400 to 270 µg C/L using 100 fg C/µm³ (scenario I) and $C = 120 \times V^{0.72}$ (scenario II), respectively. In +M, the biomass increased from 40 to 120 µg C/L in scenario I and from 140 to 210 µg C/L in scenario II. The use of a scaling factor with respect to cell size reduced the relative variations over time and the difference between –M and +M.

Bacterial carbon demand (B_C) was calculated according to

Table 15.1. Carbon Flow Scenarios in Enclosures in Lake Stigsholm[a]

Variables/substrate sources	Scenario I (100 fg C/μm^3)	Scenario II (C = 120 × $V^{0.72}$)
With macrophytes		
EOC	9	9
S_z	40	85
B_c	160	430
S_b	−110	−335
DOC_T	140	140
C_{s+p}	250	475
Without macrophytes		
EOC	80	80
S_z	22 (82)	50 (110)
B_c	94	270
S_b	8 (68)	−140 (−80)
DOC_T	−120	−120
C_s	0	20 (0)

[a]Average values of μg C/L/day calculated from values integrated over 9 days. EOC, extracellular organic release by phytoplankton; S_z, bacterial substrate from mesozooplankton grazing and substrate production by micro- and protozooplankton grazing on phytoplankton in brackets; B_c, bacterial carbon demand; S_b, surplus or deficit of planktonic substrate production for bacteria; DOC_T, changes in the concentration of DOC; C_{s+p}, carbon from sediment (s) and plants (p) to balance the demand. Assumptions used to calculate the two carbon flow scenarios:

1. Primary production (PP) in +M and −M: 100 and 900 μg C/L/day, respectively.
2. Algal respiration (R): 10% of PP.
3. EOC: 10% of (PP − R).
4. POC: 270 μg C/L/day and 0 in −M and +M, respectively.
5. DOC: 140 μg C/L/day in +M and −120 μg C/L/day in −M.
6. Zooplankton substrate recycling: 30% of ingestion (Hygum et al., 1997). The total community grazing was calculated from the production measurements compensated for changes in biomasses (e.g., the grazing on bacteria in −M was calculated as bacterial production plus the decrease in biomass and phytoplankton grazing in +M was subtracted the average increase in biomass). Phytoplankton grazing in −M was attributed to zooplankton <50 μm but allowing 50% for sedimentation. As grazing by microzooplankton was not measured, their potential carbon recycling is presented in brackets.
7. Bacterial biomass: 10 μg C/L/day in +M (both scenarios) and −10 or −18 μg C/L/day in −M, for 100 fg C/μm^3 and C = 120 × $V^{0.72}$, respectively.
8. Bacterial growth yield: 50% from bacterial regrowth experiments (Søndergaard and Theil-Nielsen, 1997).

$$B_C = Y^{-1} \cdot BB \cdot \mu$$

where Y = carbon growth yield, BB = carbon biomass, and μ = growth rate. The integrated production over the experimental period was 2.8-fold higher in scenario II than in I for both +M and −M. In both scenarios, the bacterial production was 1.7-fold higher in +M than in −M (Table 15.1).

Planktonic Carbon Flow Scenarios

The measurements of carbon pool changes, phytoplankton production, clearance rates, and planktonic bacterial production made it possible to construct a tentative budget for planktonic bacterial carbon demand and substrate sources. The values are calculated assuming carbon transfer from particles to bacterioplankton via phytoplankton extracellular release (EOC) and zooplankton grazing on both phyto- and bacterioplankton. Bacterial use of the ambient DOC pool is accounted for as the changes in DOC enter the calculations. The scenarios in Table 15.1 are based on the two "selected" calculations of bacterial biomass and a series of assumptions concerning the planktonic flow of carbon to bacteria (see Table 15.1 footnote).

The consequences of the two scenarios are quantitative differences in bacterial carbon demand, rates of grazer-recycled carbon, and EOC (Table 15.1). With macrophytes present, both scenarios resulted in a substrate "deficit" for the bacterioplankton of either 110 or 335 µg C/L/day (Table 15.1). To these deficits should be added 140 µg C/L/day caused by the increase in DOC. The sediment and the macrophyte-periphyton complex are the only sources to cover the deficit as other influxes can be considered negligible and the concentration of DOC increased.

Release of EOC and grazer-produced substrate balanced the bacterioplankton carbon demand in –M and scenario I. However, the decrease in DOC added 120 µg C/L/day, resulting in a substrate surplus (Table 15.1). In –M and scenario II, the pelagic deficit was 140 µg C/L/day, but with the decrease in DOC, only a minor substrate deficit of 20 µg C/L/day had to be accounted for by a net efflux from the sediment. Including grazing of phytoplankton by the micrograzers enhanced zooplankton organic recycling to a quantity in which the bacterioplankton carbon demand could be accounted for by the planktonic processes.

These theoretical scenarios are open to criticism. The conclusion that the bacterioplankton in +M had a much larger apparent deficit for their carbon demand than in –M is very robust as long as identical assumptions concerning carbon demand, EOC release, and organic recycling are used for both +M and –M. We have chosen low phytoplankton respiration and rather high EOC and recycling values (see Table 15.1 footnote), which all act to increase the planktonic carbon flux to bacteria and thus diminish the suggested influence from macrophytes and sediment.

Aside from bacterial biomass, the calculated bacterial carbon demand is influenced by the choice of the conversion factor for cell production per mole thymidine incorporated and the growth yield. We have no reason to expect the conversion factor to differ between –M and +M, and bacterial growth experiments with water from both +M and –M showed no difference in growth yield (Søndergaard and Theil-Nielsen, 1997).

The measured cell sizes are critical for the calculation of bacterial biomass and carbon production. If large cells have escaped our measurements in –M, the calculated differences between +M and –M will diminish and eventually disappear. Similar production values would be reached if the average cell volume for growing cells in –M was about threefold larger than the measured average. In a study on grazing-resistant bacteria in Lake Stigsholm, but using a different counting and sizing procedure than the present, Jürgens and Jeppesen (this volume, Chapter 16) reported lower abundances and generally larger cells than we have measured. Although the use of their biovolumes resulted in a diminished difference in bacterioplankton carbon demand between –M and +M, an organic input by the macrophyte-periphyton complex was still needed to establish a mass balance, and our general conclusions were not invalidated.

Conclusions

It is demonstrated that the presence of submerged macrophytes had a profound effect on the structure of the pelagic biota. The abundance of ciliates, flagellates, and bacterioplankton in +M was low compared with –M, which had a fivefold higher concentration of phytoplankton. The high mesozooplankton clearance rate in +M was one reason for the difference, although other causes should not be neglected. With respect to bacterioplankton production, a significantly ($P < .001$, paired t-test) higher growth rate in +M in six of eight measurements compensated for the lower cell abundance and resulted in similar cell production rates. Bacterioplankton production did not relate to phytoplankton biomass and production in these shallow lake enclosures. It should be emphasized that a high bacterioplankton productivity was achieved in both enclosure types, which had very different biological structures controlled by the absence or presence of a dense population of submerged macrophytes.

A high bacterioplankton production was measured in +M despite a low phytoplankton biomass, and their carbon demand could only be balanced assuming the macrophyte-periphyton complex as a substrate source. Total bacterial production in the enclosures does not only include bacterioplankton but also the bacterial production taking place on or in the sediment and in the biofilm on the macrophytes. The production in the sediments is unknown, but bacterial production on *Potamogeton pectinatus* is high (Theil-Nielsen and Søndergaard, unpublished data) and adds to the total bacterial carbon demand in +M. This reinforces the position of macrophytes as an active component in the DOC dynamics and metabolic activity of shallow lakes and their impact on the pelagic microbial food web.

Acknowledgments. The comments and suggestions by Bob Wetzel and Klaus Jürgens and the technical skills of Gitte Jacobsen and Nils Willumsen are appreciated. This study was supported by the Danish Environmental Research Programme.

References

Carlson, C.A.; Ducklow, H.W. Growth of bacterioplankton and consumption of dissolved organic carbon in the Sargasso Sea. Aquat. Microb. Ecol. 10:69–85; 1996.

Carpenter, S.R.; Kitchell, J.F., eds. The trophic cascade in lakes. Cambridge: Cambridge University Press; 1993.

Christoffersen, K.; Riemann, B.; Klysner, A.; Søndergaard, M. Potential role of zooplankton in structuring a plankton community in eutrophic lake water. Limnol. Oceanogr. 38:561–573; 1993.

Coveney, M.F.; Wetzel, R.G. Biomass, production, and specific growth rate of bacterioplankton and coupling to phytoplankton in an oligotrophic lake. Limnol. Oceanogr. 40:1187–1200; 1995.

del Giorgio, P.A.; Peters, R.H. Patterns in planktonic P:R ratios in lakes: influence of lake trophy and dissolved organic carbon. Limnol. Oceanogr. 39:772–787; 1994.

Fagerbakke, K.M.; Heldal, M.; Norland, S. Content of carbon, nitrogen, oxygen, sulfur and phosphorus in native and cultured bacteria. Aquat. Microb. Ecol. 10:15–27; 1996.

Findlay, S.; Pace, M.L.; Lints, D.; Howe, K. Bacterial metabolism of organic carbon in the tidal freshwater Hudson Estuary. Mar. Ecol. Prog. Ser. 89:147–153; 1992.

Fuhrman, J.A.; Azam, F. Bacterioplankton secondary production estimates for coastal waters of British Columbia, Antarctica, and California. Appl. Environ. Microbiol. 39: 1085–1095; 1980.

Hessen, D.O. Dissolved organic carbon in a humic lake: effects on bacterial production and respiration. Hydrobiologia 229:115–123; 1992.

Hygum, B.; Petersen, J.W.; Søndergaard, M. Dissolved organic carbon release by zooplankton grazing activity—a high quality substrate pool for bacteria. J. Plankton Res. 19:97–111; 1997.

Jeppesen, E.; Sortkjær, O.; Søndergaard, M.; Erlandsen, M. Impact of a trophic cascade on heterotrophic bacterioplankton production in two shallow fish-manipulated lakes. Arch. Hydrobiol. Beih. Ergebn. Limnol. 37:219–231; 1992.

Kairesalo, T.; Lehtovaara, A.; Saukkonen, P. Littoral-pelagial interchange and the decomposition of dissolved organic matter in a polyhumic lake. Hydrobiologia 229:199–224; 1992.

Kitchell, J.F.; Carpenter, S.R. Synthesis and new directions. In: Carpenter, S.R.; Kitchell, J.F., eds. The trophic cascade in lakes. Cambridge: Cambridge University Press; 1993: 332–350.

Norland, S. The relationship between biomass and volume of bacteria. In: Kemp, P.F.; Sherr, B.F.; Sherr, E.B.; Cole, J.J., eds. Handbook of methods in aquatic microbial ecology. Boca Raton, FL: Lewis Publ.; 1993:303–307.

Pace, M.L. Heterotrophic microbial processes. In: Carpenter, S.R.; Kitchell, J.F., eds. The trophic cascade in lakes. Cambridge: Cambridge University Press; 1993:252–277.

Porter, K.; Feig, Y.S. The use of DAPI for identifying and counting aquatic microflora. Limnol. Oceanogr. 25:943–948; 1980.

Riemann, B.; Christoffersen, K. Microbial trophodynamics in temperate lakes. Mar. Microb. Food Webs 7:69–100; 1993.

Schlüter, L.; Riemann, B.; Søndergaard, M. Nutrient limitation in relation to phytoplankton carotenoid/chlorophyll *a* ratios in freshwater mesocosms. J. Plankton Res. 19:891–906; 1997.

Simek, K.; Macek, M.; Pernthaler, J.; Straskrabova, V.; Psenner, R. Can freshwater planktonic ciliates survive on a diet of picoplankton? J. Plankton Res. 18:597–613; 1996.

Simon, M.; Azam, F. Protein content and protein synthesis rates of planktonic marine bacteria. Mar. Ecol. Prog. Ser. 51:201–213; 1989.

Smits, J.; Riemann, B. Cell production derived from [3]-H-thymidine incorporation using freshwater bacteria. Appl. Environ. Microbiol. 54:2213–2219; 1988.

Søndergaard, M.; Middelboe, M. Measurements of particulate organic carbon: a note on the use of glass fiber (GF/F) and Anodisc filters. Arch. Hydrobiol. 127:73–85; 1993.

Søndergaard, M.; Hansen, B.; Markager, S. Dynamics of dissolved organic carbon lability during a clear water phase in a eutrophic lake. Limnol. Oceanogr. 40:46–54; 1995.

Søndergaard, M.; Theil-Nielsen, J. Bacterial growth efficiency in lake water cultures. Aquat. Microb. Ecol. 12:115–122; 1997.

Timms, R.M.; Moss, B. Prevention of growth of potentially dense phytoplankton populations by zooplankton grazing, in the presence of zooplanktivorous fish, in a shallow wetland ecosystem. Limnol. Oceanogr. 29:472–486; 1984.

Tranvik, L. Allochthonous dissolved organic matter as an energy source for pelagic bacteria and the concept of the microbial loop. Hydrobiologia 229:107–114; 1992.

Wetzel, R.G. Gradient-dominated ecosystems: sources and regulatory functions of dissolved organic matter in freshwater ecosystems. Hydrobiologia 229:181–198; 1992.

Wright, R.T. Methods for evaluating the interaction of substrate and grazing as factors controlling planktonic bacteria. Arch. Hydrobiol. Beih. Ergebn. Limnol. 31:229–242; 1988.

16. Cascading Effects on Microbial Food Web Structure in a Dense Macrophyte Bed

Klaus Jürgens and Erik Jeppesen

Introduction

Heterotrophic microorganisms play a major role in the carbon and energy flow and nutrient recycling of aquatic systems. Planktonic bacteria are regulated by the supply of organic and inorganic nutrients and by predation of bacterivorous organisms. Most studies on controlling mechanisms of bacterioplankton have focused on assessing the direct effects of these factors. However, more recent studies have revealed that microbial and classic food webs have an array of interdependencies and are linked in many different ways (Turner and Roff, 1993). Therefore, the structure of the whole planktonic community must be considered for a better understanding of population dynamics at the microbial level (Pace et al., 1990).

Cascading trophic interactions are known to play an important role in the transfer of fish predation effects via zooplankton to the phytoplankton community (Carpenter et al., 1985). The trophic cascade concept has recently been examined with respect to the effects of zooplankton on the microbial food web (Pace, 1993; Jürgens et al., 1994). The capability of certain zooplankton groups (e.g., daphnids) to consume bacteria and those predation patterns on protozoans are key links between the classic and microbial food webs. Different groups of metazooplankton such as daphnids, copepods, and rotifers have strong group-specific impacts on planktonic protozoans (e.g., Arndt, 1993; Jürgens, 1994; Sanders et al., 1994). The protozoan community structure is to some extent a reflection of the metazooplankton predation regime. If and how these effects on the protozoan community are transferred to the bacterial level is not clear yet. Predatory cascades from zooplankton to bacteria could not be detected in enclosure experiments and whole-lake manipulations in an oligotrophic lake (Pace, 1993). By contrast, drastic cascading effects after zooplankton manipulation, which strongly altered bacterial biomass and community structure, were found in an enclosure experiment in a mesotrophic lake (Jürgens et al., 1994).

Submerged macrophytes have a major impact on the biological structure and water quality of shallow lakes (e.g., Scheffer et al., 1993). Dense macrophyte beds can act as a refuge for large zooplankton species, which in turn control the phytoplankton (Timms and Moss, 1984; Schriver et al., 1995; Lauridsen and Buenk, 1996), and this may affect also the structure and function of the microbial community. Microbial food web structure and dynamics were part of an integrated study of the impact of macrophytes on the biological structure in a shallow eutrophic lake (Jeppesen et al., submitted). Bacterial production and coupling to dissolved organic carbon (DOC) from the same experimental system are reported by Søndergaard et al. (this volume, Chapter 15). Here, we compare the composition and structure of protozoan and bacterial populations inside and outside submerged macrophyte beds. Results from fractionation experiments revealed the importance of zooplankton predation as a determining factor for the microbial food web structure.

Material and Methods

Lake Stigsholm is a shallow eutrophic lake in central Jutland, Denmark (mean depth, 0.8 m; mean total P concentration, 0.15 mg/L). During the summer of 1994, a large part of the littoral zone was covered by dense vegetation of submerged macrophytes, mainly *Potamogeton pectinatus*. A detailed experimental design of the enclosure system is given in Jeppesen et al. (submitted). Briefly, six 5-m diameter patches (three replicates each) were either kept free from plant vegetation (M–) or macrophytes were allowed to grow undisturbed (M+, plant infested volume 40–55%). To assess the plankton community in these patches, without interference from surrounding water, vertical polyethylene sheets were used to enclose the patches for a 2-week period. All fish inside the enclosures were removed by electrofishing and trapping. After 1 week, planktivorous perch were again added to the enclosures. Sampling of organisms began 2 days after the enclosement.

For a detailed examination of microbial food web structure and for size-fractionation experiments, we have used only two of the enclosures, in the following referred to as M– (no vegetation) and M+ (with macrophytes). Twice during the enclosement, we performed size-fractionation experiments; one (experiment 1) under fish-free conditions 3 days after enclosement and the second (experiment 2) 2 days after restocking of fish. Water from the enclosures was either left unfiltered (UF) or filtered through a 20-µm mesh net to remove both meso- and most of the microzooplankton. By comparing the population developments of the different microbial components with and without zooplankton, this design should yield information on the predation impact of the zooplankton communities in M+ and M–.

After filling the water into transparent 4.8-L Nalgene polycarbonate bottles, these were fixed to the enclosure frame, approximately 30 cm below the surface. Samples for counting of organisms were taken at least once a day. Lugol-fixed samples were used for counting of ciliates. For determination of the dominant species, some selected samples were postfixed with Bouins fixative, and ciliate species were determined after impregnation with Protargol according to Skibbe (1994). Formalin-fixed samples were used for enumeration of bacteria, hetero-trophic nanoflagellates (HNF), and autotrophic picoplankton (APP) after DAPI (4'6-diamidine-2-phenyl-indole; Sigma) staining (Porter and Feig, 1980). Volumes of bacteria and APP were determined with an image analysis system (SIS GmbH, Münster, Germany) connected to an epifluorescence microscope. We considered the following bacterial morphologies as (protozan-) grazing-resistant bacteria (GRB, see Jürgens and Güde, 1994): long filamentous bacteria (cells or chains approximately 7 μm in length) and aggregated bacteria (particles with attached bacteria or purely bacterial aggregates with a diameter of 10 μm). GRB were directly counted and measured with an ocular grid from the same DAPI prepara-tions as total bacteria.

Results and Discussion

The enclosement of littoral patches with and without macrophytes enabled a detailed analysis of the planktonic community. General composition and develop-ment of planktonic organisms proved to be very similar in the replicate enclosures both with and without macrophytes (Jeppesen et al., submitted). The separation from the surrounding water, especially from fish predation, might have led to some changes in the regulating mechanisms of the enclosed organisms. However, the relative stability of the plankton community structure during the 2-week study period (data in Jeppesen et al., submitted) indicated that the enclosure did not significantly alter the biological structure. Therefore, the experimental data char-acterize the plankton community that had previously developed within the macro-phyte bed or in areas without submerged vegetation, respectively.

The decisive factor for the structure of the microbial food web was that the zooplankton composition was different (Table 16.1): a high density of different cladocerans and cyclopoid copepods was found in M+, whereas ciliates, rotifers, and cyclopoid copepods dominated in M–. The zooplankton com-munities are the well-known result of different degrees of fish predation: large-bodied mesozooplankton were able to survive in the macrophytes probably due to the refuge effect, and small-bodied zooplankton, mainly microzoo-plankton, developed without macrophytes, indicating an intense fish predation pressure.

Table 16.1. Density of Important Planktonic Organisms in Enclosures with (M+) and without (M–) Macrophytes at the Beginning of the Size-Fractionation Experiments[a]

	Experiment 1		Experiment 2	
	M+	M–	M+	M–
Pico-/nanoplankton				
Total bacteria (10^6/ml)	2.8 ± 0.5	14.1 ± 0.5	4.0 ± 0.4	8.9 ± 0.1
Filaments[b] (10^3/ml)	13.8 ± 3.9	31.8 ± 2.8	8.5 ± 1.5	133.4 ± 13.8
Aggregates[c] (10^3/ml)	3.7 ± 0.5	4.2 ± 2.3	2.1 ± 0.7	3.9 ± 0.9
APP (10^5/ml)	0.2 ± 0.1	5.8 ± 1.6	0.7 ± 0.2	25.5 ± 0.7
HNF (10^3/ml)	0.8 ± 0.6	5.5 ± 1.4	0.9 ± 0.4	6.5 ± 0.1
Micro-/macrozooplankton				
Ciliates (/mL)	1.5 ± 0.4	245 ± 52	11.7 ± 9.0	216 ± 105
Rotifers (/L)	436 ± 26	15,140 ± 4,845	199 ± 94	20,095 ± 5,024
Daphnids (/L)	148 ± 13	1.5 ± 0.4	61 ± 30	0
Other cladocerans (/L)	213 ± 104	1.5 ± 1.0	247 ± 103	5 ± 1.7
Cyclopoid copepods (/L)	1,184 ± 331	301 ± 3	635 ± 298	418 ± 176
Phytoplankton				
Chlorophyll *a* (µg/L)	12.0 ± 5.3	60.3 ± 6.1	15.0 ± 6.1	67.3 ± 19.6

[a]Means ± SD of three replicate enclosures (zooplankton) or means ± SD of three replicate treatments from one enclosure (pico-/nanoplankton).
[b]Filamentous bacteria >6 µm.
[c]Aggregates >10 µm with attached bacteria.

Tremendous differences between M– and M+ were also visible in the structure of the microbial food web. In M+, the abundance of phytoplankton, protozoans, and bacteria was very low, whereas in M– an abundant and diverse assemblage of auto- and heterotrophic pico-, nano-, and microplankton coexisted with the meta-zooplankton community (Tables 16.1 and 16.2). Ciliates represented the most drastic difference in the microbial food web (Table 16.2). Ciliate density was approximately two orders of magnitude higher in M– compared with M+, and species composition differed. Some of the ciliates found in M+, such as the hypotrichous species, are benthic, probably macrophyte-associated species. Truly planktonic forms clearly dominated in M–. Different trophic levels and feeding modes (bacterivorous, algivorous, and carnivorous species) were present within the ciliate community. The most abundant groups were scuticociliates (mainly bacterial feeders) and small oligotrichous and prostomatid species, which feed mainly on nanophytoplankton (Müller et al., 1991). But larger predatory ciliates such as *Urotricha pelagica* and species from the Haptorida order were also present in higher numbers (up to 25 individuals/ml) in M–. From the different cell sizes, we assume that different functional groups were probably also present among the heterotrophic flagellates in M–: small bacterivorous species (3–5 µm) and larger mainly algivorous species (10–15 µm). The diverse protozoan species composition in M– must be exerting a significant grazing pressure on the bacterial as well

Table 16.2. Ciliate Species Composition in Enclosures with (M+) and without (M–) Macrophytes

Ciliate taxa	M+	M–
Total abundance[a] (per ml)	3.8 ± 3.2	282 ± 79
Prostomatida		
Urotricha cf. *furcata/farcta*	+	+
Urotricha cf. *pelagica*	–	+
Coleps sp.	+	–
Oligotrichida		
Halteria grandinella	+	+
Strobilidium lacustris	–	+
Strobilidium sp. (<25 µm)	–	+
Haptorida		
Askenasia sp.	–	+
Monodinium sp.	–	+
Lagynophria sp.	–	+
Enchylis sp.	–	+
Actinobolina sp.	+	–
Scuticociliatida		
Cyclidium sp.	+	+
Cinetochilum margaritaceum	+	+
Peritrichida		
Vorticella spp.	+	+
Vorticella mayeri	–	+
Colpodea		
Cyrtolophosis mucicola	–	+
Hypotrichia		
Aspidisca sp.	+	–
Oxytricha sp.	+	–

[a]Mean ± SD from 2 weeks.

as on the algal community. Further, many predator–prey interactions probably occur between different protozoan groups.

Differences between M+ and M– occurred at the bacterial level as well: bacterial concentrations were four- to fivefold higher in M– when compared with M+ and exhibited a more pronounced morphological diversity with a high proportion of filamentous forms and aggregated bacteria (Fig. 16.1). These complex cell morphologies are resistant to protozoan grazing and generally appear when predation pressure by protozoans is high, especially during HNF population peaks (Jürgens and Güde, 1994). APP, here exclusively chroococcid cyanobacteria, constituted another important part of the picoplankton in M– but were at a low level in M+. In M–, their contribution to total picoplankton biomass even exceeded heterotrophic bacteria in the second experiment (Table 16.3). Also,

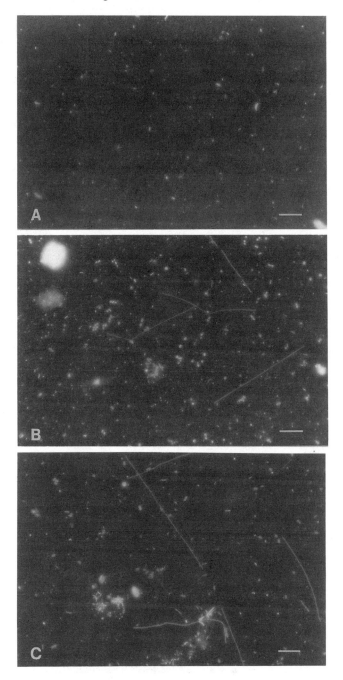

Figure 16.1. Epifluorescence microphotographs of DAPI-stained preparations (1 ml of sample on 0.2-μm Nucleporefilter). (A) M+, start of experiment 2; (B) M–, start of experiment 2; (C) M+, 20 μm filtered, after 3 days. Scale bar = 10 μm.

Table 16.3. Estimated Biovolumes (10^5 μm^3/ml) of Picoplankton in M+ and M– at the Beginning of Experiments 1 and 2[a]

	M–		M+	
	Experiment 1	Experiment 2	Experiment 1	Experiment 2
Bacteria	8.34 ± 0.28	5.34 ± 0.08	2.52 ± 0.31	3.61 ± 0.85
Filaments	0.39 ± 0.06	2.64 ± 0.27	0.13 ± 0.03	0.12 ± 0.02
APP	6.22 ± 1.25	28.05 ± 0.78	0.22 ± 0.06	0.80 ± 0.37

[a]Means ± SD of three replicate treatments. Autotrophic picoplankton (APP; mean cell volume, 1.10 μm^3), bacteria (freely suspended rods and cocci, mean cell volume 0.06 μm^3 and 0.09 μm^3 for M– and M+, respectively), and filaments (filamentous bacteria, mean cell volume 1.0–2.0 μm^3).

filamentous bacteria contributed substantially to the bacterial biomass in M– (Table 16.3).

The size-fractionation experiments revealed that the impacts of zooplankton greater than 20 μm on the microbial food web differed between M+ and M–. We considered top-down effects of zooplankton to be effective when the population levels of the various microbial components differed significantly between UF and less than 20 μm filtered water after 24 hours. The results of the first experiment are shown in Figure 16.2; the statistical results of both experiments, which showed virtually similar trends, are summarized in Table 16.4. The removal of zooplankton in M+ resulted in an immediate increase in bacteria and protozoans compared with the UF controls. Most pronounced was the response of heterotrophic bacteria and HNF; within 24 hours, their abundance increased by factors of 3 and 9, respectively. Chlorophyll *a* increased by a factor of 4 within 4 days. These findings are in accordance with the general notion that *Daphnia*, which occurred in high densities in M+, can control bacteria, algae, and a wide range of protozoans (Christoffersen et al., 1993; Jürgens, 1994).

Removal of zooplankton greater than 20 μm in M– had little or no effect on pico- and nanoplankton. Population levels remained constant or increased, which was similar to the results for the UF controls (Fig. 16.1), with the only significant increase being that of filamentous bacteria (Table 16.4). We assume that complex predatory interactions within the microbial food web took place in M–, which did not become visible in our experimental design because they prevailed in the less than 20-μm fraction. It has previously been shown by size-fractionations with smaller filter pore sizes that several trophic links can exist even among organisms less than 10 μm (Wikner and Hagström, 1988).

A pronounced microbial succession became visible in M+ after removal of zooplankton and incubation of the bottles for 5 days (Fig. 16.3). The immediate peak in bacteria (within 12 hours) was followed by an increase in HNF, which was probably responsible for the following decline in bacterial concentrations. APP increased continuously during the experiment but at a much lower rate than heterotrophic bacteria. Also ciliates showed an exponential increase during the

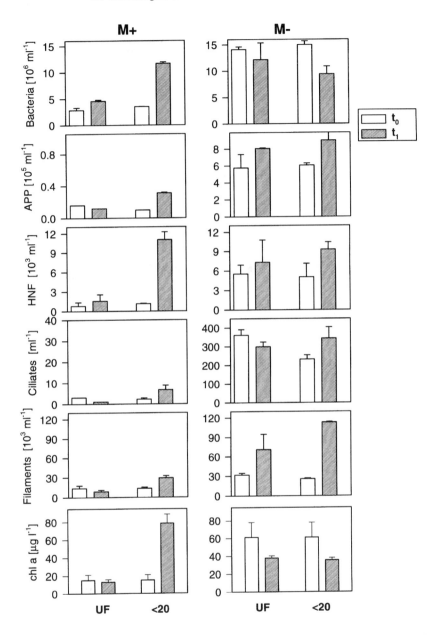

Figure 16.2. Size-fractionation (experiment 1). Development of bacteria, APP, HNF, ciliates, filamentous bacteria, and chlorophyll a in unfiltered (UF) and 20 μm filtered water from enclosure M+ and M−. t_1 is 24 hours except for chlorophyll a (chl a), where t_1 is 4 days. Note difference in scales for APP and ciliates.

Table 16.4. Statistical Comparison of the Means (after 24 hours) of Unfiltered (UF) and 20 μm Filtered Water for Enclosures M+ and M– and for Experiments 1 and 2[a]

	Experiment 1		Experiment 2	
Organisms	M+	M–	M+	M–
Bacteria	<0.0001	ns	<0.005	ns
APP	<0.05	ns	<0.05	ns
Filaments	<0.001	<0.05	<0.01	<0.05
HNF	<0.0025	ns	<0.05	ns
Ciliates	<0.05	ns	<0.05	ns

[a]Probability (*P*) levels are from unpaired two-sample *t*-tests. ns, difference of means not significant at the 0.05 level.

incubation, with *Halteria grandinella* as the dominant species (80% of total abundance). The decline in HNF might be a result of both ciliate predation and depletion of food resources. Grazing-resistant bacteria (filamentous forms and aggregates) increased as well during the incubation period. This succession sequence is very similar to the one reported by Jürgens et al. (1994) from an enclosure study in which removal of *Daphnia* resulted in an increase in phago-trophic protozoans, which shifted the bacterial assemblage toward protozoan-inedible forms.

Other aspects of microbial interactions can be deduced from this succession in fractionated water from M+. After removal of zooplankton, a huge bacterial production, which was previously controlled by mesozooplankton, became visible. Later in the experiment, it was controlled by HNF. The increase in bacteria within the first 12 hours after zooplankton removal corresponds to a bacterial doubling time of 6.6 hours. This is in agreement with direct measurements of bacterial production in the enclosures (Søndergaard et al., this volume, Chapter 15), which showed that bacteria in M+ were about as productive as in M–, despite low concentrations of phytoplankton and a substantially lower bacterial abundance.

APP seemed to have a slightly different role than heterotrophic bacteria in the microbial food web of Lake Stigsholm. Their exponential increase after zoo-plankton removal in M+ (but at a lower growth rate than bacteria) revealed the efficient top-down control by mesozooplankton. In contrast to bacteria, APP were not reduced after HNF and ciliates achieved higher numbers in the incubations (Fig. 16.3). Together with the fact that APP were present in high densities in the M– enclosures, this might indicate that they were to a lesser extent subject to protozoan predation than were heterotrophic bacteria. This supports the view that cyanobacteria have lower food quality and are probably negatively selected by protozoans (e.g., Caron et al., 1991).

The size-fractionation experiments demonstrated that zooplankton predation in M+ was the decisive factor for the observed microbial food web structure. Al-

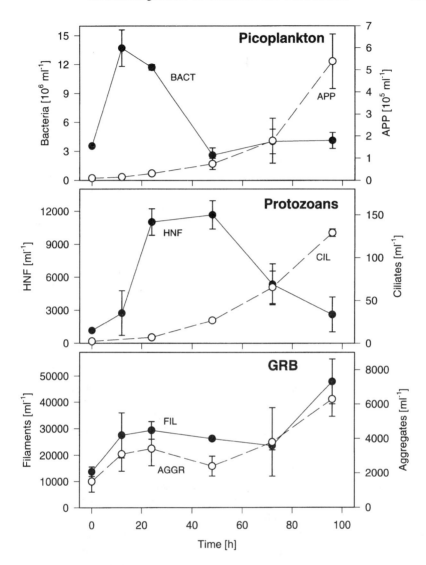

Figure 16.3. Development of picoplankton (bacteria, APP), heterotrophic protozoans (HNF, ciliates), and grazing-resistant bacteria (GRB: aggregates, filaments) in 20 μm filtered water of enclosure M+ (experiment 1).

though the microbial community in the less than 20-μm fraction is probably undergoing a transient state, it is interesting to note that after removal of zooplankton, bacteria and protozoans reached similar levels as those in M–. Also, the general morphological composition of bacteria became more similar (Fig. 16.1C).

The littoral patches, M+ and M–, were ideal for studying trophic interactions at different food web constellations as the enclosures were situated in close proximity in the littoral zone of the same lake, and nutrient levels and water chemistry were almost identical (Jeppesen et al., submitted). Although the macrophyte-periphyton complex can release bacterial substrates and thus compensate for the lack of phytoplankton in M+, this did not seem to be the crucial factor for the different microbial food web structures. Instead, different zooplankton communities and predatory cascades were of overwhelming importance.

The mesozooplankton community in M+ (dominated by cladocerans) exerted a much stronger top-down control on pico- and nano-sized organisms than the microzooplankton (and the copepods) in M–. The results are in general accordance with the view that microzooplankton assemblages are much less able to control nanophytoplankton than is a zooplankton community consisting of large filter-feeding species (e.g., Mazumder et al., 1990). This is because the latter community generally has a higher total biomass and exerts lower feeding selectivity (Lampert, 1988; Mazumder et al., 1990). Picoplankton (bacteria, APP), exceeding a certain cell size, probably find it harder to escape consumption by filter feeders such as *Daphnia* but may develop some kind of resistance against predation by protozoans (Jürgens and Güde, 1994). This is the reason why despite lower overall grazing pressure on picoplankton in M– (and higher abundance), a larger proportion consisted of protozoan-inedible or less digestible forms (filaments, aggregates, cyanobacteria).

The different zooplankton communities of M– and M+ are a result of the previous differing fish predation pressure, which was estimated to be very strong in M– but, due to the refuge effect of the macrophytes, relatively low in M+ (Jeppesen et al., submitted). Consequently, predation effects can cascade from fish to the bacterial level in these systems. It is probably a general feature of more eutrophic systems that top-down regulation is a stronger shaping factor for the plankton community than food resources (bottom-up regulation) (Jeppesen et al., 1997).

Our study and the comparable one by Jürgens et al. (1994) showed that this also holds true for planktonic bacteria, which responded immediately to an alteration in the predation regime. This is in contrast to Pace's model (1993), which assumed that the main effects of the trophic cascade on bacteria are indirect, mediated through alterations in phytoplankton biomass and productivity. We assume that this might be the case in oligotrophic systems in which bacteria are more likely to be nutrient limited, but direct predatory interactions that affect the structure and function of microbial food webs are more important in systems of higher productivity.

Acknowledgments. We are grateful to Morten Søndergaard and Kirsten Christof-fersen for their critical review of the manuscript and to Nancy Zehrbach, who improved our wording. Special thanks go to Oliver Skibbe who introduced us to Protargol staining and helped with ciliate species determination.

References

Arndt, H. Rotifers as predators on components of the microbial web (bacteria, heterotrophic flagellates, ciliates)—a review. Hydrobiologia 255/256:231–246; 1993.

Caron, D.A.; Lim, E.L.; Miceli, G.; Waterbury, J.B.; Valois, F.W. Grazing and utilization of chroococcoid cyanobacteria and heterotrophic bacteria by protozoa in laboratory cultures and a coastal plankton community. Mar. Ecol. Prog. Ser. 76:205–217; 1991.

Carpenter, S.R.; Kitchell, J.F.; Hodgson, J.R. Cascading trophic interactions and lake productivity. BioScience 35:635–639; 1985.

Christoffersen, K.; Riemann, B.; Klysner, A.; Søndergaard, M. Potential role of fish predation and natural populations of zooplankton in structuring a plankton community in eutrophic lake water. Limnol. Oceanogr. 38:561–573; 1993.

Jeppesen, E.; Jensen, J.P.; Søndergaard, M.; Lauridsen, T.; Pedersen, L.J.; Jensen, L. Top-down control in freshwater lakes: the role of nutrient state, submerged macrophytes and water depth. Hydrobiologia 342/343:151–164; 1997.

Jeppesen, E.; Søndergaard, M.; Søndergaard, M.; Christoffersen, K.; Jürgens, K.; Theil-Nielsen, J.; Schlüter, L. Cascading trophic interactions in the littoral zone of a shallow lake (submitted).

Jürgens, K. The impact of *Daphnia* on microbial food webs—a review. Mar. Microb. Food Webs 8:295–324; 1994.

Jürgens, K.; Güde, H. The potential importance of grazing-resistant bacteria in planktonic systems. Mar. Ecol. Prog. Ser. 112:169–188; 1994.

Jürgens, K.; Arndt, H.; Rothhaupt, K.O. Zooplankton-mediated changes of bacterial community structure. Microb. Ecol. 27:27–42; 1994.

Lampert, W. The relationship between zooplankton biomass and grazing: a review. Limnologica 19:11–20; 1988.

Lauridsen, T.; Buenk, I. Diel changes in the horizontal distribution of zooplankton in the littoral zone of two shallow eutrophic lakes. Arch. Hydrobiol. 137:161–176; 1996.

Mazumder, A.; McQueen, D.J.; Taylor, W.D.; Lean, D.R.S.; Dickman, M.D. Micro- and mesozooplankton grazing on natural pico- and nanoplankton in contrasting plankton communities produced by planktivore manipulation and fertilization. Arch. Hydrobiol. 118:257–282; 1990.

Müller, H.; Schöne, A.; Pinto-Coelho, R.M.; Schweizer, A.; Weisse, T. Seasonal succession of ciliates in Lake Constance. Microb. Ecol. 21:119–138; 1991.

Pace, M.L. Heterotrophic microbial processes. In: Carpenter, S.R.; Kitchell, J.F., eds. Cascading trophic interactions. Cambridge: Cambridge University Press; 1993:252–277.

Pace, M.L.; McManus, G.B.; Findlay, S.E.G. Planktonic community structure determines the fate of bacterial production in a temperate lake. Limnol. Oceanogr. 35:795–808; 1990.

Porter, K.G.; Feig, Y.S. The use of DAPI for identifying and counting aquatic microflora. Limnol. Oceanogr. 25:943–947; 1980.

Sanders, R.W.; Leeper, D.A.; King, C.H.; Porter, K.G. Grazing by rotifers and crustacean zooplankton on nanoplanktonic protists. Hydrobiologia 288:167–181; 1994.

Scheffer, M.; Hosper, S.H.; Meijer, M.-L.; Moss, B.; Jeppesen, E. Alternative equilibria in shallow lakes. Trends Ecol. Evol. 8:275–279; 1993.

Schriver, P.; Bøgestrand, J.; Jeppesen, E.; Søndergaard, M. Impact of submerged macrophytes on fish–zooplankton–phytoplankton interactions: large-scale enclosure experiments in a shallow eutrophic lake. Freshwat. Biol. 33:255–270; 1995.

Skibbe, O. An improved quantitative protargol stain for ciliates and other planktonic protists. Arch. Hydrobiol. 130:339–347; 1994.

Timms, R.M.; Moss, B. Prevention of growth of potentially dense phytoplankton populations by zooplankton grazing in the presence of zooplanktivorous fish in a shallow wetland ecosystem. Limnol. Oceanogr. 29:472–486; 1984.

Turner, J.T.; Roff, J.C. Trophic levels and trophospecies in marine plankton: lessons from the microbial food web. Mar. Microb. Food Webs 7:225–248; 1993.

Wikner, J.; Hagström, A. Evidence for a tightly coupled nanoplanktonic predator–prey link regulating the bacterivores in the marine environment. Mar. Ecol. Prog. Ser. 50:137–145; 1988.

17. Abundance, Size, and Growth of Heterotrophic Nanoflagellates in Eutrophic Lakes with Contrasting *Daphnia* and Macrophyte Densities

Kirsten Christoffersen

Introduction

Natural populations of heterotrophic nanoflagellates (HNF) can constitute an important component of the nanoplankton community because they have high growth rates and high predation rates on picoplankton (Riemann and Christoffersen, 1993). Seasonal variability in the HNF population (e.g., Carrick et al., 1992) has often been attributed to the abundance of picoplankton (especially bacteria). Berninger et al. (1991) established a predator–prey correlation based on data from numerous lakes. Several other cross-system analyses have concluded that the relationship is not strong and decreases in strength from oligotrophic to eutrophic systems (Sanders et al., 1992; Gasol and Vaqué, 1993; Gasol, 1994).

The HNF biomass in eutrophic lakes is often small, relative to the biomass of other microbes, during most of the year (Riemann and Christoffersen, 1993). Numerous studies have shown that predation by ciliates, rotifers, and metazoans can regulate HNF abundance and biomass in these ecosystems (e.g., Güde, 1988; Sanders and Porter, 1990; Weisse, 1991; Arndt, 1993, 1994; Christoffersen et al., 1993; Jürgens and Stolpe, 1995). The influence of cladocerans, in particular *Daphnia,* on HNF has previously been described (Riemann, 1985; Vaqué and Pace, 1992). The food spectrum of *Daphnia* is large, and their ability to consume nano-sized protozoans is widely recognized (Stoeckner and Capuzzo, 1990; Arndt, 1993; Jürgens, 1994). This implies that mechanisms operating on *Daphnia* populations are also indirectly acting on the protozoan community.

In a recent study, Christoffersen et al. (1993) showed that the presence of planktivorous fish changed the biomass and composition of zooplankton in a eutrophic lake, and this, in turn, affected the microbial communities. When clado-cerans dominated (*Bosmina* sp. and *D. cucullata*), they controlled the biomass of phytoplankton, HNF, rotifers, and bacteria. However, when fish reduced the cladoceran biomass, a microbial community developed despite the presence of other metazoans (mainly copepods). An increasing number of studies have recognized the key position of *Daphnia* in freshwater systems, supporting the thesis that a strong predator control is the structuring element in lakes (Jeppesen et al., 1992; Riemann and Christoffersen, 1993; Jürgens, 1994).

This study evaluates the specific interactions between populations of *Daphnia* and HNF in eutrophic lakes on a cross-system basis and includes the potential impact of macrophytes on these interactions. Because macrophytes act as refuges for zooplankton under high fish predation (Schriver et al., 1995), it can be hypothesized that accumulation of zooplankton in macrophyte stands may lead to intensive grazing activities influencing the microbial community. The data set allowed also a test of the correlation between abundance of bacteria and HNF by using the framework proposed by Gasol (1994) as well as a test of the effects of temperature and bacterial abundance on HNF growth rates.

Material and Methods

The study was carried out in four shallow and eutrophic (total phosphorus, >0.1 mg/L during summer) lakes: Frederiksborg Slotssø, Stigsholm sø, Ring sø, and Ramten sø (Table 17.1). The lakes had either naturally occurring submerged macrophytes or planted ones, except for Frederiksborg Slotssø. Differences in macrophyte density (i.e., plant-filled volume) within the specific lakes were obtained by manually adjusting the macrophytes either directly in the lake (Ring sø) or inside large enclosures (19.6 m²). The enclosures were constructed by rein-

Table 17.1. Some Characteristics of the Lakes Included in This Study

	Trophic status	Mean depth	Macrophytes (density)[a]	Dominant *Daphnia* species
Frederiksborg Slotssø	Highly eutrophic	3.1 m	None	*D. cucullata*
Stigsholm sø	Eutrophic	0.8 m	*Potamogeton* (0–50%)	*D. galeata*, *D. cucullata*
Ring sø	Highly eutrophic	2.1 m	*Potamogeton* (50–100%)	*D. magna*
Ramten sø	Eutrophic	1.0 m	*Ceratophyllum* (0–100%)	*Daphnia* sp.

[a]Given as relative plant-filled volume.

forced plastic and stainless steel or wooden poles (further details in Jeppesen et al., submitted). Each enclosure had a diameter of 5 m and a water depth of 0.6–1.2 m. All experiments were carried out between March–September during 1992–1995.

Water samples were collected daily (noon) or twice daily (noon and midnight) during the experimental periods, and subsamples for identification and enumeration of HNF, bacteria, chlorophyll, and zooplankton were taken. The sampling strategy is described in Søndergaard et al. (this volume, Chapter 15). HNF were fixed from whole-water samples with glutaraldehyde (1.5% final concentration), stained with DAPI within 24 hours and filtered onto 0.8-µm black Nuclepore or Poretics filters. The filters were kept frozen ($-20°C$), and the number of flagellates was later counted by using an epifluorescence microscope fitted with an ultra-violet filter set. Only cells of 2–10 µm in greatest dimension were recorded, and at least 50 cells were counted from each filter. The HNF biovolume was determined by measuring linear dimensions of 10–20 individuals per filter. Processing of samples for abundance of bacteria and zooplankton is described in Søndergaard et al. (this volume, Chapter 15).

Potential growth rates of HNF were measured under predator-free conditions by incubation of prescreened water. Subsamples were produced from whole-water samples by screening the water with gentle inverse filtration using a 25 L fractionator equipped with a nitex screen with a mesh size of 20 µm. The screened water was added to 1-L or 0.25-L acid-cleaned polycarbonate bottles and incubated in situ at 0.5-m depth for 24 hours. The bottles were mounted on a floating rig and were thus exposed to natural water movements. An air bubble inside each bottle ensured that the water was mixable. Calculations of growth rates were based on total changes in abundance, assuming exponential growth.

The data set, including data on the abundance of bacteria, zooplankton, and density of macrophytes, was analyzed by linear regression analysis (SAS JUMP) of log-transformed (base$_e$) data. In cases in which no *Daphnia* were found, each value was increased by 1 (n+1).

A laboratory experiment was conducted to investigate the predation pressure of *Daphnia magna* on natural HNF. Water from three stations (± macrophytes and midlake) in Ring sø was prescreened through 20-µm nets and divided into 18 subsamples of 250 ml each. Fifteen *D. magna* (1.5 mm) were then added to each of half of the samples, and all samples were incubated in the laboratory at in situ temperature (25°C) for 24 hours in dim light. The abundance of HNF was measured at the start and end of the incubation as described above. The daphnids were collected from the lake and kept in prescreened (20 µm) lake water for 2 hours before the experiment.

Results and Discussion

The abundance of HNF in the investigated lakes (Fig. 17.1) ranged from 0.1×10^2 cells/ml to 1.3×10^4/ml with marked seasonal variations. The abundance of *Daphnia* covered a range from 0 to more than 1,000/L (Fig. 17.1). Different

Figure 17.1. Relationship between the abundance of *Daphnia* sp. and heterotrophic nanoflagellates in four eutrophic and shallow lakes. Measurements in replicated enclosures or in the lakes are shown individually. Circles denote systems with macrophytes, and triangles denote systems without macrophytes. The ranges of macrophyte-filled volume in the respective lakes are given in parentheses. Five measurements at very high *Daphnia* densities (512–1,342 *Daphnia*/L; 1–2 × 10³ HNF/ml) following the general trend are omitted for clarity.

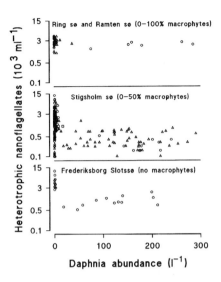

species of mainly medium-sized daphnids were present (Table 17.1), except in one case in which the population was totally dominated by large *D. magna* (Ring sø). The *Daphnia* populations were regulated by temperature and by fish predation (data not presented here). Thus, situations with low densities of daphnids as, for instance, the early spring or during intensive fish predation (typically during early summer), were associated with higher abundances of HNF than when daphnids were numerous (e.g., at high macrophyte density and/or low fish predation).

The macrophytes had no impact per se on the abundance of HNF (Fig. 17.1), but when the macrophytes functioned as a refuge for *Daphnia,* they had a negative effect on the population size of HNF (compare Fig. 17.1 middle and lower panel). This trend was not, however, found in Ring and Ramten sø (Fig. 17.1 upper panel), where the HNF abundance reached higher levels (1–3×10^3/ml) in the presence of *Daphnia* than in the other two lakes. The fact that no enclosures were established in this experiment implies that a continuous invasion of HNF from the surrounding macrophyte areas may have occurred and that this equaled the loss of HNF by *Daphnia* due to grazing.

A correlation between *Daphnia* and HNF abundance showed a significant negative relationship (Table 17.2). A general log-log model was used for this relationship. Other types of data transformation (i.e., 1/HNF and log 1/HNF) did not improve the relationship. The biovolume of individual HNF cells was not correlated with the abundance of *Daphnia* in the same manner (Table 17.2), albeit a higher frequency of small HNF was found at high *Daphnia* abundance than at low abundance. The explanation of these weak relationships is probably that a variety of other organisms feeding on nano-sized plankton can develop when *Daphnia* populations are low (Pace and Funke, 1991; Christoffersen et al., 1993; Jürgens, 1994). Consequently, a new correlation that only used *Daphnia* abundances higher than 0 individuals/L and the total HNF biovolume (i.e., abundance ×

Table 17.2. Summary of the Linear Regression Analysis on Log Transformed Data (base$_e$)[a]

	N	r^2	Slope (± SE)	P	Intercept (± SE)	P
Log HNF (per ml) vs. log *Daphnia* + 1 (per L)	251	0.172	−0.21 (0.03)	***	0.56 (0.09	***
Log HNF volume (µm³/cell) vs. log *Daphnia* + 1 (per L)	251	0.100	−0.06 (0.012)	ns	3.08 (0.04)	***
Log HNF total biovolume (µm³/ml) vs. log *Daphnia* >0 (per L)	187	0.210	−0.27 (0.03)	**	3.64 (0.10)	***
Log bacteria (per ml) vs. log *Daphnia* + 1 (per L)	212	0.160	−0.11 (0.02)	***	2.52 (0.05)	***
Log bacteria (per ml) vs. log HNF (per ml)	253	0.002	−0.022 (0.03)	ns	2.21 (0.04)	***
Log bacteria (per ml) vs. log HNF volume (µm³/cell)	253	0.000	0.0011 (0.03)	ns	2.20 (0.10)	***

[a]N, number of observations; r^2, correlation coefficient; ***$P < .001$; **$P < .01$; ns, not significant.

individual biovolume) was established (Fig. 17.2). This provided a slightly stronger correlation (Table 17.2) and allowed the calculation that 0.33 *Daphnia*/L was needed to reduce the standing stock of HNF (in terms of total biovolume) by 50%. A similar low threshold of *Daphnia* abundance was found by Jürgens (1994) when plotting the biomass of *Daphnia* versus the abundance of HNF. *Daphnia* abundances are often above 10 individuals/L in temperate eutrophic lakes during summer (Riemann and Christoffersen, 1993). This implies that the growth rate of

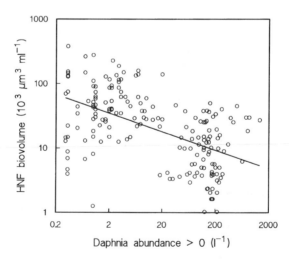

Figure 17.2. Relationship between the abundance of *Daphnia* and the total HNF biovolume. Note that both axes are on a log scale.

Figure 17.3. Laboratory experiment showing the effect of 0 and 15 *D. magna* on the growth rates of natural populations of heterotrophic nanoflagellates (from three locations in Ring sø) after a 24-hour incubation. Each column represents the average of triplicates, and the error bars denote 1 standard deviation.

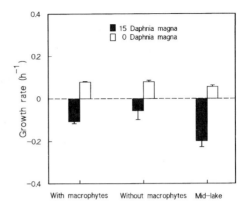

heterotrophic nanoflagellates needs to be fairly high if they are to survive as a population during intensive *Daphnia* growth.

A laboratory experiment conducted with *D. magna* and HNF populations from Ring sø indicated that the predation rate by 15 *D. magna*/L roughly equaled the growth rate of the natural HNF community as measured under predator-free conditions during 24-hour incubation (Fig. 17.3), even though the HNF population divided every 8 hours. This is, however, not a realistic picture of the instantaneous *Daphnia* predation rates because the food concentration (i.e., the HNF abundance) was not constant during the experiment. The measured predation rates were most likely underestimated, owing to the fact that *Daphnia* are not able to reduce the HNF abundance much below 1×10^2/ml (the lowest HNF abundance recorded in the entire data set was 0.7×10^2/ml).

Growth rates of the HNF community (i.e., species of 2–10 μm in diameter) were measured on a routine basis during this study under predator-free conditions (Carrick et al., 1992; Hansen and Christoffersen, 1995), implying that the water had been prescreened through 20-μm nets to ensure that no large predators were present. The measured growth rates ranged from –0.70 to 3.72/day, with an average value of 0.87/day (SD = 0.85). A few results (10 of 123 growth experiments) had to be excluded from further analysis because the growth rates were negative. Such results are possible, for example, if predators passed the 20-μm mesh net or if the enclosed HNF community was somehow damaged. Linear correlation analysis of the whole data set (base$_e$ transformed data except for temperature) revealed that HNF growth rates were neither correlated with temperature nor with bacterial abundance (data not shown). A correlation between temperature and HNF growth rates covering most of the growth season in one of the lakes (Frederiksborg Slotssø) was, however, significant ($N = 23$; $r^2 = 0.410$; $P < .05$). A calculated Q_{10} value of 2.3 seemed reasonable compared with previously published results (Weisse, 1991).

HNF abundance and individual biovolume were not correlated with the abundance of bacteria (Table 17.2). Previous analyses including a broader range of lakes have shown significant (but often weak) correlations (Berninger et al., 1991;

Sanders et al., 1992; Gasol and Vaqué, 1993; Gasol, 1994). It is generally accepted that the relationship is weakened in more eutrophic systems, because various organisms feeding on both bacteria and HNF (ciliates, rotifers, and metazoans) often increase in number along a trophic gradient (Sanders et al., 1992; Riemann and Christoffersen, 1993). It was apparent from this data set that *Daphnia* alone had a significant negative impact on the abundance of bacteria (Table 17.2).

Most of the HNF communities examined during this study increased dramatically in abundance when released from *Daphnia* predation (although alternative predators, such as ciliates, were able to prevent HNF growth in some cases), and HNF growth rates were not related to the bacterial abundance. This suggests that the HNF populations in eutrophic lakes are seldom limited by food and that they are capable of dividing several times daily on the available food.

Gasol (1994) presented a framework for the assessment of bottom-up and top-down regulation of HNF abundance. His concept is based on simple models describing the maximum attainable HNF abundance and mean realized abundance in a log-log plot of bacterial abundance versus HNF abundance. The model can, under a set of assumptions, test the relative importance of top-down or bottom-up control of HNF abundance. It was suggested that the degree of uncoupling between bacteria and HNF should be related to *Daphnia* abundance. The lack of correlation between bacteria and HNF abundance and the highly significant correlation between *Daphnia* and HNF abundance found in this study seem to follow the same trends as found by Gasol (1994). The effects of *Daphnia* predation may be more pronounced if the relationships were based on biomass instead of abundance because large and small cladocerans differ in feeding behavior (Jürgens, 1994).

From the present study, which was based on a large number of observations of bacteria, HNF, and *Daphnia* abundances, it was concluded that HNF populations in shallow and eutrophic lakes have the potential to proliferate when *Daphnia* abundance is below a few individuals per liter. This occurs typically at low temperatures (spring), low macrophyte densities, and/or high fish predation. Generally, the presence of *Daphnia* was the most important factor for regulating HNF population dynamics, whereas temperature and bacterial abundance were only marginally important. It appears that the mechanisms operating directly on the abundance and growth of *Daphnia* in eutrophic systems also have a strong but indirect effect on the microbial community. This does not necessarily mean that a one-way top-down effect is prevailing, as other factors than predation may regulate the development of *Daphnia* (e.g., food limitation or toxic cyanobacterial blooms).

Acknowledgments. I am grateful to Josep Gasol, Klaus Jürgens, and Brian Moss for valuable comments on the manuscript. Erik Jeppesen, Martin Søndergaard, Morten Søndergaard, Jon T. Nielsen, and Uffe K. Rasmussen have kindly provided data on densities of macrophytes, zooplankton, and bacteria. Nils Willumsen has provided valuable technical assistance, and Anne Mette Poulsen made linguistic corrections. Financial support was in part provided by the Strategic Environmental Research Programme—Centre for Freshwater.

References

Arndt, H. Rotifers as predators on components of the microbial web (bacteria, heterotrophic flagellates, ciliates)—a review. Hydrobiologia 255/256:231–246; 1993.

Arndt, H. Protozoen als wesentliche Komponente pelagischer Ökosysteme von Seen. Kataloge Landesmuseum N. F. 71:111–147; 1994.

Berninger, U-G.; Finlay, B.J.; Kuuppo-Leinikki, P. Protozoan control of bacterial abundance in freshwater. Limnol. Oceanogr. 36:139–147; 1991.

Carrick, H.J.; Fahnenstiel, G.L.; Taylor, W.D. Growth and production of planktonic protozoa in Lake Michigan: in situ versus in vitro comparisons and importance to food web dynamics. Limnol. Oceanogr. 37:1221–1235; 1992.

Christoffersen, K.; Riemann, B.; Klysner, A.; Søndergaard, Mo. Potential role of fish predation and natural populations of zooplankton in structuring a plankton community in eutrophic lake water. Limnol. Oceanogr. 35:1429–1436; 1993.

Gasol, J.M. A framework for the assessment of top-down vs. bottom-up control of heterotrophic nanoflagellate abundance. Mar. Ecol. Prog. Ser. 113:291–300; 1994.

Gasol, J.M.; Vaqué, D. Lack of coupling between heterotrophic nanoflagellates and bacteria: A general phenomenon across aquatic systems? Limnol. Oceanogr. 38:657–665; 1993.

Güde, H. Direct and indirect influences of crustacean zooplankton on bacterioplankton of Lake Constance. Hydrobiologia 159:63–73; 1988.

Hansen, B.; Christoffersen, K. Specific growth rates of heterotrophic plankton organisms in a eutrophic lake during a spring bloom. J. Plankton Res. 17:413–430; 1995.

Jeppesen, E.; Sortkjær, O.; Søndergaard, Ma.; Erlandsen, M. Impact of a trophic cascade on heterotrophic bacterioplankton production in two shallow fish-manipulated lakes. Arch. Hydrobiol. Beih. Ergebn. Limnol. 37:219–231; 1992.

Jeppesen, E.; Søndergaard, Ma.; Søndergaard, Mo.; Christoffersen, K.; Jürgens, K.; Theil-Nielsen, J.; Schlüter, L. Cascading trophic interactions in the littoral zone of a shallow lake (submitted).

Jürgens, K. Impact of *Daphnia* on planktonic microbial food webs—a review. Mar. Microb. Food Webs 8:295–324; 1994.

Jürgens, K.; Stolpe, G. Seasonal dynamics of crustacean zooplankton, heterotrophic nanoflagellates and bacteria in a shallow, eutrophic lake. Freshwat. Biol. 33:27–38; 1995.

Pace, M.L.; Funke, E. Regulation of planktonic microbial communities by nutrients and herbivores. Ecology 72:904–914; 1991.

Riemann, B. Potential importance of fish predation and zooplankton grazing on natural populations of freshwater bacteria. Appl. Environ. Microbiol. 50:187–193; 1985.

Riemann, B.; Christoffersen, K. Microbial trophodynamics in temperate lakes. Mar. Microb. Food Webs 7:69–100; 1993.

Sanders, R.W.; Caron, D.A.; Berninger, U-G. Relationships between bacteria and heterotrophic nanoplankton in marine and fresh waters: an inter-ecosystem comparison. Mar. Ecol. Prog. Ser. 86:1–14; 1992.

Sanders, R.W.; Porter, K.G. Bacteriovorous flagellates as food resources for the freshwater crustacean zooplankton *Daphnia ambigua*. Limnol. Oceanogr. 34:673–687; 1990.

Sanders, R.W.; Leeper, D.A.; King, C.H.; Porter, K.G. Grazing by rotifers and crustacean zooplankton on nanoplanktonic protists. Hydrobiologia 288:167–181; 1994.

Schriver, P.; Bøgestrand, J.; Jeppesen, E.; Søndergaard, Mo. Impact of submerged macrophytes on the interactions between fish, zooplankton and phytoplankton: large-scale enclosure experiments in a shallow lake. Freshwat. Biol. 33:255–270; 1995.

Stoeckner, D.K.; Capuzzo, J.M. Predation on protozoa: its importance to zooplankton. J. Plankton Res. 12:891–908; 1990.

Vaqué, D.; Pace, M.L. Grazing on bacteria by flagellates and cladocerans in lakes of contrasting food-web structure. J. Plankton Res. 14:307–321; 1992.

Weisse, T. The annual cycle of heterotrophic freshwater nanoflagellates: role of bottom-up vs. top-down control. J. Plankton Res. 13:167–185; 1991.

18. What Do Herbivore Exclusion Experiments Tell Us? An Investigation Using Black Swans (*Cygnus atratus*) and Filamentous Algae in a Shallow Lake

Robert T. Wass and Stuart F. Mitchell

Introduction

One common approach to the problems of quantifying herbivory is to determine how much plant material herbivores eat. Another is to exclude the herbivores experimentally and compare the performances of the grazed and ungrazed plant communities. Both approaches may give biased estimates of the interaction strength or effect of grazers on plant biomass increase or productivity (Mitchell and Wass, 1996a). Simple consideration of the fraction of annual plant productivity consumed ignores the effect of the time at which the material is consumed. Because grazing removes not only biomass but also the future productive potential of that biomass, consumption of small amounts of plant tissue early in a plant growth cycle has a greater effect than similar amounts of consumption later (e.g., Kiørboe, 1980). It also neglects the indirect feedback effects of herbivores, such as nutrient recycling, damage to the plants, or relief of density suppression of growth (e.g., Lodge, 1991).

Potential bias in herbivore exclusion experiments arises largely from the particular experimental period chosen. The sensitivity of plants to early grazing makes the starting time and conditions critically important. Earlier onset of density limitation in the ungrazed plants than in the grazed plants may make the time chosen to end the experiment equally important for its apparent outcome and may provide spurious indications of stimulation of productivity by grazing. Finally, results of such experiments have often been quantified with formulations that are static or otherwise inappropriate (Mitchell and Wass, 1996a).

Differences between plant biomasses in grazed and ungrazed plots are still often considered to represent grazing consumption or removal (e.g., Cyr and Pace, 1993; Cattaneo and Mousseau, 1995). We have argued that they do not (Mitchell and Wass, 1996a). In addition to the confounding effects of timing and density dependence, ungrazed plants do not experience the feedback effects of herbivores, which are part of the natural system. There is also an effect (biomass compounding) from the biomass dependence of productivity, which can be expected to cause the differences between the biomasses to overestimate consumption or removal by amounts that increase exponentially until density limitation is reached. This artifact can be removed by expressing biomass changes and grazing consumption as tissue-specific rates (e.g., Geertz-Hansen et al., 1993).

Concurrent use of the two approaches, in combination with a simple exponential model of plant growth, provides a potential means of isolating the grazing and net feedback effects of herbivores on plant biomass increase or productivity (Mitchell and Wass, 1996a). A common formulation is

$$\text{Net productivity} = B\,(B_{ig} + C + L) \tag{1}$$

where B is the initial plant biomass, and B_{ig}, C, and L are, respectively, the instantaneous rates of biomass increase in the grazed plants, grazing consumption, and losses due to decay, organic secretion, and export. This or equivalent formulations have been widely used in studies of grazing and plant production. It is unbiased. An alternative equation is

$$\text{Net productivity} = B\,(B_{iu} + L) \tag{2}$$

where B_{iu} is the instantaneous rate of ungrazed biomass increase. This equation is biased by the absence of feedback effects of the herbivores on the plants. Enumeration of the terms in the two equations may allow the net feedback effect to be quantified.

Our objectives were to test this model in a system in which feedback effects were likely to be negligible, so that the two equations become essentially equivalent and to evaluate the use of the two approaches for estimating grazing consumption and interaction strength. The investigation formed a part of wider studies on the role of black swans in shallow lake ecosystems. The study was carried out at Hawksbury Lagoon, New Zealand, where black swans are the only large deep-feeding herbivores, in 1990–1991, when the benthic vegetation consisted almost entirely of filamentous green algae. Absence or near-absence of feedback was inferred from the negligible contribution of swan excreta to nutrients in the system (Mitchell and Wass, 1995) and the lack of morphological differentiation in the algae, which eliminates the potential effects of damage to individual whole plants, selective grazing, and resource allocation by the plants. Any algae detached by the swans and removed from the control area by water currents were expected to be largely replaced by a similar influx from the surrounding waters. Another potential feedback, bioturbation by swans, appears to be insignificant for the benthic light climate (Mitchell and Wass, 1996b).

The main basin of Hawksbury Lagoon is shallow ($\bar{z}$ = 0.4–0.6 m), flat-bottomed, highly eutrophic, and brackish. It has an area of 25 ha and is located at 46° S in South Island, New Zealand. It shows irregular nonseasonal cycles of dominance by phytoplankton and by benthic algae. These cycles are closely reflected by changes in the black swan population, with birds migrating to or from the lake in response to both the long-term changes in benthic algal biomass and often also the week-to-week changes (Mitchell and Wass, 1995, 1996b). Swans show no diel migration between the lake and the surrounding land. They feed actively at night, and this population consumes little terrestrial food (McKinnon, 1989). Predation and breeding recruitment at the lake itself are insignificant.

Materials and Methods

A single rectangular swan exclosure, 20×20 m, and a similar neighboring control area were established near the center of the lake, in a region that appeared to be fairly representative of the lake in terms of both swan use and benthic algal biomass. Long-term studies have shown that swan occupancy of the region of the study area was similar to that in the whole lake. The control and experimental areas were defined by wooden stakes. Plastic mesh 0.5 m high was nailed to the stakes above the water line of the exclosure, to exclude swans but to avoid impeding the exchange of water with the lake. Duplicate algal samples were taken randomly from each of 21 sample blocks within each area by wading along defined walkways. Samples were taken with a 0.004-m^2 core sampler, collected by sieving on a 1-mm mesh, washed, dried, and weighed (McKinnon and Mitchell, 1994). Swans were counted from the shore.

Exponential rates of biomass increase (per day [i.e., g/g/day]) were calculated for the grazed and ungrazed algae for each sampling interval as $(\ln B_t - \ln B_0)/t$, where B_0 and B_t were the initial and final arithmetic mean biomasses and t is time in days. Tissue-specific grazing rates (per day) were similarly calculated from plant biomass and grazing consumption, from the feeding rate of 105 g DW/swan/ day. This figure was derived near the end of our study by using cellulose as an indigestible food marker, with fecal production being determined by two independent methods (Mitchell and Wass, 1995). Instantaneous loss rates were not determined but were assumed to be density-independent.

Results

Algal biomass increased rapidly in the exclosure from the beginning of the experiment in the southern autumn of 1990 (Fig. 18.1). The increase was highly significant (*t*-test, ln transformed biomass, $P < .01$) but brief. Control biomass remained unchanged through this period. The swan population was fairly high through winter (Fig. 18.1) but had no detectable effect on the algal biomass. A grazing effect quickly became apparent with the onset of spring growth. The

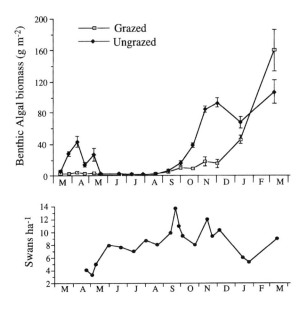

Figure 18.1. Benthic algal biomass (dry weight) in the enclosure and control areas of Hawksbury Lagoon and swan population density 1990–1991. Error bars = SE.

ungrazed biomass increased at a rate, fitted by regression, of 0.042/day ($r^2 = 0.99$, 95% confidence interval = 0.036–0.048/day) until November, when the algae apparently became density limited. By this time, they had grown to the surface, forming a thick mat across the exclosure, and there was no further significant increase in biomass. The biomass of the grazed community continued to increase, in close accordance with the exponential model, at a rate of 0.017/day until the end of the study in March ($r^2 = 0.95$, 95% confidence interval = 0.012–0.021/day). By March, the ungrazed and grazed biomasses were no longer significantly different, so that the net effect of grazing in the spring-autumn period was effectively zero. Algae identified during the study were *Enteromorpha intestinalis, Rhizoclonium* sp., and *Spirogira* sp. Trace amounts of *Ruppia megacarpa* and *Nitella hookeri* were recorded occasionally. The algal biomass was 2 orders of magnitude higher at the end of the experiment than at the beginning, a year earlier. Intermittent observations indicated that it remained high for the next 2 years, before collapsing in autumn 1993 (Mitchell and Wass, 1996b).

The difference between the growth rates of the control and experimental algae implies an average spring grazing rate of 0.025/day. The rates were, however, variable. For example, the direct estimates were as high as 0.042/day in July–August but declined progressively as the swan population density failed to keep pace with the increasing algal biomass (Table 18.1; Fig. 18.1). By the end of the study they were less than 0.001/day.

Table 18.1. Net Benthic Algal Productivity in Hawksbury Lagoon, Calculated from Equations 1 and 2[a]

Sampling interval 1990	Initial (B) and final grazed biomass (g/m²) (SE initial biomass, $n = 21$)	Grazed biomass increase (B_{ig}) (per day)	Mean grazing rate (C) (per day)	Initial and final ungrazed biomass (g/m²) (SE initial biomass, $n = 21$)	Ungrazed biomass increase (B_{iu}) (per day)	Production $B(B_{ig} + C)$ (g m⁻²/day)	Production $B \times B_{iu}$ (g m⁻²/day)
Aug 19–Sept 10	3.1– 5.6 (0.37)	0.027	0.022	2.3– 7.8 (0.32)	0.055	0.152	0.170
Sept 10–Oct 1	5.6–10.1 (0.73)	0.028	0.015	7.6–16.5 (1.05)	0.039	0.241	0.218
Oct 1–Oct 22	10.1– 9.8 (1.10)	–0.001	0.009	16.5–38.9 (3.1)	0.041	0.081	0.414
Oct 22–Nov 12	9.8–18.1 (1.51)	0.029	0.008	38.9–84.6 (2.97)	0.037	0.363	0.363

[a]Loss rates are not accounted for. Grazing rates are geometric means of initial and final direct estimates for the sampling interval. Biomass is DW.

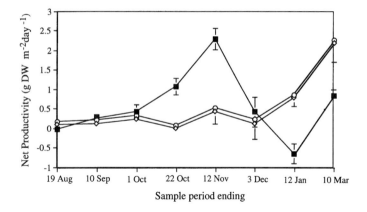

Figure 18.2. Net productivity of filamentous algae in Hawksbury Lagoon in spring-summer 1990–1991. ■ = productivity of ungrazed algae calculated from biomass changes; ○ = productivity of grazed algae, calculated from biomass changes (◊) plus amounts removed by swan grazing. Error bars = SE for biomass change.

Results of the two methods of calculating productivity agreed closely in three of the four sampling intervals for which they could be compared (Table 18.1). There was, however, a large difference in the interval from October 1–22, when the biomass of the grazed algae declined. We were unable to test this model after mid-November, as the ungrazed algae had become density limited. The productivity of the grazed algae was higher than that of the ungrazed community in two of the three subsequent samples (Fig. 18.2).

Annual net productivity (April–March), calculated from grazed biomass increases plus grazing consumption, was 185 g DW/m^2, of which 22 g DW/m^2 or 12% was removed by grazing.

Discussion

The agreement between the two independent productivity estimates in three of the four tests indicates that our approach may be useful for isolating direct consumption and feedback effects in other systems. The discrepancy in the other test might suggest that there was a large negative feedback in this interval. As feedback effects can be expected to act continuously, however, it seems more likely to represent a violation of the assumption that loss rates were density-independent. The grazed and ungrazed algae might, for example, differ in their susceptibility to being dislodged by waves. Also, the ungrazed algae that had grown toward the lake surface must have experienced a different light climate from the natural community, confined by grazing to the bottom few centimeters. Although the water was generally clear, this effect was potentially important during any resuspension events, as rates of decline in algal biomass in the lake during periods of low benthic light may be high (Mitchell and Wass, 1996b). Such effects will obviously become increasingly likely as the biomasses diverge. It is highly desirable that loss rates should be quantified in any further attempts to use this model. Alternatively, the effect of any density dependence of loss rates could be reduced by conducting sequential short-term experiments throughout a growth cycle. It is also clear that ability to detect and estimate feedback will be subject to the high sampling variation in most plant communities. At present very little is known of the magnitude of feedback effects, although Lodge (1991) among others has argued that they are potentially very important.

The final outcome of the experiment was that swans had no significant effect on the algal biomass. If we had ended the experiment earlier or begun it later, we might have reached a different conclusion. For example, on November 12 the grazed biomass was only 17% of the ungrazed biomass (Fig. 18.1). It is also predictable that if we had delayed the start of the experiment, the seasonal dynamics would have been different, with the ungrazed algae reaching the carrying capacity progressively later as the delay increased. These results reinforce the doubts that we have expressed about the interpretation of herbivore exclusion experiments conducted over arbitrary experimental periods, without reference to plant growth cycles (Mitchell and Wass, 1996a).

Our results also illustrate the reality of biomass compounding. Algal productivity estimates, uncorrected for this effect, are shown in Figure 18.2 with the estimates from equation 1. At the extreme (October 22–November 12), the difference between the ungrazed and grazed biomasses indicated a consumption or removal of 83% of the net primary productivity or biomass, whereas the correctly formulated estimate for this interval was 14%, and cumulative consumption from August 19 to November 12 was only 10% of the final ungrazed biomass. Similar differences between direct estimates of algal consumption and indirect estimates based on the crude difference between grazed and ungrazed biomasses are common in periphyton communities (Cattaneo and Mousseau, 1995). These authors attributed them to herbivore-induced losses of algae by sloughing. The level of agreement between the estimates from equations 1 and 2 confirms that this was not so in our study. Herbivores can neither remove nor eat material that has never been present in the real (grazed) system.

The higher algal productivity in the control area than in the enclosure in the last two sample intervals was not due to stimulation of productivity by the herbivores. It occurred simply because the ungrazed algae had become density limited and the grazed algae had not.

We conclude that grazing consumption cannot normally be determined without studying grazers. It can be estimated from herbivore exclusion experiments only by use of tissue-specific rates, when there are no feedback effects of the herbivores on the plants, and while plant growth remains density-independent. Interaction strengths can be understood and quantified adequately only by dynamic analysis, preferably over complete natural cycles of plant biomass increase and decline.

A potential increase in the algae was completely suppressed by swan grazing in autumn 1990. Opportunity for this effect to occur is, however, very limited. At the average spring growth rate of 0.042/day, an algal biomass of only 2 g/m² would be required for production to meet the food requirements of the 8–10 swans/ha that were present. Black swans can also play little role in the loss of benthic algae when the lake switches to phytoplankton dominance, as the final biomass represents several years' consumption by even the highest swan population ever observed there.

The high benthic algal biomass in autumn 1991 and the subsequent 2 years of benthic algal dominance of the lake are in contrast to 1994, when turbidity from phytoplankton and resuspended sediment caused *Nitella* to collapse in February–March (Mitchell and Wass, 1996b). Black swans contributed little to either outcome.

Acknowledgments. The study was supported by a University of Otago Research Grant. We are also grateful to Brian Niven and Brian Manly of the Department of Mathematics and Statistics at Otago University for statistical advice.

References

Cattaneo, A.; Mousseau, B. Empirical analysis of the removal rate of periphyton by grazers. Oecologia. 103:249–254; 1995.

Cyr, H.; Pace, M.L. Magnitude and patterns of herbivory in aquatic and terrestrial ecosystems. Nature 361:148–150; 1993.

Geertz-Hansen, O.; Sand-Jensen, K.; Hansen, D.F.; Christiansen, A. Growth and grazing control of the abundance of the marine macroalga *Ulva lactuca* L. in a eutrophic Danish estuary. Aquat. Bot. 46:101–109; 1993.

Kiørboe, T. Distribution and production of submerged macrophytes in Tipper Grund (Ringkøbing Fjord, Denmark), and the impact of waterfowl grazing. J. Appl. Ecol. 17:675–687; 1980.

Lodge, D.M. Herbivory on freshwater macrophytes. Aquat. Bot. 41:195–224; 1991.

McKinnon, S.L. The interrelationship between phytoplankton, submerged macrophytes and black swans (*Cygnus atratus*) in New Zealand lakes—test of two models. MSc thesis, University of Otago, Dunedin; 1989.

McKinnon, S.L.; Mitchell, S.F. Eutrophication and black swan (*Cygnus atratus* Latham) populations: tests of two simple relationships. Hydrobiologia 279/280:163–170; 1994.

Mitchell, S.F.; Wass, R.T. Food consumption and faecal deposition of plant nutrients by black swans (*Cygnus atratus* Latham) in a shallow New Zealand lake. Hydrobiologia 306:189–197; 1995.

Mitchell, S.F.; Wass, R.T. Quantifying herbivory-grazing consumption and interaction strength. Oikos 76:573–576; 1996a.

Mitchell, S.F.; Wass, R.T. Grazing by black swans (*Cygnus atratus* Latham), physical factors, and the growth and loss of aquatic vegetation in a shallow lake. Aquat. Bot. 55:205–215; 1996b.

19. Switches Between Clear and Turbid Water States in a Biomanipulated Lake (1986–1996): The Role of Herbivory on Macrophytes

Ellen Van Donk

Introduction

Shallow lakes may display alternate stable states over a range of nutrient concentrations, a clearwater state dominated by aquatic vegetation, and a turbid state characterized by high algal biomass (Scheffer et al., 1993). This may have important implications for the possibilities of restoring eutrophied shallow lakes.

Man-made modification of fish populations ("biomanipulation") has been applied successfully to several small shallow lakes to induce a transition from a phytoplankton-dominated state to a clearwater state with submerged macrophytes (e.g., Hanson and Butler, 1994; Lauridsen et al., 1994b; Meijer et al., 1994). These macrophytes play a key role in several mechanisms that tend to keep the system in a clearwater state at relatively high nutrient loadings (Jeppesen et al., this volume, Chapter 5). An important question underlying the use of biomanipulation as a restoration technique is its long-term effectiveness at different nutrient loadings (Jeppesen et al., 1990). Most studies published thus far lasted less than 5 years, whereas according to Frost et al. (1988), studies of fish manipulations, which involve complex interactions, should extend for longer periods.

In Lake Zwemlust, a eutrophic lake located in the middle of The Netherlands, biomanipulation measures (major changes in total fish community) were taken in March 1987 (van Donk et al., 1989). After this biomanipulation, the lake shifted several times between the turbid and the clearwater state. In this chapter, I give an overview of this switching behavior over a 10-year period (1986–1996) in relation to changes in macrophyte composition and herbivory by waterfowl (coots) and fish (rudd). An account of the period until 1994, characterized by a return of turbidity after a clear period of several years, was given previously (van Donk and Gulati, 1995). The present extension until 1996 covers an autonomous but temporary recovery of a clearwater state.

Methods

Lake Zwemlust is a small water body (area, 1.5 ha; mean depth, 1.5 m) situated in the middle of The Netherlands. The external P and N loadings to the lake are high and estimated at approximately 2 g P/m/yr and 9 g N/m/yr, respectively. Before biomanipulation in 1987, phytoplankton blooms were dominant during the whole year (Secchi depth, 0.3 m). Extensive description of the limnology of the lake, before and after biomanipulation, is given in van Donk et al. (1990, 1993) and van Donk and Gulati (1995). The biomanipulation measures are discussed at some length by van Donk et al. (1989, 1990).

Biomass and composition of submerged macrophytes in the lake were estimated according to Ozimek et al. (1990). The method of Prejs (1984) was followed to estimate the consumption of macrophytes by rudd. As for coot grazing on submerged macrophytes, a daily intake of about 45 g DW was found by Hurter (1979). The consumption of macrophytes by coots was estimated from this daily consumption per coot and the number of "birds days" (average number of birds per day multiplied by number of days).

Exclosures were used to evaluate grazing effects by fish and birds on macrophyte species composition. Cages made of an iron frame with dimensions of 4 m (length) × 1.5 m (width) × 0.6 m (height) and covered by wire netting served as exclosures for larger fish and birds. These were placed on the lake bottom at a depth of 2.0 m in May 1992, and the experiment lasted until July 1993. Initially, macrophyte characteristics (species composition and biomass) in the cages were similar to those in the lake. At the end of this experimental period, the percentages of vegetated area occupied by the different macrophyte species inside and outside the cages were determined. The design of these experiments is described in more detail in van Donk and Otte (1996).

Results and Discussion

Switches Between the Different States Related to Macrophyte Species Composition (1986–1996)

After application of biomanipulation in 1987, the lake shifted within 1 year from a turbid state dominated by phytoplankton and no submerged vegetation (state I in Fig. 19.1 and Table 19.1) to a clearwater state dominated by submerged macrophytes (state III). From 1988 until 1992, the lake remained clear during the whole

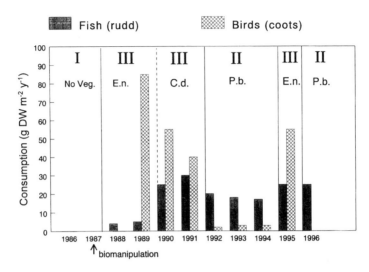

Figure 19.1. Estimates of annual macrophyte consumption by fish (rudd) and birds (coots) (macrophyte g DW/m/yr) before and after biomanipulation in Lake Zwemlust. The three different states (see Table 19.1) and the dominant submerged macrophyte species are given. State I, turbid state; state II, clear/turbid state; state III, clearwater state. No Veg, no submerged vegetation; E.n., *Elodea nuttallii;* C.d., *Ceratophyllum demersum;* P.b., *Potamogeton berchtoldii.*

year, mainly due to macrophyte-induced nitrogen limitation of the phytoplankton growth during summer and autumn (van Donk et al., 1993; van Donk and Gulati, 1995). In the first 2 years after biomanipulation (1988 and 1989), *Elodea nuttallii* was the dominant macrophyte species. *Elodea* is a perennial submerged macrophyte and rather insensitive to lower temperatures in late autumn and winter. In the subsequent period of 2 years (1990 and 1991), *Ceratophyllum demersum* became more dominant. Both *Elodea* and *Ceratophyllum* are able to compete strongly with phytoplankton for nutrients, especially for nitrogen (Best, 1977; van Donk et al., 1993).

In 1992, 1993, and 1994, the lake was turbid during late summer and autumn, but otherwise remained clear. In these years, *C. demersum* and *E. nuttallii* were nearly absent and *Potamogeton berchtoldii* became the dominant species in spring, declining to very low abundance during late summer when it started to form turions (overwintering structures). *Potamogeton* species are known to form these structures in autumn (Sastroutomo, 1981). *P. berchtoldii* was progressively covered with epiphytes from late spring to summer (van Donk and Gulati, 1995). No significant growth of epiphytes was observed on the other macrophyte species. The early collapse of *Potamogeton* was probably caused by light limitation due to epiphyte growth, inducing early formation of turions (van Donk and Gulati, 1995). Relative abundance of phytoplankton increased after the collapse of macrophytes and the rise of epiphytes. Consequently, the lake shifted to a turbid state (state II)

Table 19.1. The Three Different States and Their Ecosystem Structures as Observed in Lake Zwemlust over a 10-Year Period (1986–1996)[a]

State	Submerged macrophytes	Fish	Phytoplankton chlorophyll *a* (μg/L)	Zooplankton	Birds	Phytoplankton limitation
I (Turbid)	No	Planktiv Benthiv	80–250 whole year	Rotifers	No	Light limitation
II (Clear/turbid)	Species forming overwintering structures (e.g., *Potamogeton*)	Herbiv Pisciv	<30 (spring/summer) 30–80 (autumn)	Daphnids Rotifers	No	Zooplankton & N-limitation (spring/summer) Light limitation (autumn)
III (clear)	Perennials (e.g., *Elodea*)	Herbiv Pisciv	<30 whole year	Large daphnids	Herbiv	Zooplankton-grazing N-limitation

[a]Planktiv, planktivorous; benthiv, benthivorous; herbiv, herbivorous; pisciv, piscivorous.

but only during late summer and autumn. State II differs from state I in that a clearwater state still exists during winter and early summer (Table 19.1).

In 1995, *E. nuttallii* reappeared and became the dominant macrophyte species. The clearwater state (state III) was re-established during the whole year. In 1996, however, when *Elodea* was very scarce and *Potamogeton* abounded, the lake shifted again to the turbid state II.

Shifts in Macrophyte Species Composition Related to Herbivory

Herbivory by vertebrate and invertebrate grazers may play an important role in structuring and lowering the macrophyte biomass (e.g., Lodge, 1991; Lodge et al., this volume, Chapter 8). In Lake Zwemlust, herbivory by coots (*Fulica atra*) and rudd (*Scardinius erythrophthalmus*) was observed after re-establishment of the macrophytes (van Donk et al., 1994).

Extensive grazing of coots on submerged macrophytes occurs mainly during autumn and winter, after territories break up (Kiørboe, 1980; Perrow et al., 1997). Kiørboe (1980) stated that grazing by coots has only a minimal effect on macrophyte growth because grazing often takes place outside the growing season of the plants. Coots, however, pull out whole plants and may influence the macrophyte composition and succession by removing especially plants still present during autumn and winter. A high number of coots appeared in Lake Zwemlust when perennial macrophytes became dominant after biomanipulation (Fig. 19.1). An increase of herbivorous birds, after the restoration of submerged macrophyte communities, was also observed in various other lakes (Hanson and Butler, 1994; Hargeby et al., 1994; Lauridsen et al., 1994a; Schutten et al., 1994; Søndergaard et al., this volume, Chapter 20). High grazing pressure on *Elodea* and *Ceratophyllum* during the first years after biomanipulation probably promoted the rise of *Potamogeton* (van Donk and Gulati, 1995). *P. berchtoldii* was not negatively affected by coots because this species forms nongrazable overwintering structures. From 1992 to 1995, virtually no coots were present, probably due to absence of submerged vegetation during autumn and winter. After *Elodea* became abundant again in 1995, coots returned in winter 1995/1996 and started to graze on *Elodea*. In 1996, *Potamogeton* reappeared, followed by phytoplankton blooms during the summer.

Rudd was introduced in Lake Zwemlust in 1987 as food for pike (van Donk et al., 1989). According to Prejs (1984), only larger rudd (>1[+]) are herbivorous. The number of larger rudd was quite low until 1990 but increased in 1991 to 297 kg/ha and stabilized in 1992–1996 at 200–300 kg/ha. Rudd only graze during the growing season of the macrophytes (temp, >16°C) and had therefore probably a much lower impact on total macrophyte biomass compared with coots (van Donk et al., 1994) (Fig. 19.1). Prejs (1984) even suggested that grazing by rudd may stimulate the growth of the plants. In laboratory experiments, rudd were observed to consume mainly *Elodea* and *Potamogeton* but not *Ceratophyllum,* which is calcareous in structure and has apparently a much lower edibility (Prejs and Jackowska, 1978; Teule, 1993). In this way, rudd may, like coots, induce a change in macrophyte species composition.

In the cages designed to exclude herbivory by both larger fish and birds, *P. berchtoldii* was the dominant species at the start of the experiment in spring 1992. However, in early spring 1993, the exclosures became dominated by *E. nuttallii,* whereas in the lake, *P. berchtoldii* was still abundant. The coverage percentage of macrophytes in the exclosures was higher than that in the lake (van Donk and Otte, 1996). In cages excluding grazing by coots, Søndergaard et al. (1996) found a significantly higher macrophyte biomass than outside the exclosures, but in contrast to our findings, no shift in macrophyte species composition was reported. Further results of the exclosure experiments are described in van Donk and Otte (1996).

Conclusions

Ten years after application of biomanipulation, Lake Zwemlust has undergone several switches between turbid and clearwater states. The presence of submerged macrophytes seems to be essential in keeping this lake clear. The observations suggest that dominance of perennials such as *Elodea* and *Ceratophyllum* leads to a whole-year clearwater state, whereas species forming overwintering structures, such as *Potamogeton,* give rise to epiphyte growth in early summer and phytoplankton blooms in late summer and autumn. Herbivory, especially by coots, was probably an important factor in triggering the shift from perennials to nonperennials. The switch in 1995 from state II (turbid) back to state III (clear) was rather unexpected. Probably, the perennial *Elodea* was able to re-establish itself due to absence of coots in the previous winters. This conclusion is based on the results of the exclosure experiments. Exclosures, however, can lead to misleading conclusions because they may also change other factors such as light, climate, and water currents. Rørslett et al. (1986) observed that *Elodea canadensis* suddenly collapsed after some years of growth and then returned to its previous dominance years after, without any known external factors. In follow-up studies, not only the consumption but also the production of the macrophytes has to be considered. Also, more specific experimental research is needed to quantify the effect of herbivory on growth of different macrophyte species and consequently on stability of the clearwater state.

References

Best, E.P.H. Seasonal changes in mineral and organic components of *Ceratophyllum demersum* and *Elodea canadensis*. Aquat. Bot. 3:337–348; 1977.

Frost, T.M.; DeAngelis, D.L.; Bartell, S.M.; Hall, D.J.; Hulbert, S.H. Scale in the design and interpretation of aquatic community research. In: Carpenter, S.R., ed. Complex interactions in lake communities. New York: Springer-Verlag, 1988:229–258.

Hanson, M.A.; Butler, M.G. Responses to food web manipulation in a shallow waterfowl lake. Hydrobiologia 279/280:457–466; 1994.

Hargeby, A.; Anderssen, G.; Blindow, I.; Johansson, S. Trophic web structure in a shallow eutrophic lake during a dominance shift from phytoplankton to submerged macrophytes. Hydrobiologia 279/280:83–90; 1994.

Hurter, H. Nahrungsökologie des Blässhuhn (*Fulica atra*) an den Überwinterungsgewässern im nördlichen Alpenvorland. Der Ornitologische Beobachter 76:257–288; 1979.

Jeppesen, E.; Jensen, J.P.; Kristensen, P.; Søndergaard, M.; Mortensen, E.; Sortkjær, O.; Olrik, K. Fish manipulation as a lake restoration tool in shallow, eutrophic, temporate lakes 2: threshold levels, long-term stability and conclusions. Hydrobiologia 200/201: 219–227; 1990.

Kiørboe, T. Distribution and production of submerged macrophytes in Tripper Ground, and the impact of waterfowl grazing. J. Appl. Ecol. 17:675–687; 1980.

Lauridsen, T.L.; Jeppesen, E.; Østergaard Andersen, F. Colonization and succession of submerged macrophytes in shallow fish manipulated Lake Væng: impact of sediment composition and waterfowl grazing. Aquat. Bot. 46:1–15; 1994a.

Lauridsen, T.L.; Jeppesen, E.; Søndergaard, M. Colonization and succession of submerged macrophytes in shallow Lake Væng during the first five years following fish manipulation. Hydrobiologia 275/276:233–242; 1994b.

Lodge, D.M. Herbivory on freshwater macrophytes. Aquat. Bot. 41:195–224; 1991.

Meijer, M.-L.; Jeppesen, E.; van Donk, E.; Moss, B.; Scheffer, M.; Lammens, E.; Van Nes, E.; Faafeng, B.A.; Jensen, J.P. Long-term responses to fish-stock reduction in small shallow lakes: interpretation of five year results of four biomanipulation cases in The Netherlands and Denmark. Hydrobiologia 275/276:457–467; 1994.

Ozimek, T.; van Donk, E.; Gulati, R.D. Can macrophytes be useful in biomanipulation of lakes? The Lake Zwemlust example. Hydrobiologia 200/201:399–409; 1990.

Perrow, M.R.; Schutten, J.; Howes, J.R.; Holzer, T.; Madgwick, F.J.; Jowitt, A.J.D. Interactions between coot (*Fulica atra*) and submerged macrophytes: the role of birds in the restoration process. Hydrobiologia 342/343:241–255; 1997.

Prejs, A. Herbivory by temperate freshwater fishes and its consequences. Environ. Biol. Fish. 10:281–296; 1984.

Prejs, A.; Jackowska, H. Lake macrophytes as the food of roach (*Rutilus rutilus*) and rudd (*Scardinius erythrophthalmus* L.) I. Species composition and dominance relations in the lake and the food. Ekol. Pol. 26:429–438; 1978.

Rørslett, B.; Berge, D.; Johansen, S.W. Lake enrichment by submerged macrophytes: a Norwegian whole-lake experience with *Elodea canadensis*. Aquat. Bot. 26:325–340; 1986.

Sastroutoma, S.S. Turion formation, dormancy and germination of curly pondweed, *Potamogeton crispus* L. Aquat. Bot. 10:161–173; 1981.

Scheffer, M.; Hosper, S.H.; Meijer M.-L.; Moss, B.; Jeppesen, E. Alternative equilibria in shallow lakes. Trends Ecol. Evol. 8:275–279; 1993.

Schutten, J.; Van der Velden, A.; Smit, H. Submerged macrophytes in the recently freshened lake system Volkerak-Zoom (The Netherlands), 1987–1991. Hydrobiologia 275/276: 207–218; 1994.

Søndergaard, M.; Bruun, L.; Lauridsen, T.; Jeppesen, E.; Vindbæk Madsen, T. The impact of grazing waterfowl on submerged macrophytes: in situ experiments in a shallow eutrophic lake. Aquat. Bot. 53:73–84; 1996.

Teule, K. The effect of rudd (*Scardinius erythrophthalmus* L.) on the macrophyte biomass and species composition in Lake Zwemlust (in Dutch). Report 3062. Wageningen: Agricultural University; 1993.

van Donk, E.; Gulati, R.D. Transition of a lake to turbid state six years after biomanipulation: mechanisms and pathways. Wat. Sci. Techn. 32:197–206; 1995.

van Donk, E.; Otte, A. Effects of grazing by fish and waterfowl on the biomass and species composition of submerged macrophytes. Hydrobiologia 340:285–290; 1996.

van Donk, E.; Gulati, R.D.; Grimm, M.P. Food-web manipulation in Lake Zwemlust: positive and negative effects during the first two years. Hydrobiol. Bull. 23:19–34; 1989.

van Donk, E.; Grimm, M.P.; Gulati, R.D.; Klein Breteler, J.P.G. Whole-lake food-web manipulation as a means to study community interactions in a small ecosystem. Hydrobiologia 200/201:275–291; 1990.

van Donk, E.; Gulati, R.D.; Iedema, A.; Meulemans, J.T. Macrophyte-related shifts in the nitrogen and phosphorus contents of the different trophic levels in a biomanipulated shallow lake. Hydrobiologia 251:19–26; 1993.

van Donk, E.; De Deckere, E.; Klein Breteler, J.P.G.; Meulemans, J.T. Herbivory by waterfowl and fish on macrophytes in a biomanipulated lake: effects on long term recovery. Verh. Int. Verein. Limnol. 25:2139–2143; 1994.

20. Macrophyte–Waterfowl Interactions: Tracking a Variable Resource and the Impact of Herbivory on Plant Growth

Martin Søndergaard, Torben L. Lauridsen, Erik Jeppesen, and
Lise Bruun

Introduction

Submerged macrophytes are important for the maintenance of clear water in shallow eutrophic lakes (Jeppesen et al., 1990; Scheffer, 1990; Hargeby et al., 1994), and establishment of permanent macrophyte coverage is an important aspect of the lake restoration process after a reduction of nutrient loading. However, as macrophytes are subject to grazing by herbivores such as waterfowl (e.g., Lodge, 1991), it may be speculated whether waterfowl grazing can delay recolonization and thereby lake recovery. The impact of waterfowl grazing on macrophytes is poorly documented (Winfield, 1991), and few studies have considered it important, examples being Jupp and Spence (1977), who ascribed growth limitation of *Potamogeton filiformis* and *Potamogeton pectinatus* to wave action and waterfowl grazing, and van Donk et al. (1994), who observed that intensive coot herbivory in Lake Zwemlust (up to 120 individuals/ha) affected *Elodea* biomass and species composition.

In this study, we document large temporal variations in macrophyte and waterfowl abundance and assess the effect of waterfowl grazing on the growth of *Potamogeton crispus* in two shallow Danish lakes in the early stages of submerged macrophyte recolonization.

Methods and Study Areas

The studies were undertaken in Lake Væng and Lake Stigsholm; a detailed description of the methods applied and the two study areas can be found in Lauridsen et al. (1993, 1994) and Søndergaard et al. (1996). The main characteristics of the two lakes are summarized in Table 20.1.

Lake Væng was biomanipulated in 1986 and 1987 by removing approximately 50% of the bream (*Abramis brama*) and roach (*Rutilus rutilus*) fish stock (Søndergaard et al., 1990). After fish removal, mean summer Secchi depth increased from 0.7 m to more than 1.8 m (lake bottom, Table 20.1), primarily due to increased top-down control by large daphnids (Søndergaard et al., 1990; Table 20.1). Apart from short periods (a few weeks) with turbid water, the lake has remained clear since 1987 (Jeppesen et al., in press).

For the past decades, Lake Stigsholm has fluctuated between a macrophyte-rich clearwater state and a macrophyte-poor turbid state. This has been documented by palaeolimnological studies (B. Odgaard, unpublished observation) supported by observations of the varying number of mute swan (*Cygnus olor*), which ranged from 15 individuals/ha during the summer of 1967 to only 0.3 individuals/ha in 1969 (Skotte-Møller, 1970).

Macrophyte coverage and percentage volume infested (PVI, calculated as the product of percentage coverage and height divided by water depth) were estimated by using 9–14 transects in each lake together covering the whole lake (see Lauridsen et al., 1994, for details). Waterfowl abundance (coot and mute swan) was counted from 1991 and onward in Lake Væng and in 1990, 1994, autumn 1995, and 1996 in Lake Stigsholm.

To study the influence of grazing by coot (*Fulica atra*) and mute swan, *Potamogeton crispus* shoots were incubated at different locations (two locations in Lake Væng and seven in Lake Stigsholm). In Lake Væng, the shoots were planted

Table 20.1. Morphometric and Chemical (Mean Summer) Characteristics of Lake Væng and Lake Stigsholm

	Lake Væng	Lake Stigsholm
Area (ha)	15	21
Mean depth (m)	1.2	0.8
Max depth (m)	1.8	1.2
Hydraulic retention time (days)	21	5
Total phosphorus before 1987 (mg P/L)	0.13	—
Total phosphorus after 1987 (mg P/L)	0.08	0.15
Secchi depth (m)	0.7	—
Secchi depth (m)	1.2–1.8	1.2[a]
Chlorophyll *a* before 1987 (µg/L)	80	—
Chlorophyll *a* 1987–1995 (µg/L)	19	41

[a]Secchi depth >1.2 m in 27–73% of the samplings.

in two substrate types at each location: mud and sand. For each location and substrate type, eight pots, each with one *P. crispus* shoot, were placed in an unprotected box and in a box protected by a 2-cm mesh chicken wire fence that reached from the lake bottom to 20 cm above the lake surface. The boxes were placed 0.9 m (Lake Væng) and 0.5 m (Lake Stigsholm) below the lake surface. Mean shoot length, and in Lake Stigsholm also total shoot length, number of shoots, branches per shoot, and percentage stubble (calculated as the ratio of the number of shoots [including lateral shoots] lacking an apex to the total number of shoots × 100) per box, was measured on six to eight dates from May to August (1989) in Lake Væng and on four different dates during the period July 27 to September 3, 1990, in Lake Stigsholm. Further information can be found in Lauridsen et al. (1993) and Søndergaard et al. (1996).

Results and Discussion

The results from Lake Væng illustrate that marked seasonal and interannual variations in macrophyte coverage and species composition can take place during the recolonization phase following biomanipulation-mediated improvement in lake transparency (Fig. 20.1). Colonization started in the second and third year after biomanipulation, and by the end of the fourth year (1990), macrophyte coverage had reached 80% and PVI 54%. *P. crispus* was initially abundant, but *Elodea canadensis* soon took over and became completely dominant. *Elodea* coverage and PVI have varied widely from year to year, however. In autumn 1992 and summer 1993, *Elodea* had almost disappeared, but it regained high density again in summer 1994 and 1995, whereas it decreased again from 80% to 12% coverage in 1996.

Coot and mute swan number fluctuated with macrophyte density in the autumn and winter; high numbers were recorded only in years with high macrophyte density (Fig. 20.1). Maximum density occurred in winter 1991/1992, when up to 20 mute swans and 53 coots were recorded per hectare. Due to territorial behavior, waterfowl density was relatively low in the summer nesting season, irrespective of macrophyte abundance.

From 1989 to 1995, macrophyte coverage was relatively low in Lake Stigsholm, although Secchi depth was equivalent to water depth for long periods during the summer (Table 20.1). Macrophytes were present each year, albeit that the maximum coverage of all species (excepting macroalgae) never exceeded 30%. The dominant species were *Callitriche hermaphroditica* and *Potamogeton* (*pectinatus* and *berchtoldii*) (Fig. 20.2). In most years, filamentous algae (mainly *Enteromorpha* and *Spirogyra*) were present in large quantities for a short period during the summer. Apart from filamentous algae, which occasionally reached the surface, plant height rarely exceeded 10–20 cm, and PVI was therefore usually 25%. Macrophyte coverage during the summer was inversely correlated to chlorophyll *a*, and when chlorophyll *a* exceeded approximately 50 μg/L, macrophyte coverage was low, thus indicating that macrophyte density was closely related to

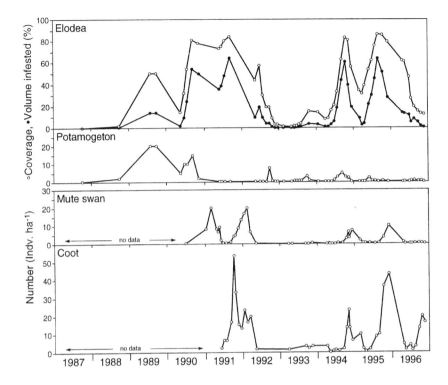

Figure 20.1. Development of submerged macrophytes (○ mean percentage coverage of total lake area and ●: mean PVI) in Lake Væng. Coot (*Fulica atra*) and mute swan (*Cygnus olor*) numbers are shown in the two lower panels.

changes in turbidity. By the end of 1995 and during 1996, macrophyte (mainly *Elodea*) coverage and PVI increased markedly (Fig. 20.2). During summer 1996, total coverage ranged between 50–70% and PVI between 18–26%. Mean summer chlorophyll *a* in 1996 was only 13 µg/L. In summer 1990 (when the grazing experiments were conducted), coot density ranged between 3–9 individuals/ha, whereas mute swan density was 0.2 individuals/ha. Corresponding with increasing macrophyte abundance, autumn densities of coot were high in 1995 and 1996 compared with the previous years, with the maximum density recorded being 38 individuals/ha in November 1996.

In the following, we give a rough estimation of waterfowl food consumption in Lake Væng. Based on direct measurements of faecal output, Mitchell and Wass (1995) calculated a mean food intake of 104 g DW/day for New Zealand black swans (mean weight, 5.6 kg) occurring at a density of 10–20 individuals/ha. Mute swan are larger, however, weighing about 10.0 kg (Kiørboe, 1980), and their food consumption is therefore presumably higher, at about 200 g DW/day. For coot, daily food intake has been estimated to be 45 g DW/day (Hurter, 1979). Applying these figures to Lake Væng waterfowl densities yields a maximum food consump-

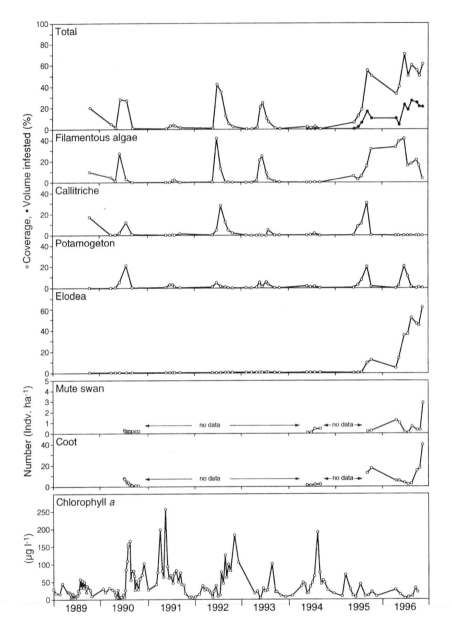

Figure 20.2. Development of submerged macrophyte coverage (○ mean percentage of total lake area and ●: PVI), chlorophyll *a*, and waterfowl abundance in Lake Stigsholm during the period 1989–1995.

tion of 4 kg DW plant material/day/ha for mute swan and 2.4 kg DW/day/ha for coot. For the whole winter period of November 1, 1991, to March 1, 1992 (from which we have the most comprehensive waterfowl data and when density was highest), total waterfowl consumption is estimated to be 440 kg DW/ha. By comparison, estimated maximum *Elodea* biomass in 1990 was 1,660 kg DW/ha (Lauridsen et al., 1994). Thus, assuming no net production by *Elodea* during the winter, the above estimates indicate that at maximum macrophyte density (e.g., as in 1990 and 1991 in Lake Væng), waterfowl grazing removes only about 25% of the macrophyte biomass. In years with low macrophyte biomass (e.g., as observed during recolonization in Lake Væng in 1990 [400 kg DW/ha in the autumn of 1989 and even lower in 1988]; see Fig. 20.1), an abundant waterfowl population is more likely to have an impact. Furthermore, the impact of waterfowl may include more than actual consumption, as both mute swan and coot tear up much larger quantities than they acutally consume (Berglund et al., 1963; Anderson and Low, 1976; van Donk et al., 1994). Finally, indirect estimation of food consumption by using bioenergetic models may yield significantly higher values, albeit that the uncertainty associated with such models is subject to debate (Mitchell and Wass, 1995). Clausen and Krause-Jensen (unpubl.) thus estimated daily consumption for mute swan and coot feeding on *Zostera* to be 530–680 g DW/day and 115–120 g DW/day, respectively. Applying these figures to Lake Væng yields a food consumption of 15–20 kg DW/day/ha at maximum waterfowl density, thus rendering a possible impact of waterfowl more likely.

The colonization pattern recorded in Lake Væng lends further support to the hypothesis that waterfowl inhibit colonization: first, colonization was delayed in relation to the potential growth areas (here defined as areas with a depth less than mean summer Secchi depth). Thus in 1988, less than 2% of the lake bottom was covered with macrophytes, despite a potential growth area of about 50%. Second, submerged macrophytes started to recolonize the wind-exposed and deepest parts of the lake and not—as had been expected—the more shallow and sheltered areas (Lauridsen et al., 1994). This colonization pattern probably reflects the impact of grazing waterfowl because waterfowl mainly inhabit and nest in sheltered and reed-covered areas during the summer and are much less abundant at wind-exposed and reed-poor locations (Lauridsen et al., 1994).

Finally, caging studies confirmed the potential grazing effect of waterfowl. In Lake Væng, grazing by waterfowl led to decreased shoot length in unprotected pots (Table 20.2), shoot length in protected pots at sheltered locations being 57.7 and 28.1 cm on mud and sand, respectively, as compared with 14.8 and 9.9 cm, respectively, in unprotected pots. The 2.8–3.9-fold difference between protected and unprotected pots at the sheltered location (compared with 1.8–2.1-fold difference at the exposed location) provides further evidence for a higher impact of waterfowl grazing in sheltered areas.

Similar results were found in Lake Stigsholm. Total shoot length, mean shoot length, and branches per shoot were significantly higher in protected pots than in unprotected pots (Table 20.3). For instance, mean total shoot length was 15.7 m in protected pots but only 3.7 m in unprotected pots. Moreover, the percentage of

Table 20.2. Mean Shoot Length ($n = 8$) of *P. crispus* Planted in Bird-Protected and Unprotected Pots in Lake Væng at Wind-Exposed and Sheltered Locations on Mud and Sand[a]

	Protected pots cm (n)	Unprotected pots cm (n)	P
Exposed, mud	67.6 (64)	38.6 (64)	.0007
Exposed, sand	48.4 (64)	22.6 (64)	.0004
Sheltered, mud	57.7 (48)	14.8 (64)	<.0001
Sheltered, sand	28.1 (56)	9.9 (64)	<.0001

[a]Mean of six–eight dates from May 22 to August 10, 1989. Significant differences between the two pot types during the period are indicated (GLM procedure using protected or unprotected pots as class variables and shoot length and date as dependent variables; SAS, 1990).

stubble, which is a direct index of grazing, was significantly higher in the unprotected pots (Table 20.3): 9.8% in protected pots versus 20.1% in unprotected pots. Furthermore, the percentage of stubble peaked at the beginning of July, when coot density was highest (Søndergaard et al., 1996).

Although grazing by fish cannot be excluded in experiments of this kind (Prejs, 1984; van Donk et al., 1994), changing the experimental set-up in Lake Stigsholm to allow fish (roach and perch) to enter the pots did not lead to any changes in macrophyte growth. The impact by fish is therefore considered to be unimportant, at least in this lake (Søndergaard et al., 1996).

It can thus be concluded that both submerged macrophyte coverage and waterfowl density are subject to marked interannual variation and that these variations are closely and positively related. The experiments undertaken in Lake Væng and Lake Stigsholm suggest that waterfowl may suppress macrophyte biomass and development in shallow lakes. The impact of waterfowl grazing is likely to be particularly important (1) in lakes in which macrophytes are in the recolonization phase after a reduction in nutrient loading, (2) in relatively small shallow lakes in

Table 20.3. *P. crispus* Shoot Characteristics in Bird-Protected and Unprotected Pots (Mean of Eight Pots, Seven Locations, and Four Dates: July 27, August 10, August 24, September 3, 1990) in Lake Stigsholm[a]

	Protected pots	Unprotected pots	P
Total shoot length (m)	15.7 (SD = 10.6)	3.7 (SD = 0.57)	<.001
Mean shoot length (cm)	10.9 (SD = 2.1)	7.2 (SD = 3.8)	<.001
Number of shoots	138 (SD = 82)	39 (SD = 51)	<.001
Branches per shoot	0.45 (SD = 0.31)	0.19 (SD = 0.35)	.004
% Stubble	9.8 (SD = 8.9)	20.1 (SD = 18.4)	.010

[a]Significant differences between the two pot types are indicated (GLM procedure as described in Table 20.2).

which the impact of waterfowl is greater due to a relatively large littoral zone and in which coot density in the absence of macrophytes is highest (Brøgger-Jensen and Jørgensen, 1992), and (3) during the autumn/winter when they are not territorial. When attempting to achieve improved lake water quality, it may therefore be a useful management strategy to protect sparsely developed macrophyte populations in the early phase of recolonization.

Acknowledgments. We wish to thank the technical staff of the National Environmental Research Institute for their assistance, in particular L. Hansen and J. Stougaard-Pedersen, B. Laustsen, J. Glargaard, K. Jensen, and L. Nørgaard. Layout and manuscript assistance was provided by K. Møgelvang, J. Jacobsen, and A.M. Poulsen, and D. Barry provided useful editorial comments.

References

Anderson, M.G.; Low, J.B. Use of sago pondweed by waterfowl on the Delta Marsh, Manitoba. J. Wildl. Manage. 20:233–242; 1976.

Berglund, B.E.; Curry-Lindahl, K.; Luther, H.; Olsson, V.; Rohde, W.; Sellerberg, G. Ecological studies on the mute swan (*Cygnus olor*) in southern Sweden. Acta Vertebratica 2:167–288; 1963.

Brøgger-Jensen, S.; Jørgensen, H.E. Vandfugles og søers miljøtilstand. Miljøprojekt 200 (in Danish). [Waterfowl versus the environmental state of lakes. Environmental Project 200]. Copenhagen: Danish Environmental Protection Agency; 1992.

Clausen, P.; Krause-Jensen, D. An annual budget of eelgrass *Zostera marina* consumption by herbivorous waterfowl in a shallow Danish estuary (in preparation).

Hargeby, A.; Andersson, G.; Blindow, I.; Johansson, S. Trophic web structure in a shallow eutrophic lake during a dominance shift from phytoplankton to submerged macrophytes. Hydrobiologia 279/280:83–90; 1994.

Hurter, H. Nahrungsökologie des Blässhuhn (*Fulica atra*) an den Überwinterungsgewässern im nördlichen Alpenvorland. Der Ornithologische Beobachter 76:257–288; 1979.

Jeppesen, E.; Jensen, J.P.; Kristensen, P.; Søndergaard, M.; Mortensen, E.; Sortkjær, O.; Olrik, K. Fish manipulation as a lake restoration tool in shallow, eutrophic, temperate lakes. 2: Threshold levels, long-term stability and conclusions. Hydrobiologia 200/201: 219–227; 1990.

Jeppesen, E.; Søndergaard, M.; Kronvang, B.; Jensen, J.P.; Svendsen, L.M.; Lauridsen, T. Lake and catchment management in Denmark. In: Harper, D.; Brierley, B.; Ferguson, A.; Phillips, G.; Madgewick, J., eds. Ecological basis for lake and reservoir management. London: J. Wiley & Sons (in press).

Jupp, B.P.; Spence, D.H.N. Limitations of macrophytes in a eutrophic lake, Loch Leven, II: Wave action, sediments and waterfowl grazing. J. Ecol. 65:431–446; 1977.

Kiørboe, T. Distribution and production of submerged macrophytes in Tipper Grund (Ringkøbing Fjord, Denmark), and the impact of waterfowl grazing. J. Appl. Ecol. 17:675–687; 1980.

Lauridsen, T.L.; Jeppesen, E.; Østergaard Andersen, F. Colonization of submerged macrophytes in shallow fish manipulated Lake Væng: impact of sediment composition and waterfowl grazing. Aquat. Bot. 46:1–15; 1993.

Lauridsen, T.L.; Jeppesen, E.; Søndergaard, M. Colonization and succession of submerged macrophytes in shallow Lake Væng during the first five years following fish manipulation. Hydrobiologia 275/276:233–242; 1994.

Lodge, D. Herbivory on freshwater macrophytes. Aquat. Bot. 41:195–224; 1991.

Mitchell, S.; Wass, R.T. Food consumption and faecal deposition of plant nutrients by black swan (*Cygnus atratus* Latham) in a shallow New Zealand lake. Hydrobiologia 306:189–197; 1995.

Prejs, A. Herbivory by temperate freshwater fishes and its consequences. Environ. Biol. Fish. 10:281–296; 1984.

SAS. SAS language version 6. SAS Institute Inc., Cary, NC, USA; 1990.

Scheffer, M. Multiplicity of stable states in freshwater systems. Hydrobiologia, 200/201: 475–487; 1990.

Skotte-Møller, H. Midtjyllands Fugle. Ornitologiske undersøgelser i Midtjylland (in Danish). [The birds of central Jutland. Ornitological investigations in central Jutland]. Meddelelse 2:34; 1970.

Søndergaard, M.; Jeppesen, E.; Mortensen, E.; Dall, E.; Kristensen, P.; Sortkjær, O. Phytoplankton biomass reduction after planktivorous fish reduction in a shallow, eutrophic lake: a combined effect of reduced internal P-loading and increased zooplankton grazing. Hydrobiologia 200/201:229–240; 1990.

Søndergaard, M.; Bruun, L.; Lauridsen, T.L.; Jeppesen, E.; Vindbæk Madsen, T. The impact of grazing waterfowl on submerged macrophytes: in situ experiments in a shallow eutrophic lake. Aquat. Bot. 53:73–84; 1996.

van Donk, E.; De Deckere, E.; Klein Breteler, J.G.P.; Meulemans, J. Herbivory by waterfowl and fish on macrophytes in a biomanipulated lake: effects on long-term recovery. Verh. Int. Verein. Limnol. 25:2139–2143; 1994.

Winfield, I.J. Fishes, waterfowl and eutrophied ecosystems: a perspective from a European vertebrate ecologist. In: Guissani, G.; Van Liere, L.; Moss, B., eds. Ecosystem research in freshwater environment recovery. Mem. Ist. Ital. Idrobiol. 48:113–126; 1991.

21. Influence of Macrophyte Structure, Nutritive Value, and Chemistry on the Feeding Choices of a Generalist Crayfish

Greg Cronin

Introduction

Herbivores are an important component of the food webs of nearly all communities that receive sunlight. The historical belief that freshwater macrophytes enter the food web only as detritus is being challenged by recent evidence suggesting direct herbivory on live macrophytes can be significant (reviewed by Lodge, 1991; Newman, 1991; Lodge et al., this volume, Chapter 8; Mitchell and Perrow, this volume, Chapter 9). Most herbivory of freshwater macrophytes is caused by generalist grazers such as aquatic insects (Newman, 1991; Jacobsen and Sand-Jensen, 1992, 1994, 1995), crustaceans (Chambers et al., 1990; Lodge, 1991; Newman et al., 1992; Creed, 1994; Lodge et al., 1994), fish (Lodge et al., this volume, Chapter 8), waterfowl (Conover and Kania, 1994; Mitchell and Perrow, this volume, Chapter 9), and mammals (Fraser et al., 1984; Doucet and Fryxell, 1993).

Little quantitative information exists about the feeding preferences of freshwater herbivores, and less is known about plant traits that determine those preferences (Lodge, 1991; Newman, 1991), although macrophyte structure (Chambers et al., 1991; Jacobsen and Sand-Jensen, 1994), nutritive value (Lorman, 1980; Fraser et al., 1984; Center and Wright, 1991; Doucet and Fryxell, 1993), and secondary metabolites (Buchsbaum et al., 1984; Newman et al., 1992) have all been implicated. As an example of our dismal mechanistic understanding of freshwater macrophyte–herbivore interactions, thousands of secondary metabolites have been characterized from terrestrial and marine plants and the effects of dozens of these compounds on the feeding behavior of terrestrial (Rosenthal and Berenbaum, 1992) and marine (Hay and Fenical, 1988; Paul, 1992) herbivores are known, but there is only one documented example of a specific plant compound defending a freshwater macrophyte from herbivores (Newman et al., 1992). Yet, there is no compelling reason to expect chemical defenses to be rarer in freshwater than in terrestrial or marine plants (Ostrovsky and Zettler, 1986).

Crayfish are very common and important freshwater omnivores that include much plant material in their diets. They can have large impacts on littoral and stream communities by reducing the abundance of macrophytes and invertebrates (Lodge and Lorman, 1987; Chambers et al., 1990; Creed, 1994; Lodge et al., 1994), yet little is known about their feeding selectivity or the effects of various plant traits on their feeding decisions. In this study, I examined the feeding preferences of the crayfish *Procambarus clarkii* among nine species of freshwater macrophytes (including macroscopic algae) and measured their responses to manipulation of the combined plant traits of morphology, toughness, and surface features (hereafter grouped as "structure") and their response to plant extracts. I also relate the preference of crayfish with measurements of plant nitrogen, protein, and polyphenolic compounds.

Methods

The Organisms

Nine species of freshwater macrophytes that represented a wide range of growth forms and taxonomic groups were collected in northern Indiana and southern Michigan (all were collected within 50 km of the Notre Dame campus) on May 1, 1995. Five of the plants were submerged, including three algae (*Chara* sp. [Characeae, Charophyta]; *Spirogyra* sp. [Zygnemataceae, Chlorophyta]; *Batrachospermum* sp. [Batrachospermaceae, Rhodophyta]) and two angiosperms (Magnoliophyta) (*Potamogeton amplifolius* [Potamogetonaceae] and *Ceratophyllum demersum* [Ceratophyllaceae]). The remaining species were angiosperms with floating leaves (*Nuphar advena* [Nymphaeaceae]) or emergent leaves (*Typha latifolia* [Typhaceae], *Iris virginica* [Iridaceae], and *Carex vesicariae* [Cyperaceae]).

The Louisiana crayfish *Procambarus clarkii* is a commercially valuable crayfish that is cultivated far outside its native range of the southeastern United States and now occurs as feral populations in Mexico, Japan, Hawaii, from coast to coast in the continental United States (Hobbs, 1972), and as far north as Lake Erie. *Procambarus* did not historically co-occur with all the macrophyte species used in this study, but their current ranges overlap. *Procambarus* used in my experiments were purchased from Alchafalaya Biological Supply Company (Louisiana) and were kept individually in 25 × 31-cm plastic tubs with about 5 cm of untreated well water. Crayfish used in these experiments had a carapace length of about 6 cm and were maintained on commercial fish food between assays.

Crayfish Preference Among Fresh Plants

Fresh pieces of the nine macrophytes were simultaneously offered to individual crayfish from May 1 to May 3, 1995, to determine their feeding preferences. Within a plant species, each piece came from a different plant shoot or clump of algae. I used pieces of plants that appeared to occupy similar volumes of water in an attempt to provide similar encounter rates to the crayfish. This procedure resulted in smaller masses of *Potamogeton* and *Carex* (about 150 mg and 200 mg, respectively) than of the other species (about 600 mg each). The wet mass of each plant piece was determined by first spinning it for about 7 seconds in a salad spinner to remove excess water and then weighing it to the nearest milligram. A weighed piece of each of the nine macrophyte species was placed in each of 20 tubs with crayfish. To control for changes in mass not due to the crayfish, 11 tubs were set up in an identical manner but without a crayfish. The pieces of plants were anchored to the bottom of the plastic tubs with rubber suction cups.

After the crayfish had foraged for 1.5 days, the macrophytes were reweighed and consumption was calculated from the equation $[(H_0 \times C_f/C_0) - H_f)]$; where H_0 and H_f were the masses of the plants in each tub with a crayfish before and after the assay, and C_0 and C_f were the mass of the macrophyte species from a randomly chosen control tub before and after the assay. These latter measures accounted for the 1–15% increase in mass unrelated to grazing activity with the following average increases: *Chara* 1%; *Ceratophyllum* 3%; *Spirogyra* 4%; *Typha* 15%; *Potamogeton* 1%; *Batrachospermum* 4%; *Nuphar* 9%; *Carex* 2%; and *Iris* 9%. Floating and emergent leaves with large air spaces *(Typha, Nuphar, Iris)* showed the largest increase in mass, presumably from absorbed water.

Crayfish Preference Among Reconstituted Plants

Additional plant tissue from each species was frozen, freeze-dried, ground into a fine powder, and stored at −20°C until used in other experiments. To determine if plant structure influenced crayfish feeding decisions, the powdered macrophytes were reconstituted into an alginate gel at natural dry mass concentration (i.e., the amount of water added to the powdered macrophytes equaled the amount of water removed by freeze-drying). To do this, a measured amount of powdered macrophyte (this amount varied among macrophyte species as percentage dry mass varied among species; Table 21.1) was mixed into a 2% solution of sodium alginate. This mixture was pressed into a mold to form a 2.6-cm-wide × 1.5-mm-thick strip on a piece of fiberglass window screening material. A solution of 0.25 M calcium chloride was poured onto the macrophyte/sodium alginate mixture to solidify it (sodium alginate is soluble in water; calcium alginate is a water-insoluble gel). The fiberglass screening material provided support for the reconstituted plants and a uniform grid that allowed me to quantify the amount eaten by counting the squares of the screen that had been cleared of reconstituted plants (see Fig. 1 in Hay et al. [1994] for an illustration of these methods). Thus, all reconstituted macrophytes were of similar structure, but the nutritive value and

Table 21.1. Tissue Properties of the Nine Macrophyte Species and Their Relation to Feeding Preferences[a]

Species	% Dry mass	Deterrency of extract (average reduction in feeding)	Protein (% WM)	Nitrogen (% WM)	Carbon (% WM)	Polyphe- nolics (% WM)
Chara	10.7	− 46	0.52	0.22	3.23	0.00
Ceratophyllum	11.4	44	0.56	0.34	4.38	0.03
Spirogyra	6.5	− 26	0.34	0.11	2.60	0.23
Typha	12.4	−109	1.36	0.42	5.20	0.03
Potamogeton	9.1	—	0.16	0.35	3.72	0.22
Batrachospermum	3.0	− 10	0.14	0.19	1.23	0.00
Nuphar	15.8	93	2.20	0.62	7.24	0.69
Carex	21.3	45	1.61	0.78	9.29	0.07
Iris	13.4	44	1.62	0.45	5.94	0.11
Pearson correlation coefficient (whole-tissue assay)	−0.391	−0.321	−0.642	−0.645	−0.530	−0.223
Pearson correlation coefficient (reconstituted tissue)	0.574	−0.336	0.265	0.422	0.446	−0.213

[a]The concentrations of soluble protein, total nitrogen, and total carbon are given as the percentage of plant wet mass. Order of species is same as in Figure 21.1.

taste of each reconstituted species were presumably little altered, although freezing and freeze-drying can affect some plant metabolites (Cronin et al., 1995).

The strips of eight reconstituted plants (*Potamogeton* powder was unavailable for this experiment) were cut into 1.2-cm sections, anchored to the bottom of tubs with rubber suction cups, and simultaneously offered to 29 separate crayfish. Because the reconstituted plants remained on the screening material in the absence of crayfish, controls for autogenic changes were unnecessary.

Effects of Macrophyte Extracts on Crayfish Feeding

The combined effects of lipophilic and water-soluble metabolites on the feeding behavior of crayfish were assessed by adding extracts to a standard palatable artificial food. The standard food consisted of freeze-dried powdered *Typha latifolia* collected from Morehead City, North Carolina, distilled water, and alginate (1 g *Typha* powder : 7.3 ml H_2O : 167 mg sodium alginate).

Four grams of powdered macrophyte were extracted first with a 1:1 mixture of methanol and ethyl acetate to remove lipophilic metabolites and then with distilled water to extract water-soluble metabolites. The lipophilic metabolites dissolved in solvent were mixed with 4 g of powdered *Typha*, and the solvents were removed by evaporation under reduced pressure. The sodium alginate was dissolved in the aqueous extract, and this solution was mixed with the *Typha* powder containing the lipophilic extract. This mixture was spread onto fiberglass screening and solidified into a gel as described above. Standard food treated in the same manner, but without added extracts, was used as control food. A parallel strip of control food was made 1 cm from the treated food on the same piece of fiberglass screen. The screen with paired treated and control food was cut into "test strips," each consisting of two rectangles of artificial food that differed only in the presence of a macrophyte crude extract in one of the pieces of food. These experiments were performed on separate days for each macrophyte species except *Potamogeton*.

Test strips were offered to individual crayfish for 1–12 hours (depending on their feeding rates), and the amount of food consumed was quantified by counting the number of squares in the screen that had been cleared of food. Replicates in which none or all of both food types were eaten were not used in data analysis. Data for each feeding assay were analyzed with a paired-sample *t*-test.

Plant Tissue Constituents

Powdered macrophytes were analyzed for total nitrogen and carbon with a Perkin Elmer CN analyzer (model 2400), soluble protein with a modified Bradford assay (Duffy and Hay, 1991), and polyphenolics with a modified Folin-Denis assay (Yates and Peckol, 1993).

For polyphenolic analyses, about 5 mg of ground, freeze-dried tissue, weighed to the nearest 0.1 mg, was extracted with 1.00 ml of 50% aqueous methanol for 1 day at 1°C with occasional mixing. Two 100-µl aliquots of the extraction solution were placed in separate test tubes and diluted with 8.4 ml of acidic 10% methanol (pH = 2). Polyvinylpolypyrrolidone (PVPP; Sigma) was added to one test tube. The Folin-Denis reagent reacts with several chemical constituents, including phenolics, proteins, amino acids, and ascorbic acid (Andersen and Todd, 1968), whereas PVPP binds specifically with polyphenolics, preventing them from reacting with the Folin-Denis reagent. The test tube without PVPP provided the traditional Folin-Denis reading that included phenolics, protein, amino acids, etc., and the +PVPP test tube provided a Folin-Denis reading for everything except polyphenolics. The concentration of polyphenolics was calculated from the difference in the two readings, using tannic acid as a standard.

Results

When plants were offered as fresh tissue, *Procambarus* consumed large amounts of *Chara*, *Ceratophyllum*, and *Spirogyra*; intermediate amounts of *Typha*; and little (if any) of *Potamogeton*, *Batrachospermum*, *Nuphar*, *Carex*, or *Iris* (Fig. 21.1A). Morphologically, the three preferred species are finely branched or filamentous. Except for *Batrachospermum*, the less preferred species have flat (*Potamogeton* and *Nuphar*) or blade-like (*Typha*, *Carex*, and *Iris*) leaves. *Batrachospermum* is very flaccid and covered in copious slime.

When plants were offered as reconstituted tissue in an alginate gel, the feeding preferences of the crayfish were altered (compare Fig. 21.1A and B). Large

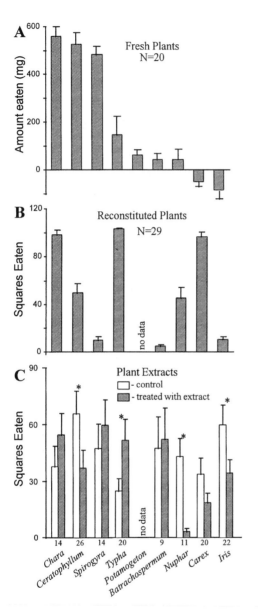

Figure 21.1. Amount of plants consumed by *Procambarus clarkii* among macrophyte species when offered simultaneously (A) as fresh tissue and (B) as freeze-dried, powdered macrophytes that had been reconstituted with an alginate gel. (C) The effects of crude extracts from the macrophyte species on feeding by the crayfish. Bars represent mean (+ 1 SE). Each pair of bars in C represents an independent feeding assay, with the sample size given below the pair. An asterisk indicates that the means in a pair differ significantly from each other at $\alpha = 0.05$.

amounts of *Chara, Typha,* and *Carex;* intermediate amounts of *Ceratophyllum* and *Nuphar;* and little *Spirogyra, Batrachospermum,* and *Iris* were consumed (Fig. 21.1B). Thus, plant structural properties are important determinants of *Procambarus* feeding decisions because altering the structure changed preferences. However, additional plant traits are also important given the crayfish fed selectively among the structurally identical reconstituted plants.

Crude extracts from plants contain both feeding stimulants and deterrents. When added to the standard food, the crude extracts from *Ceratophyllum, Nuphar,* and *Iris* significantly reduced consumption by 44%, 93%, and 44%, respectively. By contrast, the extract from *Typha* significantly stimulated feeding by 109% (Table 21.1; Fig. 21.1C). Extracts of the remaining plants did not significantly affect crayfish feeding. Thus, plant chemistry can deter or stimulate feeding by crayfish. The deterrency of the crude extracts, as determined by the average percentage reduction in feeding caused by the extract (Fig. 21.1C), was negatively related to the amount of whole tissue and reconstituted tissue consumed during feeding assays (Table 21.1).

No measured plant trait was significantly correlated with feeding preferences, either as whole plant tissue or reconstituted tissue. However, I provide correlation coefficients in Table 21.1 to indicate feeding trends with respect to tissue traits.

Polyphenolics, a class of secondary metabolites with putative defensive functions, were detected at various levels in all the angiosperms, were undetectable in *Chara* and *Batrachospermum,* and surprisingly composed 0.23% of the wet mass (=3.5% dry mass) of the green alga *Spirogyra* (Table 21.1). Nuclear Magnetic Resonance analysis of a methanol extract of *Spirogyra* revealed the presence of aromatic phenolic compounds, but these compounds differed from brown algal phlorotannins or plant tannins (W. Fenical, personal communication). Polyphenolics were negatively related to the amount of tissue consumed in both the whole-tissue assay and the reconstituted-tissue assay (Table 21.1).

Nitrogen and protein varied considerably among species and were not well correlated when expressed as percentage of dry mass (Pearson correlation coefficient = 0.044; $P = .991$; $n = 9$). However, because of the major influence of water, which constituted 78–97% of the plants wet mass, protein and nitrogen concentrations were significantly correlated with each other when expressed on a wet mass basis (Pearson correlation coefficient = 0.818; $P = .007$; $n = 9$). On a percentage dry mass basis, the concentrations of nitrogen and protein tended to be lower in the submerged plants (*Chara, Ceratophyllum, Spirogyra, Potamogeton,* and *Batrachospermum;* average 3.3% nitrogen and 4.3% protein) than in the floating (*Nuphar;* 3.9% nitrogen and 14.0% protein), emergent (*Typha;* 3.4% nitrogen and 11.0% protein), or shore (*Carex* and *Iris;* average 3.6% nitrogen and 9.9% protein) species. This pattern is exaggerated if concentrations are expressed on a percentage wet mass basis because submerged tissues have a higher water content than emergent tissues (Table 21.1). Nitrogen and protein concentrations were negatively related with the feeding preferences of crayfish during the whole-tissue assay but were positively related to the feeding preference of crayfish during the reconstituted-plant assay (Table 21.1).

Discussion

Generalist herbivores, including *Procambarus,* base their feeding decisions on multiple plant traits such as morphology, structure, chemical defenses, and nutritive value (Lodge, 1991; Newman, 1991; Lodge et al., this volume, Chapter 8). Because herbivores are generally more limited by nitrogen than carbon (Mattson, 1980), nitrogen or protein have been considered important feeding stimulants for these animals. Apparently, *Procambarus* fed on plants such as *Chara, Ceratophyllum,* and *Spirogyra* when offered as whole tissue because their finely branched or filamentous morphologies make them easier to handle, shred, and consume. When the difficulties imposed by plant structure were removed by forming all plants into a gel, the feeding preferences of *Procambarus* were altered; some plants with high concentrations of protein, nitrogen, and dry mass (i.e., nutritious) such as *Typha* and *Carex* became more popular food items while some less nutritious plants such as *Ceratophyllum* and *Spirogyra* became less popular (Fig. 21.1B).

Nutritious macrophytes that were low preference as whole and reconstituted plants (*Nuphar* and *Iris*) contained chemical defenses (Fig. 21.1C). The preferences for whole and reconstituted plants were both negatively related to the deterrency of the crude extract and concentration of polyphenolics (Table 21.1). Although some plant chemical defenses reduced crayfish feeding significantly, they were not entirely effective given that a moderate amount of reconstituted *Nuphar* was consumed (Fig. 21.1B). Additionally, *Ceratophyllum* had a deterrent crude extract, yet it was highly preferred as fresh whole tissue and moderately preferred as reconstituted tissue. Other plant traits, such as a slimy surface, may also deter crayfish: *Batrachospermum* was low preference despite a very flaccid morphology and lack of a deterrent extract, but it was extremely slimy. However, it also had little nutritive value (i.e., it was 97% water).

Procambarus will eat only plants that they can handle, shred, and ingest. Although this is self-evident, it helps explain why feeding preferences during the whole-tissue assay were negatively related to plant traits normally considered to be feeding stimulants such as protein, nitrogen, and dry mass. Plants with high concentrations of dry mass, and hence high concentrations of nutrients on a wet mass basis, also had high amounts of structural material. Only after the plants were made structurally identical were nutritive plant traits positively related to feeding preferences (Table 21.1). The importance of plant structure in determining its susceptibility to grazing has been previously noted for seaweeds; hard encrusting forms are among the least susceptible to herbivores whereas highly branched or filamentous forms are generally the most susceptible (Littler and Littler 1980). However, even seaweeds with susceptible forms can reduce grazing damage by being chemically defended (Cronin and Hay, 1996a,b). Among plants with similar levels of defense, palatability should be positively related to nutritive value (Lodge et al., this volume, Chapter 8). For the five reconstituted plants that were not chemically defended, the crayfish ate nearly all the food made from the three species (*Chara, Typha,* and *Carex*) with the highest concentration of protein and total nitrogen (wet mass basis), whereas food made from the less nutritious species (*Spirogyra* and *Batrachospermum*) was consumed less. Chambers et al. (1991) found that the feeding preferences of the crayfish *Orconectes virilis* were negatively correlated with plant nutritive value, probably because the less nutritious plants were easier to handle. Reduction in the structural integrity of macrophytes after death helps explain why macrophytes are consumed more as detritus than living tissue, an explanation that receives less attention than the "microbial conditioning" or "chemical defense leaching" explanations (Newman, 1991).

Other factors not considered above, such as cover or protection from predators afforded by the plant (Damman, 1987; Duffy and Hay, 1994), the consumer's state of hunger (Cronin and Hay, 1996b), or the consumer's prior feeding experiences (Provenza, 1995; Howard et al., 1996), will also affect herbivore feeding decisions. Although a single plant trait (e.g., thick, tough, blade-like leaf structure) can be used to accurately predict a plant's susceptibility to *Procambarus,* it may not be useful to predict a plant's susceptibility to herbivory in general because the effects of plant traits on feeding vary among herbivore species. Additionally, multiple plant traits can interact to influence herbivore feeding behavior in a mitigative, additive, or synergistic manner (Duffy and Paul, 1992; Hay et al., 1994).

It has been demonstrated that the feeding activity of generalists can have significant direct and indirect impacts on aquatic communities (Lodge and Lorman, 1987; Chambers et al., 1990; Creed, 1994; Lodge et al., 1994; Hill and Lodge, 1995; Lodge et al., this volume, Chapter 8; Mitchell and Perrow, this volume, Chapter 9). A mechanistic understanding of macrophyte–herbivore interactions will improve our ability to predict the impacts of herbivores on freshwater communities. We know that herbivores can reduce the standing stock of macrophytes, but a knowledge of herbivore feeding preferences and plant defensive traits will allow better herbivore-specific and macrophyte-specific predictions of impact.

Summary

Laboratory experiments were performed to determine the feeding preferences of the crayfish *Procambarus clarkii* among nine species of submerged, floating, emergent, and shoreline macrophytes representing a broad taxonomic range (9 families in 4 divisions). When plants were offered as whole fresh tissue, the crayfish preferred highly-branched and filamentous plants over those with thick, flat leaves. However, when the plants were freeze-dried, ground into a powder, and reconstituted with an alginate gel at the original percentage of dry mass (i.e., plant structure was equalized, but nutritive value and taste were presumably unaltered), the feeding preferences of the crayfish were greatly altered. Chemical extracts from the selected plant species that were incorporated into palatable artificial foods also altered consumption by the crayfish: crude extracts of three macrophytes deterred feeding while the extract of one macrophyte significantly stimulated feeding. Therefore, plant structure (morphology, toughness, and/or surface features) and plant chemistry are important determinants of crayfish feeding choices. Finally, total nitrogen, protein, and polyphenolics were quantified from plant tissues to relate to feeding preferences. No single plant trait could explain the feeding preferences of crayfish. Rather, crayfish apparently base feeding decisions on a variety of plant traits: *Procambarus* avoided species with structural or chemical deterrents, and preferred undefended plants high in nitrogen.

Acknowledgments. This research was funded by NSF DEB 94–08452–002 to David Lodge. I am grateful to Bill Fenical for performing NMR studies, Jill Witkowski for technical assistance, and David Lodge for insightful discussions and comments about this project and herbivory in general. Reviews by Robin Bolser, David Lodge, and two anonymous reviewers improved the manuscript.

References

Andersen, R.A.; Todd, J.R. Estimation of total tobacco phenols by their bonding to poly-vinylpyrrolidone. Tobacco Sci. 12:107–111; 1968.

Buchsbaum, R.; Valiela, I.; Swain, T. The role of phenolic compounds and other plant constituents in feeding by Canada geese in a coastal marsh. Oecologia 63:343–349; 1984.

Center, T.D.; Wright, A.D. Age and phytochemical composition of waterhyacinth (Pontederiaceae) leaves determine their acceptability to *Neochetina eichhorniae* (Coleoptera: Curculionidae). Environ. Entomol. 20:323–334; 1991.

Chambers, P.A.; Hanson, J.M.; Burke, J.M.; Prepas, E.E. The impact of the crayfish *Orconectes virilis* on aquatic macrophytes. Freshwat. Biol. 24:81–91; 1990.

Chambers, P.A.; Hanson, J.M.; Prepas, E.E. The effect of aquatic plant chemistry and morphology on feeding selectivity by the crayfish, *Orconectes virilis*. Freshwat. Biol. 25:339–348; 1991.

Conover, M.R.; Kania, G.S. Impact of interspecific aggression and herbivory by mute swans on native waterfowl and aquatic vegetation in New England. Auk 111:744–748; 1994.

Creed, R.P., Jr. Direct and indirect effects of crayfish grazing in a stream community. Ecology 75:2091–2103; 1994.

Cronin, G.; Hay, M.E. Amphipod grazing induces chemical defenses in the brown alga *Dictyota menstrualis*. Ecology 77:2287–2301; 1996a.

Cronin, G.; Hay, M.E. Seaweed–herbivore interactions depend on recent history of both the plant and animal. Ecology 77:1531–1543; 1996b.

Cronin, G.; Lindquist, N.; Hay, M.E.; Fenical, W. Effects of storage and extraction procedures on yields of lipophilic metabolites from the brown seaweeds *Dictyota ciliolata* and *D. menstrualis*. Mar. Ecol. Prog. Ser. 119:265–273; 1995.

Damman, H. Leaf quality and enemy avoidance by the larvae of a pyralid moth. Ecology 68:88–97; 1987.

Doucet, C.M.; Fryxell, J.M. The effect of nutritional quality on forage preference by beavers. Oikos 67:201–208; 1993.

Duffy, J.E.; Hay, M.E. Food and shelter as determinants of food choice by an herbivorous marine amphipod. Ecology 72:1286–1298; 1991.

Duffy, J.E.; Hay, M.E. Herbivore resistance to seaweed chemical defense: the roles of mobility and predator risk. Ecology 75:1304–1319; 1994.

Duffy, J.E.; Paul, V.J. Prey nutritional quality and the effectiveness of chemical defenses against tropical reef fishes. Oecologia 90:333–339; 1992.

Fraser, D.; Chavez, E.R.; Paloheimo, J.E. Aquatic feeding by moose: selection of plant species and feeding areas in relation to chemical composition and characteristics of lakes. Can. J. Zool. 62:80–87; 1984.

Hay, M.E.; Fenical, W. Marine plant–herbivore interactions: the ecology of chemical defense. Annu. Rev. Ecol. Syst. 19:111–145; 1988.

Hay, M.E.; Kappel, Q.E.; Fenical, W. Synergisms in plant defenses against herbivores: interactions of chemistry, calcification, and plant quality. Ecology 75:1714–1726; 1994.

Hill, A.M.; Lodge, D.M. Multi-trophic-level impact of sublethal interactions between bass and omnivorous crayfish. J. North Am. Benth. Soc. 14:306–314; 1995.

Hobbs, H.H., Jr. Crayfishes (Astacidae) of North and Middle America. U.S. Environmental Protection Agency, Washington; 1972.

Howard, J.J.; Henneman, M.L.; Cronin, G.; Fox, J.A.; Hormiga, G. Conditioning of scouts and recruits during foraging by a leaf-cutting ant, *Atta colombica*. Anim. Behav. 52: 299–306; 1996.

Jacobsen, D.; Sand-Jensen, K. Herbivory of invertebrates on submerged macrophytes from Danish freshwaters. Freshwat. Biol. 28:301–308; 1992.

Jacobsen, D.; Sand-Jensen, K. Invertebrate herbivory on the submerged macrophyte *Potamogeton perfoliatus* in a Danish stream. Freshwat. Biol. 31:43–52; 1994.

Jacobsen, D.; Sand-Jensen, K. Variability of invertebrate herbivory on the submerged macrophyte *Potamogeton perfoliatus*. Freshwat. Biol. 34:357–365; 1995.

Littler, M.M.; Littler, D.S. The evolution of thallus form and survival strategies in benthic marine macroalgae: field and laboratory tests of a functional form model. Am. Nat. 116:25–44; 1980.

Lodge, D.M. Herbivory on freshwater macrophytes. Aquat. Bot. 41:195–224; 1991.

Lodge, D.M.; Lorman, J.G. Reductions in submersed macrophyte biomass and species richness by the crayfish *Orconectes rusticus*. Can. J. Fish. Aquat. Sci. 44:591–597; 1987.

Lodge, D.M.; Kershner, M.W.; Aloi, J.E.; Covich, A.P. Direct and indirect effects of an omnivorous crayfish (*Orconectes rusticus*) on a freshwater littoral food web. Ecology 75:1265–1281; 1994.

Lorman, J.G. Ecology of the crayfish *Orconectes rusticus* in northern Wisconsin. Ph.D. thesis, University of Wisconsin, Madison; 1980.

Mattson, W.J. Herbivory in relation to plant nitrogen. Annu. Rev. Ecol. Syst. 11:119–161; 1980.

Newman, R.M. Herbivory and detritivory on freshwater macrophytes by invertebrates: a review. J. North Am. Benth. Soc. 10:89–114; 1991.

Newman, R.M.; Hanscom, Z.; Kerfoot, W.C. The watercress glucosinolate-myrosinase system: a feeding deterrent to caddisflies, snails and amphipods. Oecologia 92:1–7; 1992.

Ostrovsky, M.L.; Zettler, E.R. Chemical defenses in aquatic plants. J. Ecol. 74:279–287; 1986.

Paul, V.J., ed. Ecological roles of marine natural products. Ithaca, NY: Comstock Publishing Associates; 1992.

Provenza, F.D. Tracking variable environments: there is more than one kind of memory. J. Chem. Ecol. 21:911–924; 1995.

Rosenthal, G.A.; Berenbaum, M.R., eds. Herbivores: their interactions with secondary plant metabolites. Vol. II: Evolutionary and ecological processes. New York: Academic Press; 1992.

Yates, J.L.; Peckol, P. Effects of nutrient availability and herbivory on polyphenolics in the seaweed *Fucus vesiculosus*. Ecology 74:1757–1766; 1993.

22. Concordance of Phosphorus Limitation in Lakes: Bacterioplankton, Phytoplankton, Epiphyte–Snail Consumers, and Rooted Macrophytes

Robert E. Moeller, Robert G. Wetzel, and Craig W. Osenberg

The characterization of many unpolluted lakes as aquatic "deserts" (Whittaker, 1975) reflects the well-recognized role of phosphorus and nitrogen in controlling the abundance and growth rates of *planktonic* algae (Schindler, 1974, 1978; Kalff and Knoechel, 1978). Deficiencies or reduced availability of these elements may be less severe in the littoral zone because of marked morphological and physiological adaptations that result in metabolic and community mechanisms for nutrient recycling and retention (Wetzel, 1990a, 1993). As a result, few limiting nutrient studies have been performed in the littoral zone, and many are directed toward specific components (e.g., Fairchild and Everett, 1988; Fairchild and Sherman, 1993).

In Lawrence Lake, located in southwest Michigan, submerged macrophytes are responsible for less than one-quarter of the annual net primary productivity of 180 g C/m². The rest originates from phytoplanktonic (16%) and epiphytic (70%) algae (Burkholder and Wetzel, 1989). To evaluate the extent of phosphorus limitation in these three communities as well as the bacterioplankton, we provided local additions of phosphate separately to subsamples of each community. For comparability within the natural light and temperature regimes, all results are for vegetation and microbial communities at a depth of 2 m. Rooted vascular plants grow as deep as 6 m in this 12-m deep lake (Rich et al., 1971; Wetzel et al., 1972; Burkholder and Wetzel, 1989).

Phytoplankton assays consistently demonstrated a deficiency of phosphorus during most of the year (Wetzel, 1981). Addition of 100 µg P/L to samples of

318

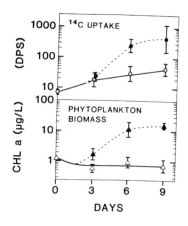

Figure 22.1. Progressive response of summer phytoplankton to P additions. Relative rates of photosynthesis (upper, as relative ^{14}C incorporation, in photosynthesis, as disintegrations per second, DPS) and biomass (lower, as µg chlorphyll a/L) for enriched (●, ▲) and control cultures (○, △). Values are means ± 95% CI for $n = 7$ dates from June 7 to September 13. Phosphate addition was 50 µg P/500 ml Pyrex flask, as K_2HPO_4. Cultures were maintained on a rotating tray at ambient lake temperatures and light (16 h light/8 h dark) (Wetzel, 1981). Chlorophyll a was corrected for phaeopigments (Wetzel and Likens, 1991). The ^{14}C activity of filtered algae from short-term labeling assays is a relative measure of photosynthesis: incubation times, volumes filtered, and specific activities were consistent throughout. Similar results were obtained with enrichments by organic phosphorus as β-glycerophosphate (Wetzel, 1981).

lakewater incubated in the laboratory caused consistent increases in phytoplankton biomass and photosynthesis over the next 9 days (Fig. 22.1). Alkaline phosphatase activity of enriched cultures increased at 9 days, after about 50% suppression of activity at 3–6 days, which suggested a return to phosphorus-limited conditions (Wetzel, 1981). The 10-fold stimulation of biomass of phytoplankton demonstrated a strongly limiting role for phosphorus under ambient planktonic conditions, which included abundant dissolved nitrate (nitrate averaged 30 µM and total dissolved P only 0.1–0.3 µM in summer at 2 m). Phytoplankton development in 1984, when the littoral vegetation was studied, was very similar to that in 1976, as had been the condition when the phytoplankton enrichments were conducted. Variations in the annual phytoplanktonic mean chlorophyll a concentrations and annual in situ rates of photosynthesis varied less than 15% over an 18-year period of continuous measurement (Wetzel, 1983).

In situ rates of productivity of bacterioplankton from the same sites were determined by incorporation rates of tritiated thymidine into DNA (Wetzel and Likens, 1991). Growth rates were also evaluated by changes in frequency of cell division. Nutrient enrichment experiments with bacterioplankton also indicated an enhanced rate of growth in response to phosphorus (Coveney and Wetzel, 1988, 1992, 1995).

A continuous and localized addition of phosphorus to the natural epiphyte community was accomplished during July–September 1984 in a 2-m-diameter site (plot A) dominated by a submerged sedge, *Scirpus subterminalis* Torr., the dominant macrophyte of the lake (Rich et al., 1971; Wetzel et al., 1972). Vertical rods (41 rods, 1.7 g P/rod positioned on July 13; 5 rods added near center on August 20) coated with resin-encapsulated pellets of calcium phosphate (Sierra Chemical Co., Milpitas, CA) were positioned 60 cm above the sediments throughout the plot; phosphate released from the pellets over 3 months was at a rate of 0.5–1% of the annual loading of phosphorus to the lake. At a nearby site (plot B), rods coated with pellets of mixed fertilizer (supplying N, P, K, S, and Ca) were embedded in the sediment (30 cm below the surface (104 rods, 0.5 g P/rod on June 12) with 60% of the rods 0.5 m from center of plots). Plot B was an attempt to stimulate phosphorus release from growing macrophyte tissue, which is known to be a direct although variable source of phosphorus for overlying epiphytes (Carignan and Kalff, 1982; Moeller et al., 1988; Burkholder and Wetzel, 1990).

A visible increase in epiphytes occurred within 5 days when phosphate was released above the sediment. After 10 weeks, we measured a 40-fold increase of epiphyte biomass at the center of plot A compared with biomass immediately outside the plot (Fig. 22.2). Epiphyte biomass was unchanged by the below-sedimentenrichment at plot B, although additional phosphorus was incorporated by *Scirpus*. Leaf phosphorus increased from 0.14 to 0.51% of dry wt within the plot ($P < .001$). Evidently, the conservative retention of phosphorus by growing macrophytes (Carignan and Kalff, 1982; Moeller et al., 1988; Burkholder and Wetzel, 1990; Wetzel, 1990a) was not significantly relaxed as excess phosphorus accumulated. In plot A, the local addition of phosphorus to the strongly phosphorus-deficient natural epiphyte vegetation led to a massive algal proliferation supported by nitrate, dissolved silica, and other nutrients continuously transported into the unenclosed site.

Epiphyte biomass increased in plot A despite an increase in the intensity of grazing by gastropods (Fig. 22.2). Snail biomass per unit area of lake bottom increased 73% (115 vs. 66 mg dry wt/0.05 m^2; $P < .05$). Snail biomass per unit of macrophyte biomass increased 46% (34 vs. 23 mg dry wt/g organic wt; $P < .1$). The response was attributable to two prosobranchs, *Amnicola limnosa* and *Valvata tricarinata*, which feed on epiphytic microflora and detritus and made up more than 80% of the total snail biomass. Both species live only 1 year and had reproduced before the enrichment started. Snail densities per unit area of lake bottom or unit of macrophyte biomass did not change ($P > .2$), indicating that neither immigration nor reproduction significantly contributed to the response. Instead, individuals of both common species were significantly larger ($P < .005$) within the zone of increased epiphyte biomass (Fig. 22.2), which demonstrated that snail growth rates responded positively to an increased food supply. Subsequent experiments replicated these results for epiphytes and snails in 2 additional years (Osenberg, 1988, 1989). Under natural conditions, therefore, food-limited grazers fed on phosphorus-limited algae. The intensity of grazing may have been sufficient to reduce algal biomass secondarily, as demonstrated elsewhere for

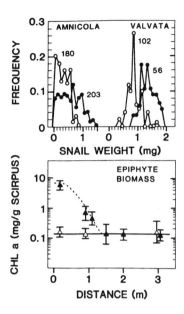

Figure 22.2. Response of epiphytic algae and snails to 10 weeks of P addition. Mean epiphyte biomass (± 95% CI, $n = 7$) increased within water-column enriched sites (▲) but not at sediment-enriched sites (△). Weight-frequency distributions of the two most common snails were shifted to larger sizes in the enriched zone (● = enriched zone <m from center of plot, o = control zone 2 m from center; $n = 56$–203 snails). Epiphytes were delicately brushed and rubbed from entire tillers of *Scirpus* collected using SCUBA, then centrifuged, lyophilized, and extracted by grinding in 90% basic acetone. Chlorophyll *a* was corrected for phaeopigment degradation products (Wetzel and Likens, 1991). For epiphytes, seven replicates were collected between September 18 and October 3 from each of nine distances from centers of plots. Snails were collected in mid-October from 0.05-m^2 quadrats of surficial sediment with overlying macrophytes; $n = 4$ enriched quadrats <m from center of plot A and 4 control quadrats >2 m from center. New growth of *Scirpus* was reduced by heavy epiphyte load inside the enrichment zone, so five–six entire stems of *Potamogeton illinoiensis* were also collected both inside and outside the enrichment zone to better compare snail biomass per unit macrophyte. Shell lengths were measured and converted to tissue dry mass based on regressions of length-mass.

littoral (Cattaneo, 1983; Cuker, 1983; Lowe and Hunter, 1988; Osenberg, 1988, 1989; Brönmark, 1989; Lodge et al., 1994) or pelagic algal communities (Carpenter and Kitchell, 1984; Wright and Shapiro, 1984; Lehman and Sandgren, 1985). Our nutrient enrichments were intentionally of sufficiently short duration that phytoplankton and epiphytes could outgrow their grazer populations, which revealed the primary underlying control by phosphorus.

Aquatic macrophytes respond more slowly than microalgae to changing nutrient availability, so we allowed a full year between initial enrichment and final determination of biomass at five sites around the lake. Each site included 10

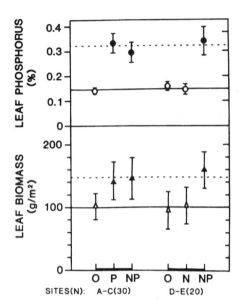

Figure 22.3. Response of the dominant submerged macrophyte *Scirpus subterminalis* to 1 year of P addition. Upper: P concentration in leaf tissue (% carbonate-free dry wt). Lower: Leaf biomass (g dry wt/m²). Means with 95% CI are compared for three sites with a parallel P-alone enrichment and for two sites with a N-alone enrichment (0 = control, NP = mixed fertilizer, N = +nitrogen, P = +phosphorus). Fertilizer rods were as in Figure 22.2, one rod/quadrat on each date; nitrogen rods carried pellets of ammonium nitrate with some ammonium sulfate (2.6 g N/rod). F-tests for site effects were insignificant ($P > .2$) for all treatments; thus, quadrats were treated as replicates.

control quadrats 0.008 m² in area, 10 quadrats receiving mixed fertilizer in November 1983 and June 1984, and 10 separate quadrats receiving either calcium phosphate or ammonium nitrate on the same dates (Fig. 22.3). Because rooted macrophytes obtain most of their phosphorus from the sediment in nutrient-poor waterbodies (Barko and Smart, 1980; Carignan & Kalff, 1980; Carignan, 1982; Wetzel, 1990b), we avoided a proliferation of epiphytes by inserting fertilizer beneath the sediment surface.

The biomass of the wintergreen leaves of *Scirpus subterminalis* as measured in late October 1984 represents much of their annual net production (Rich et al., 1971; Wetzel et al., 1972; Moeller and Wetzel, unpublished data). This biomass was homogeneous among sites within treatments. The aggregated data (Fig. 22.3) displayed a statistically significant increase caused by mixed fertilizer ($P < .01$) and by phosphorus alone ($P = .05$). Nitrogen added without phosphorus had no effect ($P > .2$). *Potamogeton illinoiensis* Morong made up 30% of total macrophyte biomass at control sites and apparently responded similarly to the mixed fertilizer. The increase from 44 to 60 g dry wt/m² for this more heterogeneously distributed plant was not statistically significant ($P > .1$).

The phosphorus additions effectively saturated plant demand for that element, as demonstrated by the accumulation of excess phosphorus in leaves (Fig. 22.3). The small size of the subsequent increase in biomass (about 50%) probably means that only the lowest leaf phosphorus concentrations encountered in Lawrence Lake (e.g., the 27% of analyses <0.12% of dry wt) truly represent phosphorus deficiency. Concentrations of nitrogen and potassium did not increase after addition of mixed or nitrogen-only fertilizers, which suggested that the enrichments may not have increased their availability. Therefore, it is possible that these elements may also have a limiting role unresolved by the less effective enrichments in these elements.

In the emergent macrophyte wetland surrounding this lake, a dominant species is the common cattail *Typha latifolia* L. Detailed analyses of tissue nutrient analyses and population and growth dynamics in response to fertilization experiments revealed that *T. latifolia* was principally phosphorus-limited in the open calcareous marshes along the eastern side of the lake (Grace and Wetzel, 1981; Dickerman and Wetzel, 1985). Populations of *T. latifolia* under shaded conditions within wooded areas of the western side of the lake were light rather than nutrient-limited (Grace and Wetzel, 1981).

Qualitatively, then, phosphorus operates concordantly as a growth-regulating factor across all biotic components examined. Quantitatively, however, phosphorus is moderately available in the anoxic calcareous sediments of Lawrence Lake that allows modest development of macrovegetation. Experiments with potted *Scirpus* grown in the lake showed that biomass can be more than doubled to 500 g dry wt/m^2 within one growing season; physical disturbance plus mixed fertilizer, not phosphorus alone, was required. The heavily developed epiphytic algal component of the littoral zone shares a phosphorus-deficient medium with the phytoplankton. Rooted macrophytes occupying the same spatial habitat as their epiphytes are functionally detached, by the nutrient geochemistry of sediments, from the aquatic desert of the water column.

Conclusions

We address the question of how pelagial and littoral habitats are integrated into functional lake ecosystems by asking if a single plant nutrient, phosphorus, can play the same biomass-limiting role for littoral submerged macrophytes and their epiphytes, as well as growth of representative consumers, that it does for the plankton. Enrichments of phosphorus to the environment of natural microbial and vegetation communities from a hardwater lake demonstrate that phosphorus can play a concordant growth-regulating role across many aquatic growth-forms. Epiphytic algae as well as phytoplankton and bacterioplankton were strongly limited by phosphorus in summer. Growth and biomass of submerged macrophytes increased significantly to phosphate released within the calcareous littoral sediments, but not to inorganic nitrogen. Phosphorus availability did not suppress growth in the rooted vegetation as severely as it did in the microbial communities. Growth rates of two dominant snail species increased rapidly in response to increased epiphytic algal food supplies under enriched conditions. These results indicate that phosphorus operates uniformly as a growth-regulating factor among many growth-forms within the aqueous portions of the ecosystem.

Acknowledgments. We gratefully acknowledge the capable technical assistance of K. Keagle, J. Sonnad, A.J. Johnson, and S.M. Ford and the critical reviews of two anonymous reviewers. R. Benson of the Sierra Chemical Co. generously provided fertilizer pellets. This work was supported by the National Science Foundation (DEB-8001190–01), a NSF predoctoral fellowship (C.W.O.), and the Department of Energy (DE-FG05–90ER60930).

References

Barko, J.W.; Smart, R.M. Mobilization of sediment phosphorus by submersed freshwater macrophytes. Freshwat. Biol. 10:229–238; 1980.

Brönmark, C. Interactions between epiphytes, macrophytes and freshwater snails: a review. J. Moll. Stud. 55:299–311; 1989.

Burkholder, J.A.; Wetzel, R.G. Epiphytic microalgae on a natural substratum in a hardwater lake: seasonal dynamics of community structure, biomass and ATP content. Arch. Hydrobiol. Suppl. 83:1–56; 1989.

Burkholder, J.M.; Wetzel, R.G. Alkaline phosphatase and algal biomass on natural and artificial plants in an oligotrophic lake: re-evaluation of the role of macrophytes as a phosphorus source for epiphytes. Limnol. Oceanogr. 35:736–747; 1990.

Carignan, R. An empirical model to estimate the relative importance of roots in phosphorus uptake by aquatic macrophytes. Can. J. Fish. Aquat. Sci. 39:243–247; 1982.

Carignan, R.; Kalff, J. Phosphorus sources for aquatic weeds: water or sediments? Science 207:987–989; 1980.

Carignan, R.; Kalff, J. Phosphorus release by submerged macrophytes: significance to epiphyton and phytoplankton. Limnol. Oceanogr. 27:419–427; 1982.

Carpenter, S.R.; Kitchell, J.F. Plankton community structure and limnetic primary production. Am. Nat. 124:159–172; 1984.

Cattaneo, A. Grazing on epiphytes. Limnol. Oceanogr. 28:124–132; 1983.

Coveney, M.F.; Wetzel, R.G. Experimental evaluation of conversion factors for the (^{3}H)thymidine incorporation assay of bacterial secondary productivity. Appl. Environ. Microbiol. 54:2018–2026; 1988.

Coveney, M.F.; Wetzel, R.G. Effects of nutrients on specific growth rate of bacterioplankton in oligotrophic lake water cultures. Appl. Environ. Microbiol. 58:150–156; 1992.

Coveney, M.F.; Wetzel, R.G. Biomass, production, and specific growth rate of bacterioplankton and coupling to phytoplankton in an oligotrophic lake. Limnol. Oceanogr. 40:1187–1200; 1995.

Cuker, B.E. Competition and coexistence among the grazing snail *Lymnaea,* chironomidae, and microcrustacea in an arctic epilithic lacustrine community. Ecology 64:10–15; 1983.

Dickerman, J.A.; Wetzel, R.G. Clonal growth in *Typha latifolia:* population dynamics and demography of the ramets. J. Ecol. 73:535–552; 1985.

Fairchild, G.W.; Everett, A.C. Effects of nutrient (N, P, C) enrichment upon periphyton standing crop, species composition and primary production in an oligotrophic softwater lake. Freshwat. Biol. 19:57–70; 1988.

Fairchild, G.W.; Sherman, J.W. Algal periphyton response to acidity and nutrients in softwater lakes: lake comparison vs. nutrient enrichment approaches. J. North Am. Benth. Soc. 12:157–167; 1993.

Grace, J.B.; Wetzel, R.G. Phenotypic and genotypic components of growth and reproduction in *Typha latifolia:* experimental studies in marshes of differing successional maturity. Ecology 62:789–801; 1981.

Kalff, J.; Knoechel, R. Phytoplankton and their dynamics in oligotrophic and eutrophic lakes. Annu. Rev. Ecol. Syst. 9:475–495; 1978.

Lehman, J.T.; Sandgren, C.D. Species-specific rates of growth and grazing loss among freshwater algae. Limnol. Oceanogr. 30:34–46; 1985.

Lodge, D.M.; Kershner, M.W.; Aloi, J.E.; Covich, A.P. Effects of an omnivorous crayfish (*Orconectes rusticus*) on a freshwater littoral food web. Ecology 75:1265–1281; 1994.

Lowe, R.L.; Hunter, R.D. Effect of grazing by *Physa integra* on periphyton community structure. J. North Am. Benth. Soc. 7:29–36; 1988.

Moeller, R.E.; Burkholder, J.M.; Wetzel, R.G. Significance of sedimentary phosphorus to a submersed freshwater macrophyte (*Najas flexilis*) and its algal epiphytes. Aquat. Bot. 32:261–281; 1988.

Osenberg, C.W. Body size and the interaction of fish predation and food limitation in a freshwater snail community. Ph.D. dissertation, Michigan State University, East Lansing; 1988.

Osenberg, C.W. Resource limitation, competition and the influence of life history in a freshwater snail community. Oecologia 79:512–519; 1989.

Rich, P.H.; Wetzel, R.G.; Thuy, N.V. Distribution, production and role of aquatic macrophytes in a southern Michigan marl lake. Freshwat. Biol. 1:3–21; 1971.

Schindler, D.W. Eutrophication and recovery in experimental lakes: implications for lake management. Science 184:897–899; 1974.

Schindler, D.W. Factors regulating phytoplankton production and standing crop in the world's freshwaters. Limnol. Oceanogr. 23:478–486; 1978.

Wetzel, R.G. Longterm dissolved and particulate alkaline phosphatase activity in a hardwater lake in relation to lake stability and phosphorus enrichments. Verh. Int. Verein. Limnol. 21:337–349; 1981.

Wetzel, R.G. Limnology. 2nd ed. Philadelphia: W.B. Saunders Co.; 1983.

Wetzel, R.G. Land–water interfaces: metabolic and limnological regulators. Verh. Int. Verein. Limnol. 24:6–24; 1990a.

Wetzel, R.G. Detritus, macrophytes and nutrient cycling in lakes. Mem. Ist. Ital. Idrobiol. 47:233–249; 1990b.

Wetzel, R.G. Microcommunities and microgradients: linking nutrient regeneration, microbial mutualism, and high sustained aquatic primary production. Netherlands J. Aquat. Ecol. 27:3–9; 1993.

Wetzel, R.G.; Likens, G.E. Limnological analyses. 2nd ed. New York: Springer-Verlag; 1991.

Wetzel, R.G.; Rich, P.H.; Miller, M.C.; Allen, H.L. Metabolism of dissolved and particulate detrital carbon in a temperate hard-water lake. Mem. Ist. Ital. Idrobiol. 29 (suppl.):185–243; 1972.

Whittaker, R.H. Communities and ecosystems. New York: Macmillan; 1975.

Wright, D.I.; Shapiro, J. Nutrient reduction by biomanipulation: an unexpected phenomenon and its possible cause. Verh. Int. Verein. Limnol. 22:518–524; 1984.

23. Sources of Organic Carbon in the Food Webs of Two Florida Lakes Indicated by Stable Isotopes

Mark V. Hoyer, Binhe Gu, and Claire L. Schelske

Introduction

Carbon cycling pathways in lacustrine systems are complex because there are often multiple sources of organic carbon available to the food webs. Among the techniques used to delineate carbon flows from organic matter to consumers, stable isotope analysis may be the most powerful one because isotope compositions of consumers reflect those of the dietary carbon assimilated and incorporated into their tissues and because no system manipulation is involved. The use of stable carbon isotopes in food web study is based on these premises: (1) there is a broad isotope range among different sources of organic matter (Peterson and Fry, 1987) and (2) consumer $\delta^{13}C$ closely resembles their diets within 1‰ (DeNiro and Epstein, 1978). Stable isotope analysis has been successfully applied to the investigations of carbon flows in some lacustrine food webs (e.g., Hecky and Hesslein, 1995).

The objective of this study was to determine carbon sources that supported the food webs of lakes Apopka and Okahumpka, Florida, phytoplankton- and macrophyte-dominated systems, respectively. We used stable carbon isotopes as natural tracers to illustrate carbon flows in each food web.

Study Area

Lakes Apopka and Okahumpka are located in central Florida and differ considerably in some major limnological characteristics (Table 23.1). Lake Apopka (28°39′

Table 23.1. Typical Limnological Characteristics of Lakes Apopka and Okahumpka, Florida

Parameter (units)	Apopka	Okahumpka
Secchi disc depth (m)	0.26	Bottom
pH	9.0	8.1
Dissolved inorganic C (mg/L)	24.7	26.7
Alkalinity (mg/L)	120	51.3
Conductivity (μS/cm^2)	340	177
Total nitrogen (mg/L)	4.03	0.95
Total phosphorus (mg/L)	0.192	0.020
Chlorophyll a (μg/L)	100.0	5.0

N; 81°39′ W) has a total surface area of 12,400 ha and an average depth of 1.7 m. This lake is hypereutrophic, with an annual net phytoplankton primary production 300 g C/m^2 (Gale and Reddy, 1994). The water column is highly turbid, as a combined result of high phytoplankton biomass and frequent resuspension of the organic sediments. Low light penetration severely restricts the growth of benthic algae and macrophytes. Lake Okahumpka (28°45′30″; 82°05′02″) has a total surface area of 271 ha and an average depth of 1.1 m. By contrast to Lake Apopka, Lake Okahumpka is a macrophyte-dominated system with eel grass (*Vallisneria americana*) covering more than 90% of the lake surface. As a result of macrophyte growth, phytoplankton biomass is extremely low. Epiphytic algae are the second important primary producer in this lake (Canfield and Hoyer, 1992).

Methods

Phytoplankton, epiphytic algae, aquatic macrophytes, invertebrates, and fish were collected in 1994 and 1995. Owing to low biomass, pure phytoplankton from Lake Okahumpka and benthic algae from Lake Apopka were not obtainable. Sample preparations and mass spectrometer determinations for isotope ratios have been described in detail by Gu et al. (1996).

^{13}C/^{12}C ratio is expressed in the conventional delta (δ) notation, defined as the per mil (‰) deviation from a standard:

$$\delta^{13}C(\text{‰}) = \left(\frac{^{13}C/^{12}C_{\text{sample}}}{^{13}C/^{12}C_{\text{standard}}} - 1 \right) \times 1,000$$

The standard is Peedee Belemnite limestone. The analytical precision of these measurements was 0.1‰.

Results and Discussion

The average $\delta^{13}C$ of consumers in Lake Apopka ranged from -15 to $-8‰$, reflecting carbon source integrated from different phytoplankton signals (-14 to $-3‰$) (Fig. 23.1). Aquatic plants and benthic algae are sparse in this lake and were unlikely to be important to the growth of fish. The small variation in $\delta^{13}C$ among the planktivores (sunfish, gizzard shad, and least killifish) was indicative of feeding similarity. Blue tilapia is known by its high feeding plasticity, and the broad range of $\delta^{13}C$ suggested that there was diet variation within the population in Lake Apopka. Although a small percentage of the individuals had the cattail signal ($-26‰$), most of them derived their carbon source from plankton production. The two benthic feeders (white catfish and brown bullhead) had the $\delta^{13}C$ resembling the phytoplankton signal. Similarly, piscivores also displayed $\delta^{13}C$ signals, reflecting a phytoplankton-based food chain.

The $\delta^{13}C$ of the consumers in Lake Okahumpka fell into a narrow range with one exception. Most of the invertebrates and all fish species (-18 to $-16‰$) showed a clear dependence on the epiphytic algal carbon (-19 to $-14‰$) (Fig. 23.2). Despite its high abundance, the eelgrass ($-8‰$) was basically unexploited by consumers. An unidentified species of snail had a $\delta^{13}C$ ($-12‰$) at the midrange between epiphytic algae and the eelgrass, suggesting that it obtained

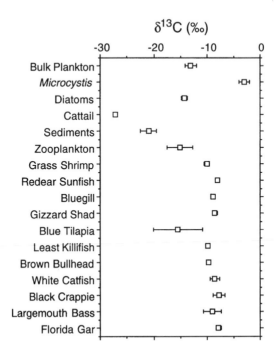

Figure 23.1. Stable carbon isotope ratios of organic matter and consumers from Lake Apopka (mean ± 1SD).

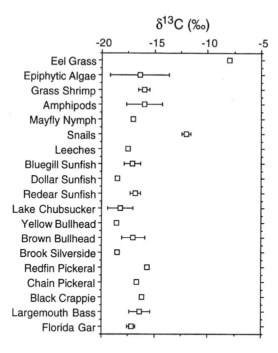

Figure 23.2. Stable carbon isotope ratios of organic matter and consumers from Lake Okahumpka (mean ± 1SD).

approximately equal amount of its dietary carbon from each source. The large difference in $\delta^{13}C$ between the carnivorous fish and snail indicated that the snail biomass was not transferred to a higher trophic level.

Our isotope data revealed two major carbon pathways from primary producers to consumers in the study lakes. Phytoplankton and epiphytic algae were the energy sources fueling the food webs in lakes Apopka and Okahumpka, respectively. The dependence of consumers on phytoplankton carbon in Lake Apopka is not surprising because phytoplankton is the only dominant primary producer in this hypereutrophic lake. Other organic sources, including terrestrial materials, macrophytes, and benthic algae, are not abundant. Only a few percentages of blue tilapia showed dependence for growth on cattail detritus.

By contrast, consumers in Lake Okahumpka showed heavy dependence on epiphytic algal carbon. It is possible that the food web may also derive some carbon from phytoplankton. However, feeding on the dilute phytoplankton is more energy expensive than feeding on the abundant epiphytic algae from the macro-phyte surface, which are also nutritionally more valuable than phytoplankton (Hecky et al., 1993). The most abundant macrophyte, eelgrass, was not used by the consumers to any great extent. The lack of trophic linkage between the consumers and aquatic macrophytes in many aquatic systems can be explained by the low nitrogen and high structural carbon contents in aquatic macrophytes (Hecky and Hesslein, 1995).

In conclusion, our results indicated that stable carbon isotope compositions of lake biota are excellent indicators of carbon flows in our study lakes. This technique provides insight into the actual carbon sources supporting the growth of consumers and has the advantage over the conventional methodologies in elucidating energy pathways in aquatic systems that receive contributions from different organic sources.

References

Canfield, D.E., Jr.; Hoyer, M.V. Aquatic macrophytes and their relation to the limnology of Florida lakes. Florida Department of Natural Resources, Tallahassee, FL; 1992.

DeNiro, M.J.; Epstein, S. Influence of diet on the distribution of carbon isotopes in animals. Geochim. Cosmochim. Acta 42:495–506; 1978.

Gale, P.M.; Reddy, K.R. Carbon flux between sediment and water column of a shallow, subtropical, hypereutrophic lake. J. Environ. Qual. 23:965–972; 1994.

Gu, B., Schelske, C.L.; Brenner, M. Relationships between sediment and plankton isotope ratios ($\delta^{13}C$ and $\delta^{15}N$) and primary productivity in Florida lakes. Can. J. Fish. Aquat. Sci. 53:875–883; 1996.

Hecky, R.E.; Hesslein, R.H. Contributions of benthic algae to lake food webs as revealed by stable isotope analysis. J. North Am. Benth. Soc. 14:631–653; 1995.

Hecky, R.E., Campbell, P.; Hendzel, L.L. The stoichiometry of carbon, nitrogen, and phosphorus in particulate matter of lakes and oceans. Limnol. Oceanogr. 38:709–724; 1993.

Peterson, B.J.; Fry, B. Stable isotopes in ecosystem studies. Annu. Rev. Ecol. Syst. 18:293–320; 1987.

24. Importance of Physical Structures in Lakes: The Case of Lake Kinneret and General Implications

Avital Gasith and Sarig Gafny

Introduction

Rock formations, plants, and woody debris are typical sources of physical structure in lakes. They determine physical complexity of littoral habitats, form the basis for the heterogeneous nature of the nearshore environment, and support metabolic (organic matter and nutrient dynamic) and nonmetabolic (structure) related functions (Wetzel and Hough, 1973; Lodge et al., 1988). Physical structures are often colonized by a diverse assemblage of microorganisms, algae, and invertebrates and attract predators (mostly fish and macroinvertebrates), which exploit this rich food resource (e.g., Lodge et al., 1988; Heck and Crowder, 1991; Diehl, 1993). Structured habitats also provide refugia for prey organisms (Heck and Crowder, 1991) and are favored as spawning sites (e.g., Goodyear et al., 1982; Gafny et al., 1992). Lake Kinneret (Israel) undergoes relatively wide water-level fluctuations, providing an opportunity to examine biotic responses to changes in littoral habitat structure in a relatively large (170 km^2), deep (43 m) lake. Here, we present selected results of our study on the effects of water-level fluctuation on habitat structure and availability, fish breeding, and community structure and discuss the importance of physical structures in lakes of different morphometry.

Results and Discussion

Effect of Water Level on Habitat Structure and Availability

Lake Kinneret water level normally fluctuates within 1.5–2 m. After drought years, the lake level may fall by 4 m (Gasith and Gafny, 1990). As water levels rise and fall, large areas along the shoreline are inundated or exposed, changing the location and structure of the littoral zone. The proportion of shores with rocky substrate declines from greater than 60% at the highest lake level to less than 10% as the water level falls by 3.5 m (Gasith and Gafny, 1990). The belt of submerged rocks also narrows markedly with falling lake level (Fig. 24.1), and stone size usually becomes smaller. Submerged macrophytes (e.g., *Potamogeton pectinatus, Myriophyllum spicatum*) develop sporadically in Lake Kinneret (e.g., Gasith and Gafny, 1990; Gafny, 1993). Emergent vegetation (mostly *Cyperus alopecuroides, Tamarix jordanensis, Typha angustata,* and *Phragmites australis*) is restricted to the supralittoral zone during periods of high lake levels. However, dense macrophyte stands develop (up to mean biomass of 1.3 kg dry weight/m^2; total lake shore biomass of ca. 1,000 metric tons ash-free dry weight) in sandy shores exposed after lake drawdown (Gasith and Gafny, 1990). Rising lake level inundates the newly developed vegetation and provides highly structured habitats for a period of a few months until the plants are uprooted or senesce and decompose (Gasith and Gafny, 1990; Gafny, 1993). In addition to affecting habitat complexity, water-level fluctuation markedly changes the availability of substrate colonized by periphyton, which, in the absence of macrophytes, form the main source of organic matter in the littoral zone (Gafny, 1993).

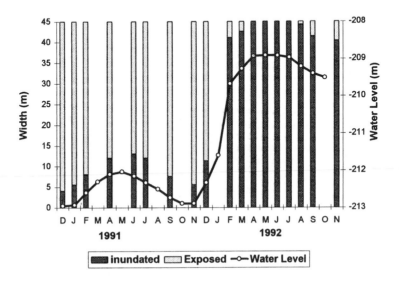

Figure 24.1. Change of the width of submerged rocky habitats in a selected littoral site (E-21) of Lake Kinneret during years of low (1991) and high (1992) lake levels.

Effect of Water-Level Fluctuation on Fish–Habitat Interaction

Of the 24 fish species extant in the lake, 15 may be found during the daytime in the shallow littoral zone (m), all at sites with boulders. Eight of these species may also be found at sites where the substrate is dominated by cobbles, and only four at sites with structurally simple, sandy substrate (Gafny, 1993). Most of these fish are small (<80 mm), either small-bodied species or juveniles of larger fish. Under conditions of high lake level, fish biomass over boulders is often an order of magnitude higher (80 g/m^2 wet weight) than over cobbles, reflecting preference of more structurally complex habitats over simple ones. During periods of low lake levels, fish biomass at both habitat types is similar and relatively low (<40 g/m^2). This reflects a decline in fish biomass over boulders and an increase over cobbles. Apparently, under low water-level conditions the fish have no choice but to use any structured habitat available (Gasith and Goren, 1995). Forcing the fish out of their preferred habitats can reduce fish survival, partly due to greater predation mortality.

Water-level fluctuation may also affect fish breeding success. During winter (November–May), large schools of the "lavnun" (*Mirogrex terraesanctae,* a keystone zooplanktivorous cyprinid) move inshore at night to spawn over rocks in very shallow waters (<50 cm). Only eggs that stick firmly to the substrate develop (Gafny et al., 1992). Winter is also the peak period of epilithon growth in Lake Kinneret (Gafny, 1993), and algae such as diatoms (e.g., *Cymbella* sp., *Gomphonema* sp., *Navicula* sp.) form a slimy covering on the stones' surface, making the substrate unsuitable for egg attachment. During periods of low lake levels, the lavnun uses the "window of opportunity" provided by the rising lake level to spawn over freshly inundated, temporarily (7–10 days) algae-free rocks found along the shoreline (Fig. 24.2). A minimal rise in lake level (ca. 30 cm) in the winter of 1988/89 produced unusual conditions in the littoral zone, and submerged rocks were overgrown by epilithon throughout most of the littoral zone. We estimated that more than 90% of the lavnun eggs were lost that winter and concluded that water-level fluctuation can strongly influence breeding success and, ultimately, recruitment of young of year (YOY) of the lavnun (Gasith and Gafny, 1990; Gafny et al., 1992; Gasith et al., 1996). This association between water level, the availability of spawning substrate, and YOY recruitment has been corroborated by findings from hydroacustic studies of fish abundance in the lake (Walline et al., 1992; Walline et al., 1994).

Water level also influences the availability of preferred spawning substrate for a blenny *(Salaria [Blennius] fluviatilis),* an important zoobenthivorous fish that spawns on rocks in the littoral zone of Lake Kinneret. Spawn density and size drop under low lake levels (Aidlin, 1995). However, the same conditions of low lake levels that deleteriously affected the above species enhance emergent macrophyte development in exposed shores. When inundated, this vegetation provides excellent breeding sites for certain cichlids (nest densities >0.8/m^2) as well as refugia and feeding grounds for the larvae and YOY. We can therefore predict higher recruitment of cichlids in littoral sites around the lake in a rainy year that follows years of low lake levels.

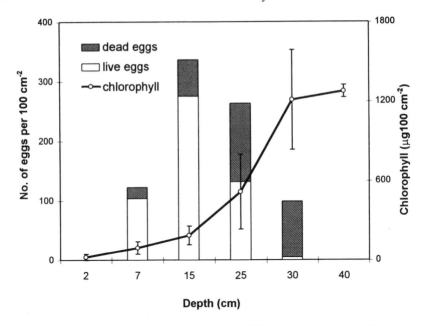

Figure 24.2. Typical relationship between lavnun (*Mirogrex terraesanctae,* Cyprinidae) egg density (mean ± SD) and survival (live and dead eggs), depth, and epilithon biomass (chlorophyll *a*) in the littoral zone of Lake Kinneret. (From Gafny et al., 1992.)

General Implications: The Role of Vegetative Versus Abiotic Formations and Effect of Lake Morphometry

Most of the knowledge on the structure and function of littoral zones comes from studies of small vegetated lakes, where the littoral region occupies a relatively large portion of the lake. In such lakes, both metabolic and structural features of the littoral zone are important and have been well recognized (e.g., Wetzel and Hough, 1973; Lodge et al., 1988). However, the role of the littoral zone in large deep lakes is less well understood (Danehy et al., 1991; Gasith, 1991). The availability of vegetated habitats in lakes is influenced by physical (e.g., wave action, light), water and sediment quality, and biotic controls (e.g., Carpenter and Lodge, 1986; Gasith and Hoyer, this volume, Chapter 29). In small and shallow lakes, temporal variability in habitat structure usually follows the seasonal cycle of macrophyte development. Macrophyte importance in lakes generally declines with increasing lake size. Duarte et al. (1986) concluded that, on average, the percentage of lake area colonized by submerged macrophytes declines with increasing lake size from about 50% in lakes of about 10 ha to about 10% in lakes of 10^4 ha and to less than 10% in larger lakes. They attributed the relatively higher importance of emergent vegetation in large lakes to a greater proportion of sheltered bays and floodplains. Large deep lakes often derive physical complexity from abiotic formations (e.g., Gasith and Gafny, 1990; Danehy et al., 1991;

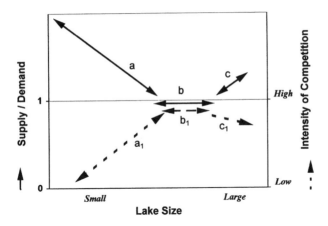

Figure 24.3. Relationship between lake size and the ratio of supply to demand of littoral resources and the corresponding intensity of competition over these resources among consumers that move into the littoral zone (for further details, see text).

Beauchamp et al., 1994). The size and form of these structures is mostly independent of water-quality conditions but is affected by wave action that sorts for the larger and heavier particles in the high-energy, nearshore zone and for finer ones in deeper water (e.g., Walter, 1985). Like plants, the surface of abiotic substrates is often seasonally modified by growth of periphyton. The availability of littoral habitats with vegetative or abiotic structure may be altered by water-level fluctuations.

The portion of the lake occupied by the littoral zone is inversely related to basin slope (Duarte and Kalff, 1986) and to the degree of shoreline regularity, being smallest in large deep lakes with low shoreline development index. Gasith (1991) proposed that physical structures and habitat complexity are unique littoral zone features, and thus littoral resources can be limiting, particularly in large and deep lakes where the littoral zone occupies a small proportion of the lake area. Similarly, Danehy et al. (1991) stated that "the relatively few regions of highly structured habitats [in the Great Lakes] may be more important to the ecology of fish populations than has been previously recognized." If this is true, we can predict that in small lakes, where the supply-to-demand ratio of littoral resources is high (Fig. 24.3, a), the intensity of competitive interactions among consumers that move from the pelagic zone to use littoral resources should be relatively low (Fig. 24.3, a_1). The cross boundary movers are typically adult or large fish that move between deep and shallow water for feeding, for cover (e.g., at night time; Gasith, Goren, and Gafny, unpublished data), and seasonally, for breeding. These fish may use different littoral resources than those used by the "resident," mostly young or small-bodied littoral species. Higher intensity of competition over littoral resources may be expected in larger deeper lakes (Fig. 24.3, b_1), where littoral resources are relatively limited (Fig. 24.3, b). The underlying assumption is that, other factors being equal, larger lakes with larger pelagic region can sustain larger consumer populations, some of which at times or at some part of their life history move back and forth between the pelagic and littoral zones. Beyond a certain size of a water body, the dependency on littoral resources is expected to diminish due to increased cost of energy expenditure in moving large distances across habitat boundary, and interhabitat links weaken (Lodge et al., 1988). For example, most lacustrine fish use nearshore habitats for spawning and nursery, whereas open-water spawning is usually typical to large deep lakes (Goodyear et al., 1982) and is characteristic of ocean-breeding fish. In such a situation, supply of littoral resources may again exceed the demand (Fig. 24.3, c), lowering the intensity of competition among cross-boundary movers over these resources (Fig. 24.3, c_1). Competition for food among resident littoral populations may be as intense in large and small lakes, depending on the availability of cover, which reduces predation mortality and increases density-dependent food limitation. High competition for food has been reported in small lakes (Mittelbach, 1988), mostly under excessive growth of foliated plants species such as milfoil (Heck and Crowder, 1991; Carpenter et al., this volume, Chapter 11).

Conclusions

Water-level fluctuations markedly modify the structure and availability of littoral zone habitats in Lake Kinneret and can influence fish community structure in a multiple mechanistic way. During periods of low lake levels, the littoral zone of Lake Kinneret supports fewer fish. Small fish and YOY are probably most affected by a reduction in the availability of structured habitats. Low water level is detrimental for fish breeding over rocky substrate but may be beneficial for other species that exploit the increased availability of vegetated habitats. Two hypotheses on the role of the littoral zone in lakes, and of physical structure in particular, are evoked from Lake Kinneret study. (1) Structural complexity is unique to littoral zones making it a potentially limiting factor, particularly in large deep lakes. In small shallow lakes, the availability of littoral resources may exceed the demand placed on them by interhabitat consumers. Thus, intensity of competitive interactions over littoral zone resources is expected to increase with increasing lake size and diminishing proportion of littoral zone area. (2) Under constant lake levels, abiotic formations may constitute a stable source of structural complexity of littoral habitats, allowing more temporal leeway in resource utilization. In vegetated lakes, littoral structure and complexity are amenable to biological feedback and usually follow seasonal plant cycle. The organisms using these resources are forced to synchronize with the window of opportunity provided by macrophyte growth. Competitive interactions over vegetative resources is expected to be highest where the period of plant growth is shortest.

    In conclusion, the case of Lake Kinneret underscores the importance of water level fluctuation as a major environmental factor that can strongly influence habitat structure and related interactions. Due to global climate changes, this factor may become relevant in lakes that presently exhibit stable water levels.

Acknowledgments. We thank Larry B. Crowder for constructive remarks and Merav Bing and Susan Gilman for assistance in manuscript preparation. The study was supported by the German-Israel Research and Development Fund (BMFT-DISUM 00016-GR 00984) and the Kurt Lion Fund for Scientific Cooperation between the University of Konstanz (Germany) and Tel-Aviv University.

References

Aidlin, M. Biological and ecological aspects of *Salaria fluviatilis* in the littoral zone of Lake Kinneret. M.Sc. thesis, Tel-Aviv Univ., Israel (Hebrew, English summary); 1995.

Beauchamp, D.A.; Byron, E.R.; Wurtsbaug, W.A. Summer habitat use by littoral-zone fishes in Lake Tahoe and the effects of shoreline structures. North Am. J. Fish. Manage. 14:385–394; 1994.

Carpenter, S.R.; Lodge, D.M. Effects of submersed macrophytes on ecosystem processes. Aquat. Bot. 26:341–370; 1986.

Danehy, R.J.; Ringler, N.H.; Gannon, J.E. Influence of nearshore structure on growth and diets of yellow perch *(Perca flavescens)* and white perch *(Morone americana)* in Mexico Bay, Lake Ontario. J. Great Lakes Res. 17:183–193; 1991.

Diehl, S. Effects of habitat structure on resource availability, diet and growth of benthivorous perch, *Perca fluviatilis.* Oikos 67:403–414; 1993.

Duarte, C.M.; Kalff, J. Littoral slope as predictor of the maximum biomass of submerged macrophyte communities. Limnol. Oceanogr. 31:1072–1080; 1986.

Duarte, C.M.; Kalff, J.; Peters, R.H. Pattern in biomass and cover of aquatic macrophytes in lakes. Can. J. Fish. Aquat. Sci. 43:1900–1908; 1986.

Gafny, S. The effect of substrate type on the structure and function of the littoral zone in Lake Kinneret, Israel. Ph.D. dissertation. Tel-Aviv Univ., Israel (Hebrew, English summary); 1993.

Gafny, S.; Gasith, A.; Goren, M. Effect of water level fluctuation on shore spawning of *Mirogrex terraesanctae* (Steinitz), (Cyprinidae) in Lake Kinneret, Israel. J. Fish Biol. 41:863–871; 1992.

Gasith, A. Can littoral resources influence ecosystem processes in large, deep lakes? Verh. Int. Verein. Limnol. 24:1073–1076; 1991.

Gasith, A.; Gafny, S. Effects of water level fluctuation on the structure and function of the littoral zone. In: Tilzer, M.M.; Serruya, C., eds. Large lakes, ecological structure and function. New York: Springer-Verlag; 1990:156–171.

Gasith, A.; Goren, M. Fish community of the littoral zone of large lakes: the Lake Kinneret and Lake Conctance experience. Part I—Lake Kinneret. Joint German-Israeli Research Projects. Final report INCR Tel-Aviv Univ., Israel; 1995.

Gasith, A.; Goren, M.; Gafny, S. Ecological consequences of lowering Lake Kinneret water level: effect on breeding success of the "Kinneret sardine." In: Steinberger, Y., ed. Preservation of our world in the wake of change. Jerusalem: ISEEQS VIB:569–573; 1996.

Goodyear, C.D.; Edsall, T.A.; Ormsby Dempsey, D.M.; Moss, G.D.; Polanski, P.E. Atlas of the spawning and nursery areas of Great Lakes fishes. Washington, DC: Fish and Wildlife Service, U.S. Department of the Interior; 1982.

Heck, K.L.; Crowder, L.B. Habitat structure and predator-prey interactions in vegetated aquatic systems. In: Bell, S.S.; McCoy, E.D.; Mushinsky, H.R., eds. Habitat structure, the physical arrangement of objects in space. London: Chapman & Hall; 1991:281–299.

Lodge, D.M.; Barko, J.W.; Strayer, D.; Melack, J.M.; Mittlebach, G.G.; Howarty, R.W.; Menge, B.; Titus, J.E. Spatial heterogeneity and habitat interactions in lake communities. In: Carpenter, S.R., ed. Complex interactions in lake communities. New York: Springer-Verlag; 1988:181–208.

Mittelbach, G.G. Competition among refuging sunfishes and effect of fish density on littoral zone invertebrates. Ecology 68:614–623; 1988.

Walline, P.; Pisanty, S.; Lindem, T. Acoustic assessment of the number of pelagic fish in Lake Kinneret, Israel. Hydrobiologia 232:153–163; 1992.

Walline, P.; Kalichman, Y.; Ostrovsky, I. Effect of water level fluctuation on the increase of the "lavnun" population size. In: Zohary, T.; Hambright, K.D., eds. Preliminary assessment of potential impacts of lowering Lake Kinneret water levels to –214 altitude. Kinneret Limnol. Lab: Special Report T24/94 ILOR (Hebrew); 1994:42–45.

Walter, R.A. Benthic macroinvertebrates. In: Likens, G.E., ed. An ecosystem approach to aquatic ecology: Mirror Lake and its environment. New York: Springer-Verlag; 1985: 280–288.

Wetzel, R.G.; Hough, R.A. Productivity and the role of aquatic macrophytes. An assessment. Pol. Arch. Hydrobiol. 20:9–19; 1973.

25. Clear Water Associated with a Dense *Chara* Vegetation in the Shallow and Turbid Lake Veluwemeer, The Netherlands

Marcel S. Van den Berg, Hugo Coops, Marie-Louise Meijer, Marten Scheffer, and Jan Simons

Introduction

The presence of submerged aquatic macrophytes in lakes is affected by the underwater light climate. Lakes with clear water can show abundant macrophyte vegetation, whereas lakes with turbid water usually have a poor submerged vegetation (Moss, 1990; Scheffer et al., 1993). Moreover, macrophytes improve their own light climate by enhancing the water transparency.

Various mechanisms have been proposed to explain the effect of macrophytes on water transparency. Plants can reduce the concentration of inorganic macronutrients by uptake from the water column, preventing excessive phytoplankton growth (Ozimek et al., 1990; Kufel and Ozimek, 1994). Furthermore, they may act as a refugium for zooplankton against predators (Timms and Moss, 1984; Schriver et al., 1995). As a consequence, a high grazing pressure inside macrophyte beds decreases the amount of phytoplanktonic algae. In addition, it has been found in laboratory studies that macrophytes reduce phytoplankton growth by the release of allelopathic substances (Wium-Andersen et al., 1982; Hootsmans and Blindow, 1994). Apart from these biological interactions, macrophytes may affect sedimentation and resuspension characteristics of the bottom sediment in favor of clear water (James and Barko, 1990; Petticrew and Kalff, 1992). In Lake Veluwemeer in The Netherlands, large areas of clear water have been observed associated with dense meadows of *Chara*. These clear areas are in strong contrast to the turbid nonvegetated parts of the lake (Scheffer et al., 1994). In 1995, the effect of *Chara* on the water transparency in the lake was studied by monitoring some physico-chemical and some biological parameters at sites inside and outside the *Chara* vegetation.

Study Area

Lake Veluwemeer (3,350 ha; mean depth, 1.45 m) is a large shallow lake in the center of The Netherlands (Fig. 25.1). The shallow part of the lake is colonized by charophytes (depth between 30–80 cm). In 1995, *Chara* vegetation (dominated by *C. aspera*) was present in one-third of the lake area, and about 420 ha were covered by a very dense *Chara* meadow (maximum biomass, about 500 g DW/m²).

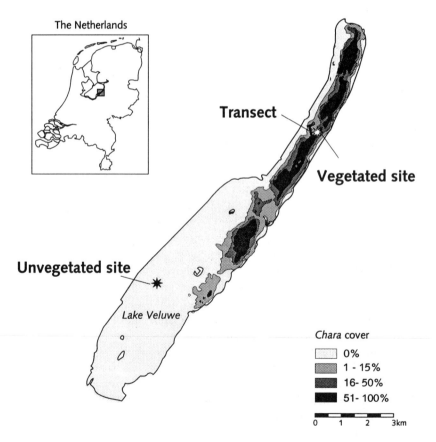

Figure 25.1. Map of Lake Veluwemeer: sample sites and the area covered by *Chara* in 1995 are indicated.

Materials and Methods

Comparison Between Two Sites in 1995

Light attenuation, chlorophyll *a,* inorganic suspended solids, and detritus were sampled at two sites in Lake Veluwemeer (Fig. 25.1), an unvegetated and a *Chara*-vegetated site at weekly or fortnightly intervals between March–November 1995. The *Chara* cover at the vegetated site was determined by placing a tube (surface area, 100 cm^2) in the water layer at 10 randomly chosen sites, whereby the bottom covered area of *Chara* was estimated by eye. The vertical downward attenuation coefficient of light (K_d, PAR: 400–700 nm) was measured directly below the water surface and at 25-cm depth. Water samples (25 L) were taken by using a tube sampler (perspex tube: length, 1 m; diameter, 5 cm). Chlorophyll *a* was measured after ethanol extraction. Suspended solids were measured as the dried (105°C) fraction remaining after filtering over a 0.45-μm membrane filter. Inorganic suspended solids were measured as the remaining residue after ignition at 600°C. Detritus concentration was calculated as total suspended solids minus inorganic suspended solids and algal biomass, assuming that 1 μg chlorophyll *a* corresponded to 0.07 mg algal biomass (Van Duin, 1992).

Transect Measurements 1995

On July 4, July 18, and August 1, 1995, several physical, chemical, and biological parameters were sampled along a transect crossing the lake from inside to outside a dense *Chara*-vegetated area. *Chara* cover, suspended solids, chlorophyll *a,* light attenuation, and gross sedimentation rate were determined as mentioned above at nine points along the transect. Sedimentation was measured by placing sediment traps (PVC; length, 0.5 m; diameter, 0.020 m) 25 cm into the sediment. After 1 or 2 weeks, the content of the traps was collected. The dry weight was determined as described above. Concentrations of phosphorus (PO_4–P), nitrogen (NO_3–N, NH_4–N), and bicarbonate (HCO_3) were measured according to International Standards (ISO).

On two dates (July 4 and July 18), zooplankton and phytoplankton species composition as well as their density were determined at seven points along the transect. To determine phytoplankton composition and density, 1 L of water was fixed with Lugol's solution. To determine microcrustacean zooplankton composition and density, 25 L water was filtered through a 120-μm filter, whereafter the remaining zooplankton was fixed with 96% ethanol. Biomass of the dominant zooplankton groups (*Daphnia* sp. and *Bosmina* sp.) was estimated by using the length/weight relationships of Culver et al. (1985). A potential grazing pressure index (percentage grazed phytoplankton biomass in 1 day [%/day]) was calculated by assuming that the algal biomass grazed by *Daphnia* sp. and *Bosmina* sp. is equal to their own biomass (Schriver et al., 1995). On July 4 and 18, the density as well as the particle sizes (divided in four fractions 1–5 μm, 5–30 μm, 30–150 μm, and >150 μm) was determined by using flow cytometrical analysis (Jonker et al., 1995).

Analysis of the Light Climate in Lake Veluwemeer

In Lake Veluwemeer, the contribution of humic acids and water to the vertical attenuation coefficient has been estimated at 0.55/m (Buiteveld, personal communication). The measured attenuation coefficient, chlorophyll a, detritus, and inorganic suspended solids concentrations were used to estimate the contribution of these fractions to the light attenuation by multiple regression (Blom et al., 1994), resulting in the model

$$K_d = 0.55 + 0.036 * [\text{inorganic suspended solids}] + 0.128 * [\text{detritus}] + 0.016 * [\text{chlorophyll } a]$$

where K_d = light attenuation in m^{-1}. Concentrations of inorganic suspended solids and detritus are in mg/L and the concentration of chlorophyll a in μg/L. The model explained 85% of total variance in K_d. The relative contribution of the components to K_d was calculated from the model.

Results

Comparison Between Two Sites

At the vegetated location, *Chara* vegetation emerged between April 10 and 25 (Fig. 25.2). Complete cover by the vegetation was reached in July. After August, the vegetation cover decreased due to grazing by water birds. In November, shortly before the lake became ice covered, an average cover of 60% was left.

Comparison of the light attenuation coefficients (K_d) between both sites showed small differences in March–June (Fig. 25.2). From May to June, the K_d values at both sites were considerably lower compared with March–April. In July–August, large differences between the two sites were found: K_d values of the vegetated site were very low, whereas K_d values of the unvegetated site were high. In September–November, differences between the sites were small again. The high transparency at the vegetated location was due to a low contribution of inorganic suspended solids, detritus, and chlorophyll a. On average, the contribution of inorganic suspended solids, detritus, and chlorophyll a to K_d at the vegetated site in July and August decreased with a factor 46, 6, and 13, respectively, compared with the unvegetated site.

Transect Measurements

The cover of charophytes inside the meadow was 100%, and at the border of the meadow, the cover decreased over a short distance to 0%. Along the transect from outside to inside the vegetation, the water transparency improved with increasing *Chara* cover (Fig. 25.3). On July 4 and 18, transparency increased already at a low cover of *Chara*. Possibly movement of water from inside to outside the vegetation caused this phenomenon, because the course of water transparency was closely

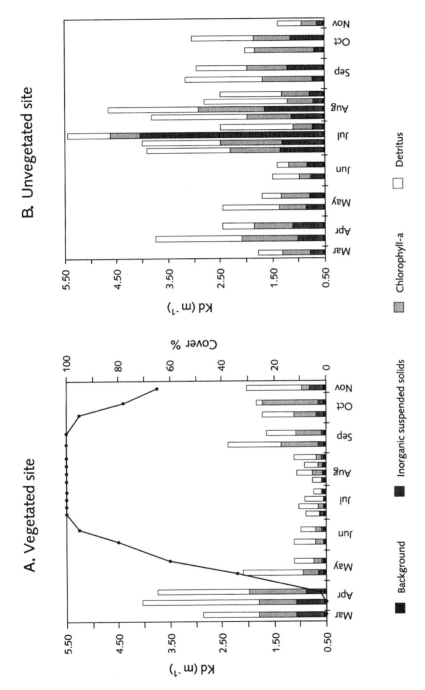

Figure 25.2. Seasonal course of light extinction (K_d) in Lake Veluwemeer during 1995, and the contribution of background attenuation, inorganic suspended solids, chlorophyll a, and detritus to the total light attenuation (A) vegetated site and (B) unvegetated site. The course of *Chara* cover is indicated by a line.

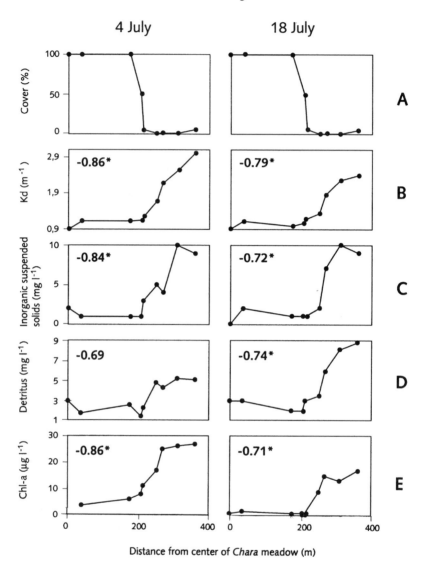

Distance from center of *Chara* meadow (m)

Figure 25.3. Vegetation cover (A), light attenuation (B), inorganic suspended solids (C), detritus (D), chlorophyll *a* (E), bicarbonate (F), phosphate-P (G), ammonium-N (H), and nitrate-N (I) measured over a transect starting inside and ending outside *Chara*-vegetated area on July 4 and 18, 1995. Spearmann correlation coefficients with *Chara* cover are given; significant correlations are shown; * *P* .05, ** *P* .001.

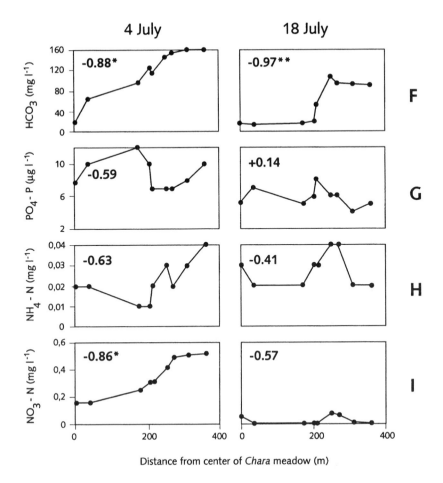

Figure 25.3. (*Continued*).

related to the course of ions (i.e., Ca and HCO₃) deposited on charophytes. The improvement of water transparency was related to both lower chlorophyll *a*, inorganic suspended solids, and detritus concentrations (Fig. 25.3). Results found on July 18 were similar to those found on August 1.

Gross sedimentation increased much more along the gradient from vegetation to the water layer than the suspended matter concentration (Fig. 25.4). Within the vegetation, the density of large-sized particles decreased more significantly than small-sized particles (Fig. 25.5). Along the transect, the density of large particles was very low compared with that of small-sized particles. Phytoplankton (excluding cyanobacteria) showed a comparable pattern, but cyanobacteria showed no differences in the size distribution between high and low *Chara* cover. A comparable pattern was found on July 18 (not in figure).

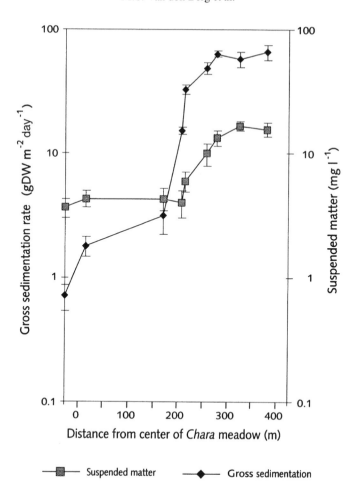

Figure 25.4. Suspended matter concentration and gross sedimentation rate over the transect starting inside and ending outside *Chara* vegetation, three data (July 4, July 18, and August 3) averaged. Error bars indicate standard error.

On July 18, chlorophyll *a* concentrations were extremely low at the vegetated part of the transect (Fig. 25.3D). Also, the relative contribution of cyanobacteria and green algae to the total density was lower in the vegetation, whereas the share of flagellates (*Cryptomonas* sp. and *Rhodomonas* sp.) was higher (Fig. 25.6A, Spearman correlation test, $P < .05$). On July 4, this shift in relative abundances occurred only in the center of the vegetation. The estimated grazing pressure of zooplankton on July 4 increased over the transect to the *Chara*-covered part (Fig. 25.6B, Spearman correlation test, $P < .05$). On July 18, the grazing pressure tended to be higher at the border of the *Chara* meadow. On both dates, the estimated potential grazing pressure exceeded by at least two times phytoplankton biomass.

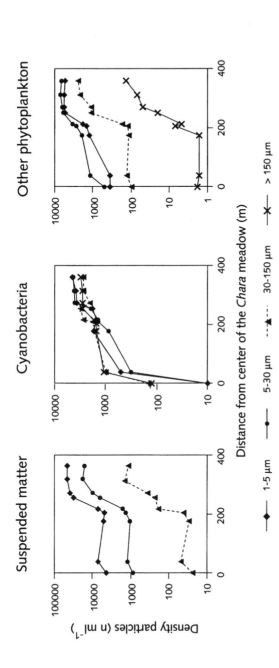

Figure 25.5. Densities of suspended particles along the transect with size fractions 1–5 μm, 5–30 μm, 30–150 μm, and >150 μm on July 4, 1995. The fraction 150 μm of suspended matter was below the detection limit.

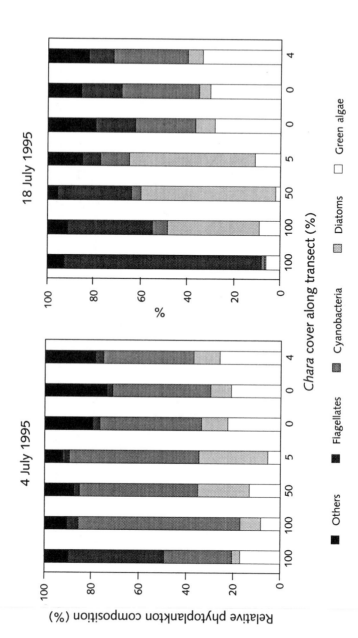

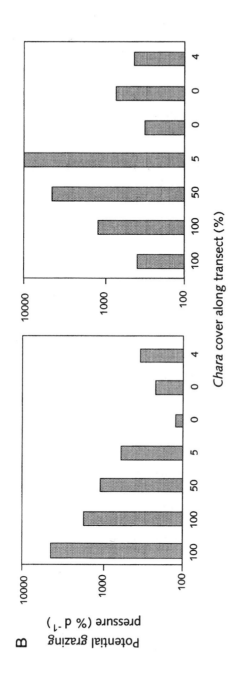

Figure 25.6. Composition of plankton over the transect starting inside and ending outside the *Chara* vegetation on July 4 and 18, 1995: (A) relative composition of phytoplankton groups and (B) potential zooplankton grazing pressure.

Ammonium and orthophosphate were not correlated to *Chara* cover, but on July 4 nitrate and on both dates bicarbonate were strongly negatively correlated with cover (Fig. 25.3). The pattern of the macronutrient concentrations on July 18 was similar to August 1.

Discussion

The effect of *Chara* on biotic and abiotic components of the water layer in Lake Veluwemeer is pronounced. The clear water above the *Chara* meadows was relatively stable, and the border between turbid and clear water was sharp compared with the size of the lake. This suggests that the water exchange between these two parts was low relative to the rate of the processes that increase water transparency. As a consequence, during the summer Lake Veluwemeer effectively consists of two separated parts: one with clear water covered by *Chara* and a turbid part characterized by pelagic algae.

The improvement of the light climate inside the vegetation was explained by a lower concentration of inorganic suspended solids and chlorophyll *a* and, in less extent, to lower detritus concentrations. The lower contribution of detritus and inorganic suspended solids to the light attenuation may be due to the reduction of resuspension above *Chara* meadows. The light climate in Lake Veluwemeer is strongly influenced by wind-induced resuspension (Blom et al., 1994). The dense charophyte vegetation is likely to restrict the resuspension of bottom sediment almost completely, whereas sedimentation of particles from the water layer to the bottom continues. In 1994, it was observed that during high wind velocities the water above the Characeae became turbid (unpublished data). After stormy weather conditions (June 21, 1994) the water cleared within 1 day, indicating that rapid sedimentation within the *Chara* meadows may largely explain the difference in transparency. Moreover, low densities of large-sized particles and a low ratio between the gross sedimentation rate and suspended matter concentration inside the vegetation give further support for the hypothesis that the restriction of resuspension is an important mechanism. However, it remains unsolved as to what extent phytoplankton density was reduced due to sedimentation. Larger algae and algae without flagella or floating vesicles in particular show net sinking in the water layer (Reynholds, 1984; Sommer, 1995). The sedimentation of phytoplankton may be increased by higher pH values inside the vegetation, because a higher rate of calcite incrustation on algae might occur at a high pH (Koschel et al., 1983). Indeed, the species composition of phytoplankton shifted to flagellates, which can escape from sedimentation (Reynholds, 1984). Despite the relatively high densities of flagellates, however, the absolute density was very low, indicating that mechanisms other than sedimentation are involved in reducing algal biomass. The decrease of algal densities inside the vegetation may be related to competition with macrophytes for nutrients, the release of allelopathic substances, or an increase of zooplankton within the macrophyte stands. In the measurements, it was observed that during summer no significant difference of dissolved phosphorus or nitrogen concentration occurred between the site inside and those outside the *Chara* vegetation. The uptake of nutrients from the water layer by rooted macrophytes becomes higher when the ratio between nutrient concentration in the water layer and in the sediment increases (Carignan, 1982; Granéli and Solander, 1988). Nutrients may be taken up from the water layer by charophytes in early summer, which was indicated by a decrease in the nitrate concentration on July 4. On July 18, nitrate and ammonium concentrations along the transect were very low. Hence, a nitrogen limitation of pelagic algae induced by macrophytes might occur (Ozimek et al., 1990).

The toxic effect of allelopathic substances by charophytes has been demonstrated for green algae (Hootsmans and Blindow, 1994) and for cyanobacteria (Jasser, 1995), but the role of allelopathic substances released by macrophytes under field conditions is still speculative (Forsberg et al., 1990). The importance of macrophytes as refuge for zooplankton has often been stressed (Schriver et al., 1995). Indeed, the grazing pressure of zooplankton was positively correlated to *Chara* cover on July 4 and tended to be higher at the border of the charophyte meadow on July 18. However, the relatively high grazing pressure of zooplankton outside the vegetation (two to five times the phytoplankton biomass) remains unexplained but is probably overestimated. The food availability outside the vegetation may be higher because detritus can be used as a food source for zooplankton (Jeppesen et al., 1997). Moreover, the relatively high concentration of small inorganic suspended matter outside the vegetation might negatively affect the effective grazing rate (Kirk and Gilbert, 1990).

Obviously, reduced resuspension inside dense *Chara* vegetation explains at least partly the clearwater patches associated with the charophytes, but other mechanisms such as nutrient uptake by charophytes and zooplankton grazing may also be important. To estimate the quantitative contribution of the mechanisms involved, further experiments are needed.

Acknowledgments. This study was partly financed by Rijkswaterstaat, Directie IJsselmeergebied. We thank Irmgard Blindow and Christer Brönmark for critically reviewing the manuscript.

References

Blom, G.; Van Duin, E.H.R.; Vermaat, J.E. Factors contributing to light attenuation in Lake Veluwe. In: van Vierssen, W.; Hootsmans, M.J.M; Vermaat, J. Lake Veluwe, a macrophyte dominated system under eutrophication stress. Dordrecht: Kluwer Academic Publishers; 1994:158–174.

Carignan, R. An empirical model to estimate the relative importance of roots in phosphorus uptake by aquatic macrophytes. Can. J. Fish. Aquat. Sci. 39:243–247; 1982.

Culver, D.A.; Boucherle, M.M.; Bean, D.J.; Fletcher, J.W. Biomass of freshwater Crustacean zooplankton from length weight regressions. Can. J. Fish. Aquat. Sci. 42:1380–1390; 1985.

Forsberg, C.; Kleiven, S.; Willén, T. Absence of allelopathic effects of *Chara* on phytoplankton in situ. Aquat. Bot. 38:289–294; 1990.

Granéli, W.; Solander, D. Influence of aquatic macrophytes on phosphorus cycling in lakes. Hydrobiologia 170:245–266; 1988.

Hootsmans, M.J.M.; Blindow, I. Allelopathic limitation of algal growth by macrophytes. In: van Vierssen, W.; Hootsmans, M.J.M.: Vermaat, J., Lake Veluwe, a macrophyte dominated system under eutrophication stress. Dordrecht: Kluwer Academic Publishers; 1994:175–192.

James, W.F.; Barko, J.W. Macrophyte influence on the zonation of sediment accretion and composition in a north-temperate reservoir. Arch. Hydrobiol. 120:129–142; 1990.

Jasser, I. The influence of macrophytes on a phytoplankton community in experimental conditions. Hydrobiologia 306:21–32; 1995.

Jeppesen, E.; Jensen, J.P.; Søndergaard, M.; Lauridsen, T.; Pedersen, L.J.; Jensen, L. Top down control in freshwater lakes: the role of nutrient state, submerged macrophytes and water depth. Hydrobiologia 342/343:151–164; 1997.

Jonker, R.R.; Meulemans, J.T.; Dubelaar, G.B.J.; Wilkins, M.F.; Ringelberg, J. Flowcytometry: a powerful tool in analysis of biomass distributions in phytoplankton. Wat. Sci. Techn. 32:177–182; 1995.

Kirk, K.L.; Gilbert, J.J. Suspended clay and the population dynamics of planktonic rotifers and Cladocerans. Ecology 71:1741–1755; 1990.

Koschel, R.; Benndorf, J.; Proft, G.; Recknagel, F. Calcite precipitation as a natural control mechanism of eutrophication. Arch. Hydrobiol. 98:380–408; 1983.

Kufel, L.; Ozimek, T. Can *Chara* control phosphorus cycling in Lake Luknajno (Poland)? Hydrobiologia 275/276:277–283; 1994.

Moss, B. Engineering and biological approaches to the restoration from eutrophication in which aquatic plant communities are important components. Hydrobiologia 275/276: 367–377; 1990.

Ozimek, T.; Gulati, R.D.; van Donk, E. Can macrophytes be useful in biomanipulation of lakes, the Lake Zwemlust example. Hydrobiologia 200/201:399–407; 1990.

Petticrew, E.L.; Kalff, J. Water flow and clay retention in submerged macrophyte beds. Can. J. Fish. Aquat. Sci. 49:2483–2489; 1992.

Reynholds, C.S. The ecology of freshwater phytoplankton. London: Cambridge University Press; 1984.

Scheffer, M.; Hosper, S.H.; Meijer, M.-L.; Moss, B.; Jeppesen, E. Alternative equilibria in shallow lakes. Trends Ecol. Evol. 8:276–279; 1993.

Scheffer, M.; Van den Berg, M.; Breukelaar, A.; Breukers, C.; Coops, H.; Doef, R.; Meijer, M.-L. Vegetated areas with clear water in turbid shallow lakes. Aquat. Bot. 193–196; 1994.

Schriver, P.; Bøgestrand, J.; Jeppesen, E.; Søndergaard, M. Impact of submerged macrophytes on fish-zooplankton-phytoplankton interactions: large scale enclosure experiments in a shallow eutrophic lake. Freshwat. Biol. 33:255–270; 1995.

Sommer, U. Planktologie. Berlin: Springer-Verlag; 1995.

Timms, R.M.; Moss, B. Prevention of growth of potentially dense phytoplankton populations by zooplankton grazing, in the presence of zooplanktivorous fish, in a shallow wetland ecosystem. Limnol. Oceanogr. 29:472–486; 1984.

Van Duin, E.H.S. Sediment transport, light and algal growth in the Markermeer. Thesis, Agriculture University, Wageningen; 1992.

Wium-Andersen, S.; Anthoni, U.C.; Christophersen, C.; Houen, G. Allelopathic effects on phytoplankton by substances isolated from aquatic macrophytes (Charales). Oikos 39: 187–190; 1982.

26. Alternative Stable States in Shallow Lakes: What Causes a Shift?

Irmgard Blindow, Anders Hargeby, and Gunnar Andersson

Introduction

Lake Tåkern and Lake Krankesjön, two shallow, moderately eutrophic, calcium-rich lakes in southern Sweden have shifted between turbid and clearwater states several times during the past decades (Fig. 26.1). Lake Krankesjön shifted from a clearwater state with abundant submerged vegetation to a turbid state with sparse vegetation during the mid-1970s (Karlsson et al., 1976) and back to a clearwater state during 1985. Today, the lake is in the clearwater state, with abundant submerged vegetation dominated by Charophyta (Blindow et al., 1993; Fig. 26.2). Both shifts coincided with deviations from the average water level. During the mid-1970s, the water level during spring and summer was about 15 cm higher than average, whereas it was about 10 cm lower than average during 1983–1985 (Blindow, 1992).

In Lake Tåkern, the submerged vegetation disappeared at least twice during the beginning of the century due to catastrophic events (extremely low water level causing dry-out during summer and damage by ice during winter, respectively), but it soon recovered. During the 1950s, a similar disappearance of submerged plants occurred due to dry-out, causing a shift to a turbid state. In the mid-1960s, the lake shifted back to a clearwater state with abundant submerged vegetation dominated by Charophyta. This shift took place after the application of a new water regime with lower amplitudes in water level. Since then, the lake has been in a clearwater state (Ekstam, 1975; Blindow et al., 1993). However, during the

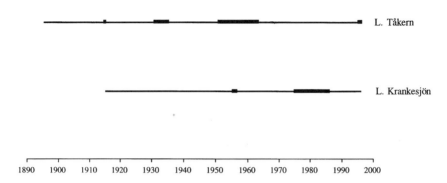

Figure 26.1. Long-term shifts between clearwater (thin lines) and turbid (thick lines) states in Lake Tåkern and Lake Krankesjön, schematically.

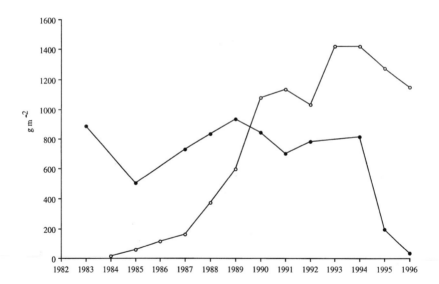

Figure 26.2. Average biomass (fresh weight) of submerged macrophytes per lake surface outside the reed belts of Lake Tåkern (black symbols) and Lake Krankesjön (white symbols). Biomass values were calculated from data on distribution of submerged macrophytes (obtained by investigation from boat, combined with air photographs during several years) multiplied with biomass values for different macrophyte species (sampled with Plexiglas corer). Values for 1995 and 1996 plant distribution are rough estimates as the investigation of plant distribution was hampered by high water turbidity.

past two summers the vegetation developed slowly, simultaneously with extended periods of turbid water. The monitoring undertaken during these years may give information regarding shifts between clearwater and turbid states.

Recent Observations in Lake Tåkern

During 1994–1996, the development of submerged vegetation was monitored in Lake Tåkern during spring and summer. Samples for biomass of submerged vegetation were taken about three times per month with a Plexiglas core sampler in a stand of *Chara tomentosa*. This species is one of the dominant submerged macrophytes in Lake Tåkern and hibernates as a green plant (Blindow, 1992). The samples were frozen, dried at 105°C (24 hours) and 550°C (2 hours) for determination of dry weight and ash-free dry weight, respectively. Whole-column water samples were taken with a Plexiglas core in an unvegetated area close to the site where the plant samples were taken. A part of the sample was frozen and analyzed later for suspended material (GF/C-filtered sample). For analysis of chlorophyll$_a$, water was filtered immediately (GF/C). The filters were deep frozen for later analysis according to standard procedures by using extraction with methanol.

During 1994, the biomass of *Chara tomentosa* decreased during the end of April and increased by the end of May. Simultaneously with this increase, water turbidity decreased (Fig. 26.3). During the summer, the plants reached the water surface, and estimated values for overall biomass of submerged macrophytes (including *Chara tomentosa*) were similar to values obtained during previous years (Fig. 26.2).

Also during 1995, the biomass of *Chara tomentosa* decreased by the end of April. In the opposite to 1994, however, only a minor increase of plant biomass was observed during spring and summer, and the water was turbid throughout the whole period of investigation (Fig. 26.3). In contrast to all previous years when the submerged vegetation has been investigated (yearly since 1983, except 1993), neither *Chara tomentosa* nor any other species of submerged plants reached the water surface during summer 1995. Consistently, estimated biomass for submerged macrophytes was lower than in previous years (Fig. 26.2). First during the autumn, submerged plants reached water surface in limited areas of the lake (L. Gezelius, personal communication). Values for chlorophyll taken during the summer of 1995 were almost twice as high as the years before (Fig. 26.4).

During 1996, the biomass of *Chara tomentosa* was low at the beginning of the season and remained low throughout the summer, with a minor increase during the autumn. The turbidity was low during the beginning of the season but increased continuously (Fig. 26.3). In the opposite to 1994 and 1995, the vegetation grew patchily, and most of the areas previously covered with dense stands of *Chara tomentosa* were vegetation-free. Estimates in July indicated that the total biomass of submerged macrophytes was even lower than during 1995 (Fig. 26.2). Values for summer chlorophyll increased further compared with 1995 (Fig. 26.4).

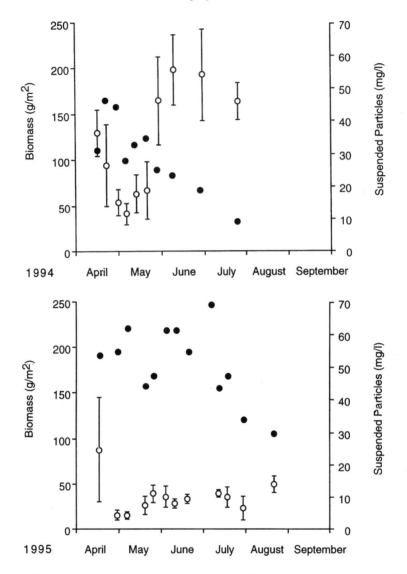

Figure 26.3. Biomass of submerged vegetation (*Chara tomentosa;* white circles, mean ± standard error) and turbidity (black circles) in Lake Tåkern during 1994 (above), 1995 (middle), and 1996 (next page).

Summer densities of Cladocera (including *Daphnia*) were low during all 3 years. Spring peak densities of both total number of Cladocera and *Daphnia* were lower during 1994 than in the years of turbid water (1995 and 1996; Table 26.1). The few data that exist on fish assemblage are hard to evaluate in terms of biomass for the whole lake. Catches with survey nets in July in a restricted area

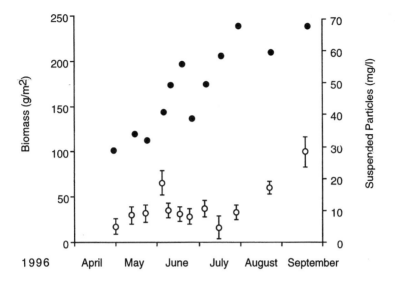

Figure 26.3. (*Continued*).

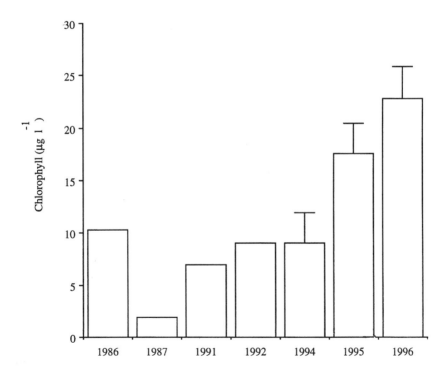

Figure 26.4. Chlorophyll concentration in Lake Tåkern during different years. Mean values of samples taken from June to August (only single samples taken during 1986–1992). Standard error is given for samples taken during 1994 to 1996.

Table 26.1. Numbers of Cladocera and *Daphnia* During Spring (Peak Values) and Summer (Average Values for Samples Taken During July and August) in Lake Tåkern

Year	Spring peak numbers, all Cladocera[a]	Spring peak numbers, *Daphnia* spp.[b]	Summer average, all Cladocera	Summer average, *Daphnia*
1994	62	16	11 ($n = 2$)	0
1995	304	136	11 ($n = 5$)	0
1996	430	54	13 ($n = 5$)	2

[a]Dominant species: *Bosmina longirostris*.
[b]Dominant species: *Daphnia longispina*.

(about 2 km^2) in the center of the lake, however, do not indicate drastic changes in abundance or species composition (Table 26.2) between 1990 and 1996.

Discussion

Shifts between the turbid and the clearwater state in shallow lakes are often caused by human interference. Increased nutrient loading to shallow lakes has in many cases led to disappearance of submerged vegetation and a drastic decrease of water transparency (e.g., Scheffer et al., 1994), whereas reduction of the fish stock has been applied in several cases to attain a switch back to the clearwater state (e.g., van Donk et al., 1990; Meijer et al., 1994). Similar to Lake Tomahawk Lagoon in New Zealand (Mitchell et al., 1988; Mitchell 1989), both Lake Tåkern and Lake Krankesjön several times switched "spontaneously" between the two states, without any obvious influence or manipulation from "outside." Furthermore, the higher densities of Cladocera (including *Daphnia*) during 1995 and 1996 compared with 1994 suggest that the observed change in Lake Tåkern was not caused by top-down mechanisms. Instead, the recent results from Lake Tåkern presented above support our earlier suggestion (Blindow et al., 1993) that water-level fluc-

Table 26.2. Fish Catches (kg) in Survey Nets (4–100-mm Mesh) in an Area with Sparse Vegetation (*Myriophyllum spicatum* in 1985 and 1986, Sparsely Occurring and Low-Grown *Chara tomentosa* 1990 and 1996)

	n	Pike[a]	Roach[a]	Rudd[a]	Tench[a]	Crucian carp[a]	Perch[a]	Ruffe[a]
1985	2	1.1 ± 1.6	2.4 ± 0.5	0.1 ± 0.1	3.7 ± 0.7	0.6 ± 0.2	2.1 ± 0.2	2.1 ± 1.2
1986	3	0.3 ± 0.5	1.5 ± 0.5	0.2 ± 0.2	2.8 ± 1.8	0.2 ± 0.4	0.4 ± 0.2	0.01 ± 0.01
1990	3	0.5 ± 0.8	0.6 ± 0.2	0.1 ± 0.2	1.2 ± 1.5	0.0	0.5 ± 0.4	0.01 ± 0.01
1996	5	0.9 ± 2.0	1.8 ± 0.5	0.01 ± 0.03	1.5 ± 1.2	0.0	0.5 ± 0.2	0.02 ± 0.03

[a] Means ± SD.

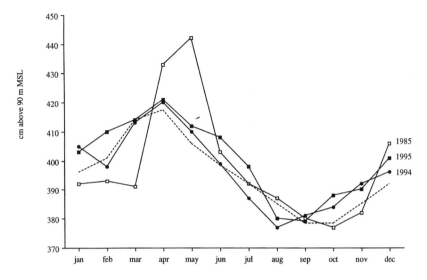

Figure 26.5. Water level (monthly average of daily readings) in Lake Tåkern during 1985 (white squares), 1994 (black circles), and 1995 (black squares). Data obtained from the Swedish Meteorological and Hydrological Institute. Dotted line: water level aimed at according to court decision.

tuations are an important factor affecting the submerged vegetation and eventually causing a switch. High spring water level in Lake Tåkern during 1985 was the most probable cause of the relatively low biomass of submerged vegetation registered in the summer of that year (Fig. 26.5). However, this disturbance was not sufficient to cause a switch to the turbid state (Blindow, 1992). Also during 1995, the spring water level was relatively high (Fig. 26.5). Compared with 1985, the deviation from normal water level was lower but occurred during a longer time period and coincided with low temperature during April to mid-May (unpublished data). High precipitation during this period probably caused a somewhat higher nutrient loading to the lake due to runoff from the agricultural area surrounding the lake. This is indicated by the fact that both water turbidity and chlorophyll concentrations at the beginning of the growing season were higher during 1995 than during 1994. We suggest that the combination of these factors—high spring water level, high turbidity, and low temperature—was the reason for the substantial decline in submerged macrophytes during 1995.

Despite lower turbidity in the beginning of the growing season, the transition toward a turbid state continued during 1996 in Lake Tåkern. The reason for this development may be the low spring temperature also that year, in combination with reduced biomass of hibernating *Chara tomentosa* and possibly other submerged macrophytes. Thus, the unfavorable conditions in 1995 may have affected the development of submerged plants the following year, weakening the feedback mechanisms that stabilize the clearwater stable state (Scheffer et al., 1993).

Acknowledgment. This study was financially supported by the World Wide Fund of Nature, Sweden ("Tåkernfonden").

References

Blindow, I. Long- and short-term dynamics of submerged macrophytes in two shallow eutrophic lakes. Freshwat. Biol. 28:15–27; 1992.

Blindow, I.; Andersson, G.; Hargeby, A.; Johansson, S. Long-term pattern of alternative stable states in two shallow eutrophic lakes. Freshwat. Biol. 30:159–167; 1993.

Ekstam, U. Förändringar av fågelfauna och miljö i och vid Tåkern 1850–1974 (Changes of the avifauna and the nature and environment of Lake Tåkern in 1850–1974). Vår Fågelvärld 34:268–282 (in Swedish with English summary); 1975.

Karlsson, J.; Lindgren, A.; Rudebeck, G. Drastiska förändringar i vegetation och fågelfauna i Krankesjön och Björkesåkrasjön 1973–1976 (Drastic changes in vegetation and bird fauna in Lake Krankesjön and Lake Björkesåkrasjön, South Sweden, in 1973–1976). Anser 15:165–184 (in Swedish with English summary); 1976.

Meijer, M.-L.; van Nes, E.H.; Lammens, E.H.R.R.; Gulati, R.D.; Grimm, M.P.; Backx, J.; Hollebeek, P.; Blaauw, E.M.; Breukelaar, A.W. The consequences of a drastic fish stock reduction in the large and shallow Lake Wolderwijd, The Netherlands. Can we understand what happened? Hydrobiologia 275/276:31–42; 1994.

Mitchell, S.F. Primary production in a shallow eutrophic lake dominated alternately by phytoplankton and by submerged macrophytes. Aquat. Bot. 33:101–110; 1989.

Mitchell, S.F.; Hamilton, D.P.; MacGibbon, W.S.; Nayar, P.K.B.; Reynolds, R.N. Interrelations between phytoplankton, submerged macrophytes, black swans (*Cygnus atratus*) and zooplankton in a shallow New Zealand lake. Int. Rev. Ges. Hydrobiol. 73:145–170; 1988.

Scheffer, M.; Hosper, S.H.; Meijer, M.-L.; Moss, B.; Jeppesen, E. Alternative equilibria in shallow lakes. Trends Ecol. Evol. 8:275–279; 1993.

Scheffer, M.; van den Berg, M.; Breukelaar, A.; Breukers, C.; Coops, H.; Doef, R.; Meijer, M.-L. Vegetated areas with clear water in turbid shallow lakes. Aquat. Bot. 49:193–196; 1994.

van Donk, E.; Grimm, M.P.; Gulati, R.D.; Klein Breteler, J.P.G. Whole-lake food web manipulation as a means to study community interactions in a small ecosystem. Hydrobiologia 200/201:275–291; 1990.

27. Clear and Turbid Water in Shallow Norwegian Lakes Related to Submerged Vegetation

Bjørn A. Faafeng and Marit Mjelde

Introduction

Timms and Moss (1984) suggested that fertile shallow lakes may have alternative stable states, a clearwater state with dense vegetation and a turbid water state dominated by phytoplankton and with little submerged and floating-leaved vegetation. This phenomenon has also been observed and discussed by several other authors (e.g., Irvine et al., 1990; Jeppesen et al., 1990; van Donk et al., 1990; Blindow et al., 1993; Scheffer et al., 1993). According to the model of Scheffer (1989, 1990), the main controlling factor for the two alternative states is the turbidity of water regulating the vertical light penetration. When the nutrient level increases, phytoplankton growth is often stimulated, which in turn increases the turbidity. This leads to increased light attenuation and a reduction of the maximum growth depth of submerged vegetation. Increased nutrient concentrations will therefore reduce the bottom area covered by submerged vegetation until it is virtually absent. In shallow lakes with most bottom areas at similar depths, the change from high plant cover to plantless bottoms may be abrupt. Uncovered sediments in shallow lakes are much more vulnerable to resuspension and more easily give rise to turbid water during periods of wind and wave action than in lakes with plant-covered sediments. Benthivorous fish may be favored under these circumstances and add to the turbidity by foraging on the sediment. A more-or-less stable turbid state may be the result.

On the contrary, when dense submerged vegetation covers a large part of the bottom area, these plants may successfully remain with sparse development of phytoplankton even at high nutrient concentrations (Timms and Moss, 1984; van Donk et al., 1993; Mjelde and Faafeng, 1997). The water adjacent to these macrophyte stands may be clear, also inside patches of vegetation (Scheffer et al., 1994). Several mechanisms connected with the submerged macrophytes support the stability of the clearwater stage: competition with phytoplankton for available nutrients (van Donk et al., 1993) and with epiphytes for light (Phillips et al., 1978; Sand-Jensen and Søndergaard, 1981), release of allelopathic substances (Wium-Andersen, 1987), and their acting as refuge for *Daphnia* and piscivorous fish (Timms and Moss, 1984; Grimm, 1989; Schriver et al., 1995). These mechanisms add to the turbidity effect.

In oligotrophic lakes, the two alternative stable states will not occur, as the phytoplankton biomass is not sufficiently high. Jeppesen et al. (1990) observed the two alternative states in Danish lakes with a total P concentration between approximately 50 and 125 μg total P/L, and the clearwater state has also been observed in very small lakes at much higher concentrations (Balls et al., 1985; Jeppesen et al., 1990).

The transition between a "clear" and a "turbid" state is so far, however, highly subjective. In lakes with a total P concentration of 20 μg P/L, the average concentration of chlorophyll (Chla) throughout the growth season may vary at least between 5 and 20 μg Chla/L (Faafeng and Hessen, 1993). It is obvious that in lakes with a total P of 20 μg P/L, the growth potential is normally much lower and the expected minimum transparency caused by phytoplankton much higher than at total P levels of, for instance, 200–300 μg P/L. Therefore, although the low light intensity needed to shade out the macrophytes is the same, we suggest that in this respect the limit between clear and turbid should relate to their total P levels. In a previous paper (Mjelde and Faafeng, 1997) we demonstrated that the clearwater state appeared in the investigated lakes with mean depths less than 1.9 m and when the bottom area covered with vegetation was greater than 50%. We also suggested that the average Chla/total P ratio over the growth season (May–September) indicates whether the water is clear (close to 1:10) or turbid (close to 1:1). The focus of this chapter is to give a more detailed study of the clear and turbid states over high and low total P concentrations by using the ratios Chla/total P and total P/transparency on a data set from shallow Norwegian lakes.

Materials and Methods

This study includes 10 small (<1 km^2) and shallow (mean depth, <4 m) lakes with an average total P greater than 20 μg P/L. The lakes are situated in northern and southern parts of Norway and cover latitudes between 58–69°N. All lakes were shallow enough to allow growth of vegetation over a major part of the bottom area under favorable light conditions. Integrated water samples were taken from the phototrophic layer (twice the Secchi depth or at most down to 0.5 m above the lake bottom) at least four times during May–September for analysis of water chemistry and quantitative phytoplankton and zooplankton. Total P was analyzed with a Technicon autoanalyzer after persulfate digestion, and chlorophyll *a* was measured spectrophotometrically after acetone extraction. Between 1988–1993, at least one growth season was studied in each lake.

The distribution of rooted submerged vegetation toward depth in lakes is normally limited by light, preventing vegetation cover in deeper areas (Vant et al., 1986). When lakes are shallow and clear, submerged and floating-leaved plants may cover the whole bottom area. In our survey, we used Secchi disc transparency as a measure for the light conditions. Transparency is often a realistic alternative to measurement of vertical light attenuation and is a useful approximation (Canfield et al., 1985; Chambers and Kalff, 1985).

Major phytoplankton blooms may occur in the spring (May) before the submerged vegetation has established, and this may seriously affect the average growth-season values of nutrients, chlorophyll, and transparency. This phenomenon was especially prominent in some of the lakes with conspicuous unrooted growth of *Ceratophyllum demersum* from early June. To avoid this problem, we therefore use "average late summer" values (ALS) calculated as an average of the values from July, August, and September.

The submerged vegetation was studied once in each lake (in late July–September) during the years 1992–1996 by using a hydroscope and by dredging from a boat. The species distribution was recorded and the bottom cover estimated. The methods are discussed in further detail in Mjelde and Faafeng (1997).

Results

The lakes had ALS total P concentrations ranging from 23 to more than 500 µg P/L (Table 27.1), whereas average transparency ranged between 0.3–3.8 m. In three of

Table 27.1. Late Summer Average Values (July–September) of Phosphorus, Chlorophyll *a,* and Secchi Disc Transparency for Each Year in the 10 Lakes

Lake	Year	Total P (µg P/L)	Chl$_a$ (µg/L)	Transparency (m)
Søylandsvatnet	1992	594	23	1.7[a]
Østensjøvann	1988	223	82	1.0
Østensjøvann	1992	228	94	0.6
Østensjøvann	1993	307	109	0.6
Hellesjøvann	1988	188	139	0.3
Hellesjøvann	1992	206	111	0.3
Hellesjøvann	1993	165	79	0.5
Lille Gleinsvatn	1993	100	43	0.9
Smokkevatn	1992	86	13	2.0
Kringelvatn	1992	61	17	1.9[a]
Mosvatn	1992	52	16	1.7
Altervatn	1992	30	3	3.3[a]
Haversvatn	1992	27	7	2.4
Stavsengvatn	1993	23	3	3.8

[a]Estimated values (see text).

Table 27.2. Lake Morphometry and Bottom Cover of Submerged and Floating-Leaved Vegetation

Lake	Lake area (km²)	Mean depth (m)	Bottom cover (%)
Søylandsvatn	0.65	0.4ᵃ	100
Østensjøvann	0.31	1.9	20
Hellesjøvann	0.53	1.3	32
Lille Gleinsvatn	0.10	4.0	28
Smokkevatn	0.14	1.2	74
Kringelvatn	0.09	1.1	74
Mosvatn	0.50	1.6	64
Altervatn	0.08	0.5ᵃ	100
Haversvatn	0.15	2.4	17
Stavsengvatn	0.06	1.9	61

ᵃEstimated values.

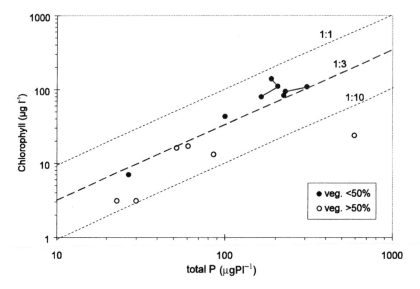

Figure 27.1. Total P versus chlorophyll *a* in the 10 lakes (late summer average values). Lakes with high bottom cover of vegetation are shown as white symbols whereas lakes with low bottom cover are shown with black symbols. 1:1 and 1:10 lines are indicated. Connected points are several years from the same lake. All lakes with high plant cover are perceived as "clear," whereas lakes with low plant cover, except one (Haversvatn: lower left black symbol), are "turbid."

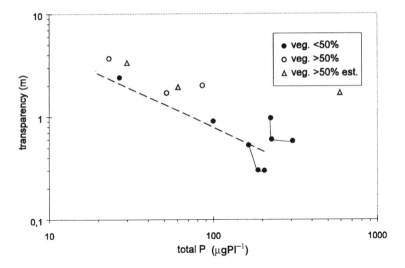

Figure 27.2. Concentration of total P versus transparency in lakes with high bottom cover of vegetation (white symbols) and with low bottom cover (black symbols). In lakes with transparency higher than the actual maximum depth and with high bottom cover, transparency has been estimated (white triangles). All values are late summer averages. Connected points are several years from the same lake. A linear regression line from 76 lake-years from deep and shallow Norwegian lakes is included (see text).

the lakes, one or more of the Secchi disc readings were higher than the maximum depth of the lake, or the vegetation was too dense to allow proper transparency measurements. For these lakes, transparency was estimated from the Chla versus transparency linear regression line of the other seven lakes covering 11 lake-years: (transparency = $4.225 - 0.813*\ln(\text{Chla})$, $r^2 = 0.947$, $P < .001$). We found no particular reason to expect a severe deviation from this line for the three lakes in question. In 6 of the 10 lakes, the submerged vegetation covered 50% of the lake bottom area or more (Table 27.2), and in 2 of the lakes, the whole bottom area was covered with vegetation, *Ceratophyllum demersum* being the dominant species.

In a plot of average growth season total P versus Chla, we previously found that most lakes with greater than 50% bottom cover of vegetation had a Chla/total P ratio close to 1:10, whereas most lakes with less than 50% cover had a Chla/total P ratio greater than close to 1:1 (Mjelde and Faafeng, 1997; Fig. 27.1). By using ALS values instead of "growth season averages," a better distinction between the lakes with high versus low cover of submerged vegetation was found.

The total P versus transparency plot (Fig. 27.2) shows a clear trend toward higher transparency in lakes with high vegetation cover than in lakes with a lower cover, both at high and low P concentrations. In this figure, a regression line from 76 lake-years is also given from both deep and shallow Norwegian lakes with total

P concentrations ranging between 20–200 µg P/L; ln(transparency) = 3.306 – 0.770*ln(total P), r^2 = 0.60, P <.01 (Faafeng, unpubl.).

Discussion

The shallow lakes with a high bottom cover of vegetation were obviously among those with the highest transparency at a certain total P concentration compared with other deep and shallow lakes as indicated by the regression line (Fig. 27.2). This is due to a lower Chla concentration as shown in Figure 27.1. This fact is even more prominent when taking into account the general trend for shallow un-stratified lakes to have a higher phytoplankton yield per unit total P than deeper lakes (Riley and Prepas, 1985; Mazumder, 1994).

Unfortunately, we have no shallow lakes in our investigation representing the turbid state in the lower end of the total P range. Lake Haversvatn, with a submerged bottom cover of only 17% and represented by a black symbol in the figures, does not produce the same phytoplankton yield as might be expected from its total P concentration (Faafeng and Hessen, 1993), and it is consequently not classified as turbid. Concentrations less than detection limits of available inor-ganic nitrogen and a constantly small population of *Daphnia* (<11 individuals/L) during May–September support this discrepancy and indicate N limitation. The lake with the highest total P concentration, Lake Søylandsvatn, is probably also N limited throughout the growth season (Mjelde and Faafeng, 1997) and therefore deviates considerably from the general trend of both Figures 27.1 and 27.2. It is a drawback to our analysis that we have related our discussion to total P only. However, at lower total P concentrations, which prevail in Norwegian lakes, total P is often a fair approximation of the growth potential of the phytoplankton.

Timms and Moss (1984) referred to "very clear water (<20 µg Chla/L)" in some lakes in the Norfolk Broads with dense plant communities "despite extremely high concentrations of phosphate and nitrogen." This fits well with the observations of van Donk and Gulati (1995), who distinguish between the "clear" and the "turbid" states in Lake Zwemlust at average May–September chlorophyll concentrations less than or greater than 30 µg Chla/L. Lake Zwemlust had very high average total P concentrations ranging from 1,863 to 2,697 µg total P/L in the years 1986, 1989, and 1992, and the lake phytoplankton was obviously not P limited during the investigated years. In years with dense submerged vegetation, most of the N was stored in the macrophytes and the phytoplankton became N limited. Their Table 1 shows that the corresponding average transparency of clear and turbid states was greater than 1.9 m or less than 1.5 m, respectively. All the "clear" lakes in our survey also had ALS chlorophyll concentrations less than 20–30 µg Chla/L (Fig. 27.1) and a transparency greater than 1.5–2 m (Fig. 27.2). This fits well also with Hargeby et al. (1994), who observed "turbid" water in Lake Krankesjön. During June–September, the average Chla concentrations in Krankesjön ranged, for ex-ample, between 25–30 µg Chla/L in the years when the cover of submerged vegetation was lowest, whereas a shift occurred to a clearwater state in the following years with Chla concentrations less than 20 μg Chla/L. The average Secchi depth during the growth season increased from less than 0.6 m in the turbid phase to 1.2–2.5 m in the clearwater phase (Blindow et al., 1993). Turbidity simultaneously decreased from greater than 22 to less than 8 JTU units. During these years, total P spontaneously changed from 50–70 μg P/L before the reinvasion of submerged vegetation to less than 35 μg P/L afterward. This gives a Chla/total P ratio of 0.49 and 0.46 during the turbid state and 0.35, 0.43, 0.46, and 0.36 in the clear state. This indicates that the Chla/total P ratio is less useful before the phytoplankton biomass and nutrient concentrations have stabilized in a new state.

It may be argued that water with 20–30 μg Chla/L can be perceived as turbid when assessing lakes with total P concentrations as low as to 20–30 μg P/L. In fact, these are the maximum Chla levels expected in such lakes (Faafeng and Hessen, 1993). As evidenced by Figures 27.1 and 27.2, clear lakes may contain Chla concentrations between one-third and one-tenth of the maximum concentrations (e.g., 2–10 μg Chla/L), at this P level.

More investigations in shallow lakes are needed to obtain a better statistical relationship between total P, chlorophyll, and transparency, especially in meso-eutrophic lakes with low vegetation cover. Turbidity caused by factors other than phytoplankton biomass should also be taken into consideration.

Acknowledgments. The authors thank Hanne Edvardsen, Bodø, and a number of colleagues at NIVA for their assistance and support during macrophyte registration and sampling and analysis of water samples. Identification and enumeration of phytoplankton were carried out by P. Brettum, and the zooplankton was analyzed by D.O. Hessen and J.E. Løvik. Sincere thanks are also due to Ingrid Blindow for valuable comments. The National Eutrophication Survey of Norwegian Lakes is financed by the Norwegian State Pollution Authority.

References

Balls, H.; Moss, B.; Irvine, K. The effects of high nutrient loading on interactions between aquatic plants and phytoplankton. Verh. Int. Verein. Limnol. 22:2912–2915; 1985.

Blindow, I.; Andersson, G.; Hargeby, A.; Johansson, S. Long-term pattern of alternative stable states in two shallow eutrophic lakes. Freshwat. Biol. 30:159–167; 1993.

Canfield, D.E., Jr.; Langeland, K.A.; Linda, S.B.; Haller, W.T. Relations between water transparency and maximal depth of macrophyte colonization in lakes. J. Aquat. Plant Manage. 23:25–28; 1985.

Chambers, P.A.; Kalff, J. Depth distribution and biomass of submersed aquatic macrophyte communities in relation to Secchi depth. Can. J. Fish. Aquat. Sci. 42:701–709; 1985.

Faafeng, B.A.; Hessen, D.O. Nitrogen and phosphorus concentrations and N:P ratios in Norwegian lakes: perspectives on nutrient limitations. Verh. Int. Verein. Limnol. 25(1): 465–469; 1993.

Grimm, M.P. Northern pike (*Esox lucius* L.) and aquatic vegetation, tools in the management of fisheries and water quality in shallow waters. Hydrobiol. Bull. 23:61–67; 1989.

Hargeby, A.; Andersson, G.; Blindow, I.; Johansson, S. Trophic web structure in a shallow lake during dominance shift from phytoplankton to submerged macrophytes. Hydrobiologia 279/280:83–90; 1994.

Irvine, K.; Balls, H.; Moss, B. The enterostracan and rotifer communities associated with submerged plants in the Norfolk Broadland—effects of plants and species composition. Int. Rev. Ges. Hydrobiol. 75:121–141; 1990.

Jeppesen, E.; Jensen, J.P.; Kristensen, P.; Søndergaard, M.; Mortensen, E.; Sortkjær, O.; Olrik, K. Fish manipulation as a lake restoration tool in shallow, eutrophic, temperate lakes. 2: Threshold levels, long-term stability and conclusions. Hydrobiologia 200/201: 219–227; 1990.

Mazumder, A. Phosphorus–chlorophyll relationships under contrasting herbivory and thermal stratification: predictions and patterns. Can. J. Fish. Aquat. Sci. 51:390–400; 1994.

Mjelde, M.; Faafeng, B. *Ceratophyllum demersum* (L.) hampers phytoplankton development in some small Norwegian lakes over a wide range of phosphorus level and geographic latitude. Freshwat. Biol. 37:355–365; 1997.

Phillips, G.L.; Eminson, D.F.; Moss, B. A mechanism to account for macrophyte decline in progressively eutrophicated freshwaters. Aquat. Bot. 4:103–126; 1978.

Riley, E.T.; Prepas, E.E. Comparison of the phosphorus–chlorophyll relationships in mixed and stratified lakes. Can. J. Fish. Aquat. Sci. 42:831–835; 1985.

Sand-Jensen, K.; Søndergaard, M. Phytoplankton and epiphyte development and their shading effect on submerged macrophytes in lakes of different nutrient status. Int. Rev. Ges. Hydrobiol. 66:529–552; 1981.

Scheffer, M. Alternative stable states in eutrophic shallow fresh water systems: a minimal model. Hydrobiol. Bull. 23:73–84; 1989.

Scheffer, M. Multiplicity of stable states in freshwater systems. Hydrobiologia 200/201: 474–486; 1990.

Scheffer, M.; Hosper, S.H.; Meijer, M.-L.; Moss, B.; Jeppesen, E. Alternate equilibria in shallow lakes. Trends Ecol. Evol. 8:275–279; 1993.

Scheffer, M.; Van den Berg, M.; Breukelaar, A.; Breukers, C.; Coops, H.; Doef, R.; Meijer, A.-L. Vegetated areas in turbid shallow lakes. Aquat. Bot. 49:193–196; 1994.

Schriver, P.; Bøgestrand, J.; Jeppesen, E.; Søndergaard, M. Impact of submerged macrophytes on fish–zooplankton–phytoplankton interactions: large-scale enclosure experiments in a shallow eutrophic lake. Freshwat. Biol. 33:255–270; 1995.

Timms, R.M.; Moss, B. Prevention of growth of potentially dense phytoplankton populations by zooplankton grazing, in the presence of zooplanktivorous fish, in a freshwater wetland ecosystem. Limnol. Oceanogr. 29:472–486; 1984.

van Donk, E.; Gulati, R.D. Transition of a lake to turbid state six years after biomanipulation: mechanisms and pathways. Wat. Sci. Techn. 32:197–206; 1995.

van Donk, E.; Grimm, M.P.; Gulati, R.D.; Heuts, G.M.; de Kloet, W.A.; van Liere, L. First attempt to apply whole lake food-web manipulation on a large scale in the Netherlands. Hydrobiologia 200/201:291–301; 1990.

van Donk, E.; Gulati, R.D.; Iedema, A.; Meulemans, J.T. Macrophyte-related shifts in the nitrogen and phosphorus content in the different trophic levels in a biomanipulated shallow lake. Hydrobiologia 251:19–26; 1993.

Vant, W.N.; Davies-Colley, R.J.; Clayton, J.S.; Coffey, B.T. Macrophyte depth limits in North Island (New Zealand) lakes of different clarity. Hydrobiologia 137:55–60; 1986.

Wium-Andersen, S. Allelopathy among aquatic plants. Erg. Limnol. 27:167–172; 1987.

28. Macrophytes and Turbidity in Brackish Lakes with Special Emphasis on the Role of Top-Down Control

Erik Jeppesen, Martin Søndergaard, Jens Peder Jensen,
Eva Kanstrup, and Birgitte Petersen

Introduction

Evidence from both empirical studies (Canfield et al., 1984; Jeppesen et al., 1990; Faafeng and Mjelde, this volume, Chapter 27) and numerous experimental field studies (see e.g., Gulati et al., 1990; van Donk et al., 1990; Mortensen et al., 1994) indicates that in freshwater lakes extensive growth of submerged macrophytes may lead to clearwater conditions, even at high nutrient concentrations. Several factors seem to be involved, including both increased zooplankton grazer control and nutrient constraint on phytoplankton, alterations in the physical environment that result in less wind-induced and fish-induced resuspension, and possibly also allelophatic effects (Jeppesen et al., 1990; Moss, 1990; Scheffer et al., 1993). A cross-analysis of survey data from 35 Danish brackish lakes revealed a significant decrease in Secchi depth with increasing concentrations of total phosphorus (TP); in contrast to freshwater lakes, however, transparency was independent of whether submerged macrophytes were present at high density (Jeppesen et al., 1994). Similarly, Moss (1994) found that nutrient-rich brackish lakes with extensive growth of submerged macrophytes tend to be in a turbid state. By using both empirical data and field experiments conducted in several brackish and freshwater shallow Danish lakes, we examine here how differences in top-down control may influence the turbidity of freshwater and brackish lakes in the macrophyte state.

Materials and Methods

The analysis is based on survey data from 50–100 freshwater lakes and 35 brackish lakes. Fish population estimates are based on fish caught overnight in gill nets (3×1.5-m sections, 14 different mesh sizes from 6.25 to 75 mm) expressed as catch per unit effort (CPUE = fish/net/19 h). Most of the sampling procedures and methods are described by Jeppesen et al. (1994) and Aaser et al. (1995), and only additional methods are presented here. *Leptodora kindti* was counted on zooplankton samples taken at equidistant intervals from the surface to the bottom at one–three stations in the pelagic zone. *Chaoborus* spp. density was estimated from their abundance (n/m^2) in sediment samples collected during the day in autumn or spring, and we assumed that they were evenly distributed in the pelagic zone at night. Between 3–10 samples were taken with a Kajak sampler (diameter, 5.2 cm) in each lake on one–five occasions during winter or spring. The estimate is thus conservative, as summer densities of *Chaoborus* are higher (e.g., Christoffersen, 1990).

Sampling of *Neomysis integer* in shallow Lake Ørslevkloster (40 ha; mean depth, about 2 m; max depth, about 4 m; salinity, 2–4‰) was conducted at 16 stations by vertical hauling with a 0.5-mm net (diameter, 0.5 m) according to Aaser et al. (1995). Fish sampling in this lake was conducted by using 1.5×32-m sinking gill nets (eight 4-m sections; each including 1-m sections of 6.25-, 8-, 10-, and 12.5-mm mesh size, respectively), four nets being placed overnight in the littoral zone running parallel to the shore and two in the pelagic at a mid-lake station. Physico-chemical data were obtained by sampling at mid-lake stations (pooled samples from the entire water column).

Results and Discussion

A plot of TP versus Secchi depth in Danish lakes shows that nutrient-rich brackish lakes are turbid even when macrophyte densities are high (Fig. 28.1). Chlorophyll *a* was significantly linearly related to TP and unrelated to submerged macrophyte coverage (Fig. 28.2). These results indicate that zooplankton grazing on phytoplankton is unaffected by the presence or absence of macrophytes in brackish lakes. In eutrophic freshwater lakes, by contrast, the macrophyte state is generally associated with high transparency (Fig. 28.1) and usually, but not always (Meijer et al., 1994), with a high zooplankton/phytoplankton biomass ratio and hence a potentially high grazing pressure on phytoplankton (Moss et al., 1994; Jeppesen et al., 1997; Van den Berg et al., this volume, Chapter 25). Lower zooplankton grazing in brackish lakes may partly reflect differences in zooplankton community structure. Although large-bodied *Daphnia* (e.g., *D. magna),* which play a key role in grazer control of phytoplankton in freshwater lakes (Carpenter and Kitchell, 1993), may become dominant in slightly brackish lakes (Jürgens and Stolpe, 1995), they most frequently disappear above salinities of 2–4‰ (Jeppesen et al., 1994; Moss, 1994). Instead, the lakes are dominated by the copepods, *Eurytemora* spp. and

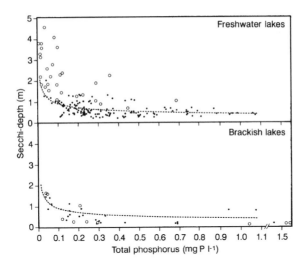

Figure 28.1. Secchi depth in relation to lake water total phosphorus in shallow freshwater (upper panel) and brackish lakes (lower panel). ○ lakes with more than 30% submerged macrophyte coverage; ● lakes with a low (< 30%) or unknown submerged macrophyte coverage. Each point represents one lake and is a time-weighted average of all data collected between May 1 and October 1. The broken line indicates an exponential curve developed by Kristensen et al. (1991) on the basis of data from freshwater and brackish lakes with low submerged macrophyte coverage. (From Jeppesen et al., 1994. Published with permission from Kluwer Academic Publishers.)

Acartia spp., and occasionally by rotifers (Jeppesen et al., 1994), which are probably less efficient in controlling phytoplankton than large-bodied cladocerans. In addition, the zooplankton/phytoplankton ratio is low in brackish lakes (Jeppesen et al., 1994).

Lake Ørslevkloster is an example of how a salinity-mediated shift in trophic structure may reduce grazer control on phytoplankton in macrophyte-rich brackish lakes. The lake shifted from a brackish state (1–3‰) dominated by *Eurytemora affinis* and rotifers to a slightly brackish state (0.5–1‰) dominated by *Daphnia galeata* (Fig. 28.3). Chlorophyll *a* was 2.5–3.5-fold higher in the brackish state

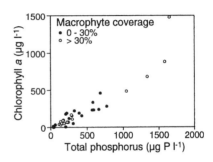

Figure 28.2. Summer mean chlorophyll *a* versus total phosphorus in some Danish shallow brackish lakes with submerged macrophyte coverage in the range 0–30% (●) or greater than 30% (○).

(1993 and 1994) during summer than in the slightly brackish state and 17–23-fold higher during autumn. Correspondingly, Secchi depth was 20–50% and 10–12% lower during summer and autumn, respectively (Fig. 28.4). Despite the fact that external loading did not change (Viborg County, 1995), lake water TP was highest in the brackish state, which is probably due to internal loading caused by FeS formation and a resultant release of iron-bound phosphorus as demonstrated in the nearby Hjarbæk Fjord (H. Jensen, unpublished results). The higher chlorophyll *a* in the brackish state may therefore partly reflect the higher P concentration. However, the chlorophyll *a*/TP ratio during summer and autumn also tended to be higher, which may indicate lower zooplankton grazing on phytoplankton. This is supported by the substantially lower zooplankton/phytoplankton biomass ratio in the brackish state during summer (0.02–0.06 versus 0.49 in the slightly brackish state) (Fig. 28.3). The shifts in turbidity and zooplankton/phytoplankton biomass ratios in Lake Ørslev-kloster thus follow the predictions of the established empirical relations, but we cannot, however, exclude the possibility that the shift is caused by other factors. Changes in the recruitment of fish unrelated to changes in salinity may, for instance, also have played a role, but sufficient data to elucidate this are not available.

The low zooplankton/phytoplankton biomass ratio in the brackish state in Lake Ørslevkloster (Fig. 28.3) and in other Danish brackish lakes (Jeppesen et al., 1994) cannot simply be explained by lack of edible phytoplankton. Green algae and diatoms, which are a common food source for the dominant zooplankton (*Eury-temora affinis* and rotifers), are abundant in most eutrophic brackish lakes (Balls et al., 1993; J.P. Jensen et al., unpublished observation), as well as in Lake Ørslevkloster (Nielsen, 1995). Another explanation is top-down control via inver-tebrates and fish. In North European eutrophic brackish lakes, *Neomysis integer* is a major invertebrate predator (Irvine et al., 1990; Moss, 1994; Aaser et al., 1995), and mysid density increases with increasing TP, particularly above 400 µg P/L (Fig. 28.5), reaching densities as high as 13/L (Jeppesen et al., 1994). The marked increase in mysid density coincides with a shift in the fish community from dominance by roach (*Rutilus rutilus*), perch (*Perca fluviatilis*), whitefish (*Core-gonus* spp.), smelt (*Osmerus* spp.), etc., to exclusive dominance by small-sized sticklebacks (*Gasterosteus* spp.) (Jeppesen et al., 1994). The latter coexist with *N. integer,* probably because unlike the larger fish, sticklebacks prey selectively on smaller stages of mysids rather than on ovigorous females (Jeppesen et al., 1994; Kanstrup, 1996). By contrast, the maximum density of the pelagic invertebrate predators in freshwater lakes (*Leptodora kindti* and *Chaoborus* spp.) occurs at 100–200 µg P/L, and they almost disappear at high TP (Fig. 28.5). This is probably due to increased fish predation because large-sized fish such as roach and bream (*Abramis brama*) dominate in north European hypertrophic lakes (Persson et al., 1988; Jeppesen et al., 1990). Although *N. integer* is omnivorous (Arndt and Jansen, 1986) as opposed to the more strict carnivorous *Chaoborus* and *Leptodora* found in freshwater lakes, the predation pressure on zooplankton by pelagic invertebrate predators is probably higher in hypertrophic brackish lakes than in comparable freshwater lakes because of the very high predator density that *Neomysis* may reach in the brackish lakes.

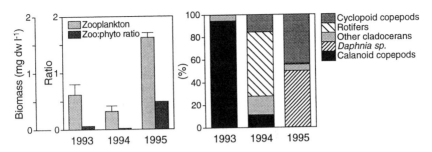

Figure 28.3. Zooplankton biomass and zooplankton/phytoplankton biomass ratio (A) and percentage of biomass accounted for by the various zooplankton groups (B) in shallow Lake Ørslevkloster in 1993, 1994, and 1995. No quantitative data are available for 1986, but high density of *Daphnia hyalina* was observed in littoral fauna samples (Viborg County, 1988), thus indicating that the lake was then in a cladoceran state.

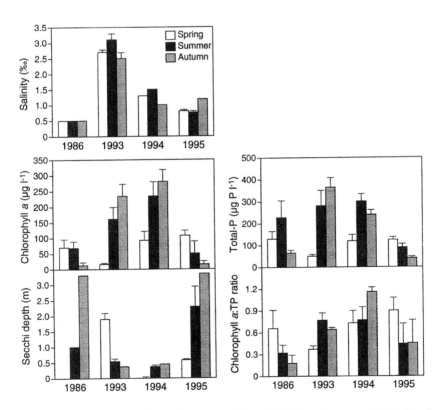

Figure 28.4. Spring (Jan 1–May 1), summer (May 1–Oct 1) and autumn (Oct 1–Jan 1) mean (±SE) values of chlorophyll *a*, Secchi depth, total phosphorus and chlorophyll *a*/total phosphorus ratio in Lake Ørslevkloster during 4 years differing in salinity (upper panel).

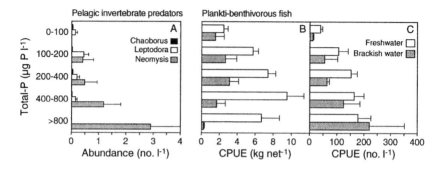

Figure 28.5. Summer (May 1–Oct 1) mean abundance of some invertebrate predators in Danish freshwater and brackish lakes (A). CPUE of planktivorous fish in terms of biomass (B) and number (C) (multiple mesh size gill netting in late July–August, 14 different mesh sizes, 6.25–75 mm), each versus summer mean lake water total phosphorus.

Planktivorous fish significantly affect zooplankton abundance, composition, and the zooplankton/phytoplankton biomass ratio in Danish shallow eutrophic freshwater lakes (Jeppesen et al., 1997). The biomass of planktivorous fish caught in multiple mesh-sized gill nets in brackish lakes is lower than in freshwater lakes, particularly at higher TP, when small-sized sticklebacks dominate (Fig. 28.5; for discussion about net selectivity, see Jeppesen et al., 1994). In terms of numbers, however, CPUE was not lower at high TP concentrations. The data from the freshwater lakes therefore suggest that the predation pressure on zooplankton should be high also in eutrophic brackish lakes (Jeppesen et al., 1994). In addition, sticklebacks produce offspring several times during the summer and autumn. Predation pressure on zooplankton by fish fry is particularly high (e.g., He and Wright, 1992; Søndergaard et al., 1997; Jeppesen et al., 1997), suggesting that there is more likely a continuously high fish predation pressure on zooplankton in eutrophic brackish lakes than in comparable freshwater lakes. This idea of higher invertebrate and fish predation is further supported by the lower zooplankton/phytoplankton biomass ratio in brackish lakes (Fig. 28.3; Jeppesen et al., 1994).

Aggregation of *G. aculeatus* and *N. integer* in the littoral zone (Arndt and Jensen, 1986) may be a contributory factor to the higher turbidity of macrophyte-rich brackish lakes, as it may diminish the ability of the pelagic zooplankton to use macrophytes as a daytime refuge. In Lake Ørslevkloster, the gill net catch of sticklebacks was about sevenfold higher in the littoral zone than in the open water during November–August, and 10–25-fold higher during the summer (Kanstrup, 1996; Fig. 28.6). After August, the pattern changed, however, with the number of stickleback caught being highest in the open water. Correspondingly, the annual mean concentration of *N. integer* was 120-fold higher in the littoral zone than in the pelagic zone (Fig. 28.6), or 30-fold higher per unit area. An experiment involving partial harvesting of macrophytes in the littoral zone of the same lake

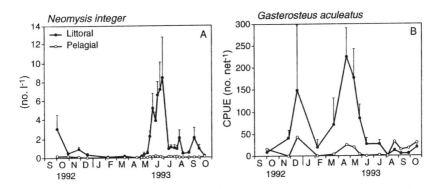

Figure 28.6. Seasonal variation in the density (±SE) of *Neomysis integer* (A) and gill net CPUE of three-spined stickleback (*Gasterosteus aculeatus*) (B) in the littoral and pelagic zone of Lake Ørslevkloster in 1992/93.

showed an 80% higher mysid density inside the plant beds than at similar depths outside (Petersen, 1994). The suggested low refuge effect of macrophytes in brackish lakes may contribute to the low zooplankton control of phytoplankton and hence to the high turbidity of macrophyte-rich lakes, as the possibility of seeking daytime refuge in the vegetation has been shown to be a key factor for the survival of pelagic cladocerans in macrophyte-rich freshwater lakes with a high density of planktivorous fish (Jeppesen et al., this volume, Chapter 5; Lauridsen et al., this volume, Chapter 13).

Nutrient release by *N. integer* may also contribute to a different response of the two lake types. Experiments conducted in two shallow brackish lakes thus showed a considerably higher TP in enclosures containing mysids than in controls devoid of mysids (Aaser et al., 1995; Nielsen, 1995). The results indicated that mysids enhance nutrient release from the sediment, perhaps because some of the nutrients ingested when feeding on sediment detritus and benthic invertebrates are subsequently released to the pelagic. This, in turn, may stimulate phytoplankton growth, thereby contributing to the low Secchi depth in brackish lakes.

Although we are beginning to understand the mechanisms behind the high turbidity of eutrophic macrophyte-rich brackish lakes, more research is needed before any firm conclusions can be drawn. Further studies are important not only from a basic science point of view but also with regard to lake management. Thus, the difference in trophic structure and dynamics of the two different lake types has important implications when transferring the ecotechnological restoration methods known from freshwater lakes to brackish lakes (Jeppesen et al., 1994; Moss, 1994).

Acknowledgments. We thank the Danish Counties for providing access to some of the data used in the analyses. The assistance of the technical staff of the National Environmental Research Institute, Silkeborg, is gratefully acknowledged. We also thank Mark Hoyer and Marten Scheffer for valuable comments. The study was supported by the Centre for Freshwater Environmental Research.

References

Aaser, H.F.; Jeppesen, E.; Søndergaard, M. Seasonal dynamics of the mysid *Neomysis integer* and its predation on the copepod *Eurytemora affinis* in a shallow hypertrophic brackish lake. Mar. Ecol. Prog. Ser. 127:47–56; 1995.

Arndt, E.A.; Jansen, W. *Neomysis integer* (Leach) in the Chain of Boddens south of Darss/Zingst (Western Baltic). Ecophysiology and population dynamics. Ophelia 4:1–15; 1986.

Balls, M.; Moss, B.; Phillips, G.L.; Irvine, K.; Stansfield, H. The changing ecosystem of a shallow, brackish lake, Hickling Broad, Norfolk II. Long-term trends in water chemistry and ecology and their implications for restoration of the lake. Freshwat. Biol. 29:141–165; 1993.

Canfield, D.E.; Shireman, J.V.; Colle, D.E.; Haller, W.T.; Watkins, C.E.; Maceina, M.J. Prediction of chlorophyll *a* concentrations in Florida lakes: importance of aquatic macrophytes. Can. J. Fish. Aquat. Sci. 41:497–501; 1984.

Carpenter, S.R.; Kitchell, J.F., eds. The trophic cascade in lakes. New York: Cambridge University Press; 1993.

Christoffersen, K. Evaluation of *Chaoborus* predation on natural populations of herbivorous zooplankton in a eutrophic lake. Hydrobiologia 200/201:459–466; 1990.

Gulati, R.D.; Lammens, E.H.R.R.; Meijer, M.-L.; van Donk, E. Biomanipulation, tool for water management. Hydrobiologia 200/201:1–628; 1990.

He, X.; Wright, R. An experimental study of piscivore–planktivore interactions: population and community responses to predation. Can. J. Fish. Aquat. Sci. 49:1176–1185; 1992.

Irvine, K.; Bales, M.T.; Moss, B.; Stansfield, J.H.; Snook, D. Trophic relations in Hickling Broad—a shallow and brackish eutrophic lake. Verh. Int. Verein. Limnol. 24:576–579; 1990.

Jeppesen, E.; Jensen, J.P.; Kristensen, P.; Søndergaard, M.; Mortensen, E.; Sortkjær, O.; Olrik, K. Fish manipulation as a lake restoration tool in shallow, eutrophic, temperate lakes 2: Threshold levels, long-term stability and conclusions. Hydrobiologia 200/201: 219–227; 1990.

Jeppesen, E.; Søndergaard, M.; Kanstrup, E.; Petersen, B.; Eriksen, R.B.; Hammershøj, M.; Mortensen, E.; Jensen, J.P.; Have, A. Does the impact of nutrients on the biological structure and function of brackish and freshwater lakes differ? Hydrobiologia 275/276: 15–30; 1994.

Jeppesen, E.; Jensen, J.P.; Søndergaard, M.; Lauridsen, T.L.; Junge Pedersen, L.; Jensen, L. Top-down control in freshwater lakes: the role of nutrient state, submerged macrophytes and water depth. Hydrobiologia 342/343:151–164; 1997.

Jürgens, K.; Stolpe, G. Seasonal dynamics of crustacean zooplankton, heterotrophic flagellates and bacteria in a shallow eutrophic lake. Freshwat. Biol. 33:27–38; 1995.

Kanstrup, E. Trepigget hundestejles *Gasterosteus aculeatus* L. betydning for de biologiske interaktioner i en lavvandet, eutrof brakvandssø (in Danish). [The influence of three-spined stickleback *Gasterosteus aculeatus* on the biological interactions in a shallow eutrophic brackish lake.] MSc thesis, National Environmental Research Institute and the University of Aarhus, Aarhus; 1996.

Kristensen, P.; Jensen, J.P.; Jeppesen, E. Simple empirical lake models. In: Nitrogen and phosphorus in fresh and marine water. Danish Environmental Protection Agency Abstracts, Copenhagen; 125–145; 1991.

Meijer, M-L.; Jeppesen, E.; van Donk, E.; Moss, B.; Scheffer, M.; Lammens, E.; van Nes, E.; van Berkum, J.A.; de Jong, G.J.; Faafeng, B.A.; Jensen, J.P. Long-term response to fish-stock reduction in small shallow lakes: interpretation of five-year results of four biomanipulation cases in the Netherlands and Denmark. Hydrobiologia 275/276:457–466; 1994.

Mortensen, E.; Jeppesen, E.; Søndergaard, M.; Kamp Nielsen, L., eds. Nutrient dynamics and biological structure in shallow freshwater and brackish lakes. Hydrobiologia 275/276:1–507; 1994.

Moss, B. Engineering and biological approaches to the restoration from eutrophication of shallow lakes in which aquatic plant communities are important components. Hydrobiologia 200/201:367–378; 1990.

Moss, B. Brackish and freshwater lakes—different systems or variations on the same theme? Hydrobiologia 275/276:367–378; 1994.

Moss, B.; McGowan, S.; Carvalho, L. Determination of phytoplankton crops by top-down and bottom-up mechanisms in a group of English lakes, the West Midland meres. Limnol. Oceanogr. 39:1020–1029; 1994.

Nielsen, F. Græsning af copepoder *Eurytemora affinis* i to eutrofe søer (in Danish). [*Eurytemora affinis* grazing in two eutrophic lakes.] MSc thesis, National Environmental Research Institute, Silkeborg, and the University of Aarhus, Aarhus; 1995.

Persson, L.; Anderson, G.; Hamrin, S.F.; Johansson, L. Predation regulation and primary production along the productivity gradient of temperate lake ecosystems. In: Carpenter, S.R., ed. Complex interactions in lake communities. New York: Springer-Verlag; 1988: 45–65.

Petersen, B. *Neomysis integers* økologiske rolle i en lavvandet, eutrof brakvandssø (in Danish). [The ecological role of *Neomysis integer* in a shallow eutrophic brackish lake.] MSc thesis, National Environmental Research Institute, Silkeborg, and the University of Aarhus, Aarhus; 1994.

Scheffer, M.; Hosper, S.H.; Meijer, M.-L.; Moss, B.; Jeppesen, E. Alternative equilibria in shallow lakes. Trends Ecol. Evol. 8:275–279; 1993.

Søndergaard, M.; Jeppesen, E.; Berg, S. Pike (*Esox lucius* L.) stocking as a biomanipulation tool. 2. Effects on lower trophic levels in Lake Lyng (Denmark). Hydrobiologia 342/343:319–325; 1997.

van Donk, E.; Grimm, M.P.; Gulati, R.D.; Klein, J.P.G. Whole-lake food-web manipulation as a means to study community interactions in a small ecosystem. Hydrobiologia 200/201:275–289; 1990.

Viborg County. Miljøtilstand i Ørslevkloster Sø 1986–1987 (in Danish). [The environmental state of Lake Ørslevkloster 1986–1987.] Viborg, Denmark; 1988.

Viborg County. Ørslevkloster Sø 1994. Belastning, fysisk-kemiske forhold, vegetation samt fiskebestand (in Danish). [Lake Ørslevkloster 1994. Loading, physico-chemical interactions, vegetation and fish stock.] Viborg, Denmark; 1995.

3. Interdisciplinary Discussions

29. Structuring Role of Macrophytes in Lakes: Changing Influence Along Lake Size and Depth Gradients

Avital Gasith and Mark V. Hoyer

Introduction

Emergent, floating-leaved, and submergent macrophytes grow in the littoral region of most lakes. These aquatic macrophytes are influenced by geomorphology, environmental conditions, and biotic interactions (Sculthorpe, 1967; Hutchinson, 1975), while exerting their own influence on the lake environment and biota (Carpenter and Lodge, 1986; Engel, 1988). The capacity of macrophytes to provide a substrate for colonization of algae and invertebrates (Sozska, 1975; Cattaneo and Kalff, 1980; Dvorak and Best, 1982; Cattaneo, 1983; Morin, 1986; Schram et al., 1987; Miller et al., 1989), to affect water and sediment chemistry as well as other limnological conditions (Carpenter and Gasith, 1978; Prentki et al., 1979; Jaynes and Carpenter, 1986), and to influence biogeochemical cycles and productivity (Wetzel and Hough, 1973; Godshalk and Wetzel, 1978; Wetzel, 1979; Carpenter, 1980; Cattaneo and Kalff, 1980; Carpenter, 1983; Wetzel, 1990) and biotic interactions (Crowder and Cooper, 1982; Heck and Crowder, 1991; Schriver et al., 1995; see also this volume) is well recognized. The understanding of the role of macrophytes in lacustrine systems is based mostly on process studies, small-scale investigations (ponds, test plots), observations in small lakes, and modeling (Carpenter and Lodge, 1986). It is intuitively obvious that the influence of macrophytes in most small or shallow aquatic systems is proportional to their abundance (density, biomass, or extent of cover) and productivity. Little is known about the role of macrophytes in situations in which they are less conspicuous, as in large deep lakes. Danehy et al. (1991), Gasith (1991), and Gasith and Gafny (this volume, Chapter 24) argue that the potential influence of littoral resources, including those provided by macrophytes to the biotic functioning of large deep lakes, has been overlooked. Both abundance and productivity of macrophytes vary about two orders of magnitude among lakes of different trophic levels (Carpenter, 1983), regardless of lake size. It is less clear, however, how the role of macrophytes varies in lakes of similar trophic status that differ in size and depth. The purpose of this discussion is to assess how the potential structuring role of macrophytes can change along lake size (surface area) and depth gradients.

We first point out the inherent difficulty in the terminology used to describe a lake size; we then consider the factors that interact with lake size and depth and affect plant growth; and finally, we assess the changing role of macrophytes along lake size and depth gradients.

The macrophyte-epiphyte complex is functionally inseparable. Whenever we generally use the term *macrophytes,* it is inclusive of their epiflora. For sake of the required brevity, we also fail to distinguish among the different macrophyte types and growth forms, despite evidence for possible type or growth form–specific effects as well as effects of mixed plant associations (Emery, 1978; Guillory et al., 1979; Eadie and Keast, 1984; Conrow et al., 1990; Dionne and Folt, 1991; Lillie and Budd, 1992; Chick and Mclvor, 1994).

Large Versus Small and Deep Versus Shallow

Lakes are commonly categorized as small or large and shallow or deep despite lack of clear-cut morphological definitions. Generally, large lakes tend to be deeper and have longer retention times than small lakes. Only large deep lakes have truly pelagic communities that are usually more important in the overall cycling and production processes than the littoral zone and bottom communities (Tilzer, 1990). The term *shallow* is often associated with lakes that do not thermally stratify and where continuous sediment–water interaction makes internal nutrients cycling more efficient than in deeper lakes that stratify. This definition ignores the important presence of aquatic macrophytes. Thus, a definition more pertinent to the aim of this discussion is that shallow lakes are those whose bottom is significantly covered by submerged macrophytes (Moss, 1995). In general, large deep lakes have less aquatic macrophytes than small shallow lakes, with the exception that highly turbid, shallow lakes may be devoid of submerged vegetation.

Factors Affecting Plant Growth: Interaction with Lake Size and Depth

Here, we consider lake size (surface area) and depth on a relative scale in connection to the potential growth of aquatic macrophytes. Unless stated otherwise, we assume similar growth conditions for the lakes compared, except for those arising from the gradients in surface area and depth.

Comparison of the role of macrophytes along lake size and depth gradients is complicated because plant development is variable even in lakes of similar morphometery (Sculthorpe, 1967; Hutchinson, 1975). Several site-specific environmental factors affecting the abundance and distribution of aquatic macrophytes in lakes have been identified. These include climatic factors such as irradiance, temperature, wave action generated by winds, size and edaphic features of the catchment basin that affect nutrient loading and general water chemistry, and biotic factors of grazing by invertebrates, fish, and birds. We limit our discussion to those factors that interact with lake size and depth.

Light availability and wave action are directly and indirectly influenced by morphometric features. Due to exponential light attenuation in water, depth is one of the most critical environmental factors determining the lakeward growth of macrophytes and their species richness (Hutchinson, 1975; Chambers and Kalff, 1985; Duarte et al., 1986). As a general rule of thumb, submerged macrophytes will grow to a depth of two to three times the Secchi depth (Canfield et al., 1985; Chambers and Kalff, 1985). Thus, macrophyte growth will be limited in lakes with small or large surface areas where the majority of lake bottom exceeds the above Secchi depth. Additionally, even if a lake is physically shallow and does not thermally stratify (i.e., 1–2 m, mean depth), if the Secchi depth is less than 0.5 m there is a strong probability that submerged aquatic macrophytes will be absent. With some exceptions, a depth range between 10 and 15 m appears to be a limit for most angiosperms. Lakes in which most of the basin is deeper than 10–15 m are not expected to have abundant submerged aquatic macrophytes. Emergent and floating-leaved aquatic macrophytes seldom grow in waters exceeding a depth of 3 m (Canfield and Hoyer, 1992). Climatic differences associated with lake latitude appear to have a strong influence on the relationship between depth distribution of submerged plants and water transparency (Duarte and Kalff, 1987). At low latitudes, angiosperms colonize deeper and reach maximum biomass at greater depth than those growing in lakes of similar transparency at higher latitudes. Warmer water, greater irradiance, and longer growing period in lower latitude lakes may account for the difference.

Basin slope (square root of the area divided by mean depth; Hakanson, 1981), surface area, and basin configuration are among the most important morphological features that influence the potential development of macrophytes in lakes (Pearsall, 1917; Spence, 1982; Duarte and Kalff, 1986). These factors interact directly and indirectly with other environmental factors such as light, nutrient availability, substrate characteristics, and wind-generated erosion to determine the site-specific extent of plant development and macrophyte types.

Maximum biomass of submerged macrophytes is inversely related to slope (Duarte and Kalff, 1986). The probable reasons for this relation is the difference in the relative area suitable for plant growth and in sediment stability and quality between gently and steeply sloped littoral zones. The area of littoral zone available for emergent growth declines with increasing slope of the basin. In addition, steep-sided basins are areas of erosion and sediment transport (Pearsall, 1917; Hakanson, 1977), whereas nearshore regions of gently sloped basins are sites of accretion of fine, relatively more stable, and nutrient-richer sediment, where macrophytes can become established. Pearsall (1920) demonstrated that the variation in the quantity and quality of silts largely controls the distribution of submerged vegetation. Thus, irrespective of lake size, steep-sided lakes will have lower cover and biomass of submerged macrophytes than lakes with gently sloped basins.

A large lake has a long fetch and a greater wave energy than a smaller lake. Exposure to waves can directly and indirectly affect plant distribution and abundance in lakes (Keddy, 1983; Chambers, 1987; Coops et al., 1991). Wave action and currents also affect sediment transport and distribution in lakes (Davidson-Arnottand Pollard, 1980; Keddy, 1982), concomitantly affecting the distribution of aquatic plants (Spence, 1982). Unless physically protected, points and shallows where wave energy is highest tend to be swept clean of fine sediments (Lorang and Stanford, 1993) and have little or no growth of macrophytes. Bays and areas below the wave-mixed depth tend to silt in providing more stable sediments, suitable for the establishment of macrophytes (Pearsall, 1929). Waves and strong currents can also retard vegetation growth by exerting a mechanical stress on the plants (Hutchinson, 1975; Coops et al., 1991). High concentration of suspended solids generated by wind mixing of bottom sediments (Kristensen et al., 1992) can limit light for plant growth particularly in large, shallow, unstratified lakes, whereas in stratified lakes suspended particles tend to settle out of the mixed layer (Osgood, 1988). High wave energy, currents, and turbidity in the shallows often restrict macrophyte growth in large deep lakes to protected bays and coves (Duarte et al., 1986). Overall, lakes with large surface areas and longer fetch are expected to have fewer vegetated littoral regions in relation to the amount of open water than smaller lakes (Rounsefell, 1946; Spence, 1982).

Lakes with a large surface area tend to be deeper than smaller lakes (a positive correlation exists between lake area and mean depth; Duarte et al., 1986). The cover and biomass of submerged macrophytes are expected to decline with increasing lake size if only for the reason that larger lakes have greater proportion of area below the compensation depth for macrophytes. In analyzing 139 lakes, Duarte et al. (1986) indeed found that the percentage surface area covered by submerged plants is not a constant proportion of the lake area but tends to be smaller in bigger lakes. Rather surprising, however, was their finding that emergent macrophytes colonized on average a constant proportion (7%) of the lake area regardless of the size of the lake. A similar relation was reported for Polish lakes, showing that emergents covered a relatively narrow range of lake surface areas (9.3–12.3%; Planter, 1973). This contradicts the expectation of declining growth of macrophytes with increasing fetch and greater wave action in the littoral zone (Spence, 1982). Duarte et al. (1986) suggested that a greater number of sheltered bays and floodplains in larger lakes where macrophytes can grow compensates for decreases in vegetation caused by greater wave action. If this is indeed so, it is apparently sufficient to compensate for the lower growth of emergents in shoals of large lakes but not of submerged macrophytes. Duarte et al.

(1986) concluded that on average submerged macrophytes are more important in small lakes and emergent plants will become more important with increasing lake size. It should be pointed out, however, that an opposite trend of a transition from submergents' dominance to that of emergent vegetation is part of the natural process of lake succession, which is most accelerated in small shallow productive lakes. The accumulation of refractory macrophyte detritus further limits growth of submerged macrophytes and hastens the transition to emergent vegetation that is more tolerant of organic rich sediments (Wetzel, 1979; Carpenter, 1981; Barko and Smart, 1983).

The proportion of littoral zone areas in a lake declines with increasing depth and lake size (Gasith, 1991) and increases with increasing shoreline irregularity (high shore development figure). Therefore, highly irregular lakes may have a higher proportion of vegetation zones compared with lakes of similar area but with a more regular shoreline.

The trophic status of lakes is inversely related to mean depth (Vollenweider, 1975; Canfield and Bachmann, 1981). Deep lakes tend to be more oligotrophic and support lesser growth of aquatic macrophytes than shallow lakes. A study by Canfield and Hoyer (1992) shows that oligotrophic and mesotrophic lakes rarely have aquatic macrophyte abundance exceeding 20% volume infested (PVI), whereas eutrophic and hypereutrophic lakes have the potential to reach 100 PVI. High turbidity may limit growth of submerged macrophytes in these lakes even though nutrients availability can support extensive growth.

Structuring Role of Macrophytes: Changing Importance Along Size and Depth Gradients

When established in a lake, aquatic macrophytes can influence the lake ecosystem in multiple ways (reviewed in Carpenter and Lodge, 1986) and mediate biotic interactions (Crowder and Cooper, 1982; Savino and Stein, 1982; Diehl, 1988; see also this volume). The structuring role of macrophytes in a lake ecosystem falls into three main categories: (1) limnological effects related to changes in physical and chemical conditions in the water and sediment; (2) metabolic effects related to production and processing of organic matter and nutrient cycling; and (3) effect on biotic interactions and community structure related to the role of macrophytes in providing a structured habitat.

It may be useful to approach the question of how the role of macrophytes changes along lake size and depth gradients by considering each of the above categories separately. We suggest that the limnological and metabolic effects of macrophytes in lakes diminish with increasing depth and lake size faster than their importance in providing structured habitats (Fig. 29.1). This implies that, by providing structure, macrophytes may still play a role affecting biotic interactions in situations in which they may have no significant effect on water-quality, nor are they important for nutrient cycling, nor as a source of organic matter.

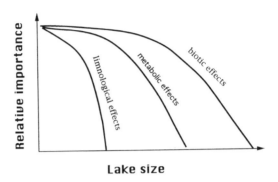

Figure 29.1. Comparison of the changing relative importance of limnological, metabolic and biotic effects of macrophytes along increasing lake size gradient. A positive relation between size (surface area) and depth is assumed.

It is reasonable to assume that a PVI exceeding 40% is required for macrophytes to be able to change the water-quality conditions of an entire lake ecosystem (Canfield and Jones, 1984); this would be a situation more typical of marshes and shallow eutrophic lakes (Canfield and Hoyer, 1992). Large-scale oxygen depletion, for example, is most likely to occur following rapid senescence of dense macrophyte stands in warm, poorly circulated waters (Carpenter and Greenlee, 1981). Indeed, the fish community of densely vegetated wetlands is composed of the most tolerant species that are able to function in dense vegetation and survive periodic high temperatures and low dissolved oxygen levels (Mac-Crimmon, 1980; Johnson, 1989). The effects of macrophytes on sediment and water quality (Carpenter and Greenlee, 1981; Carpenter and Lodge, 1986) are expected to be restricted to the plant bed, particularly in large and deep water bodies. It is possible, however, that biotic changes in the littoral zone in response to chemical-physical gradients (e.g., change in composition and abundance of prey organisms) will be carried across habitat boundary and influence limnetic communities using littoral resources.

Organic matter originating from the littoral zone may have metabolic importance especially in small shallow lakes where macrophytes are highly productive (Sculthrope, 1967; Wetzel and Hough 1973; Wetzel, 1979; Carpenter, 1983; Carpenter and Lodge, 1986). In large and deep lakes, the proportion occupied by the littoral region is often less than 10% of the total lake area (Gasith, 1991). In the Great Lakes, for example, the important spawning and nursery areas of most fish species are in littoral water less than 10 m deep (Goodyear et al., 1982; O'Gorman, 1983). In addition, large lakes are often more oligotrophic and support a lesser growth of aquatic macrophytes. In these and in highly eutrophic lakes, phytoplankton may dominate the production of organic matter and control the recycling of nutrients (Carpenter, 1983; Hough et al., 1989; Tilzer, 1990).

The effect of decreasing plant abundance with increasing lake size on biotic interactions is unclear. Biotic interactions can be influenced over a wide range of plant abundance. In the absence of alternative sources for physical structure, even sparse vegetation or isolated patches of macrophyte beds can be important in providing substrate for colonization, refuge, feeding, and spawning grounds. For example, in Orange Lake (Florida) young bluegills were found primarily in small isolated islands of panic grass (*Panicum* spp.), which constituted less than 2% of the lake's area (Conrow et al., 1990). Bluegills have been reported to prefer lateral concealment (Casterlin and Reynolds, 1978) and probably favored panic grass, which provided both protection from predators as well as access to open-water zooplankton (Conrow et al., 1990). In another case, Danehy (1984) and Danehy et al. (1991) found greater diversity and abundance of fish at relatively isolated cobbles and rubble sites than at sandy sites in Lake Ontario. Moreover, Danehy et al. (1991) found that yellow perch captured at the structured sites grew faster than those collected from the sandy sites. They attributed this difference in growth to lower energy expenditure associated with greater cover and lower predation risk as well as to higher food availability at the structured sites. At the sandy sites, individuals may have been required to "commute" more in search of cover and food. This led them to conclude that even small structured habitats may be important to local fish populations. The significance that this may have in the context of the whole lake ecosystem is yet to be evaluated. Another example in which a relatively limited plant structure can be important in a large lake situation is illustrated by the evidence that although spawning on macrophytes is unusual for salmonids, at least a portion of the population of lake trout (*Salvelinus namaycush*) in Lake Tahoe spawns in deepwater mounds (40–60 m deep) over beds of *Chara* (Beauchamp et al., 1992). No evidence of spawning was found over rocky formations that exist at various depths in the lake. Apparently, the *Chara* mounds are favored as they provide the basic requirements for successful egg incubation by anchoring the eggs against currents and providing protection from effective invertebrate and small vertebrate egg predators (Beauchamp et al., 1992). Similarly, it has been suggested that macrophyte beds provide cover for predation-vulnerable grazers such as large herbivorous zooplankton (Timms and Moss, 1984; Davies, 1985; Jeppesen et al., 1991; Moss et al., 1994; Lauridsen and Lodge, 1996; Jeppesen et al., this volume, Chapter 5; Lauridsen et al., this volume, Chapter 13). Survival of herbivorous zooplankton even in limited macrophyte coverage may accelerate establishment of larger populations (Lauridsen et al., 1996) that may, in turn, play a role in the switch from algae dominance to macrophytes (Scheffer et al., 1993; Hargeby et al., 1994; Jeppesen et al., this volume, Chapter 28). Due to their limited capacity for horizontal movement relative to fish, zooplankton would probably benefit less from scattered isolated plant beds than would fish. Restricted plant cover may therefore be expected to provide more effective refuge for zooplankton populations in small rather than large lake situations. Fish, however, are probably able to exploit isolated plant beds over a wider lake size range.

Freshwater fish use vegetation for cover (Crowder and Cooper, 1982; Tabor and Wurtsbaugh, 1991), foraging on benthos, epifauna, and prey organisms in the water among the vegetation (Fairchild, 1982; Mittelbach, 1984; Heck and Crowder, 1991; Diehl and Kornijów, this volume, Chapter 2) directly as food (Prejs,

1984) and as spawning and nursery sites (Goodyear et al., 1982; O'Gorman, 1983; Beauchamp et al., 1992). Most of the information on the use of structured habitats by fish is based on daytime studies. There is evidence, however, of much higher fish density in littoral habitats at night (Beauchamp et al., 1994) as well as a difference in size distribution of the fish between the day and night-time littoral zone assemblages (Gasith, Gafny, and Goren, unpublished data, Lake Kinneret). Further studies are needed to assess the importance of diurnal shifts in fish abundance, size, and species composition of littoral habitats.

As lake size and depth increase, macrophyte abundance declines, and structured habitats and associated resources may become in short supply (Gasith, 1991; Beauchamp et al., 1994). Consequently, competition over littoral resources (Mittelbach, 1988), particularly among species moving from the limnetic zone into the littoral region, is expected to increase with increasing lake size and depth (Gasith and Gafny, this volume, Chapter 24). In addition, unlike abiotic structures (e.g., rocky formations) macrophytes undergo temporal and spatial variations. In lakes where physical structure is provided mostly by macrophytes, organisms using littoral resources are forced to synchronize with the "window of opportunity" provided by macrophyte growth. If this is indeed so, competitive interactions over macrophyte-supported resources should be highest in large deep lakes where the abundance of macrophytes is low and in lakes where the period of macrophyte growth is shortest (e.g., high latitudes).

Conclusion

The changing influence of macrophytes along lake size and depth gradients is currently mostly speculative. Generally, the importance of macrophytes is expected to be proportional to their abundance in the water body, and thus their influence will decline with increasing lake size and depth. Existing information suggests that macrophytes can affect biotic interactions in situations in which they have no more limnological or metabolic significance. We therefore may conclude that only in shallow and small lakes can macrophytes potentially have significant effects on the physical-chemical condition in the water and sediment, on internal nutrient loading, and on lake productivity as well as on biotic interactions. In large deep lakes, macrophyte influence on lake ecosystem diminishes and is probably limited to some effect on biotic interactions.

Relatively small and isolated plant beds may have greater importance than have so far been assumed. In this connection, it is possible that cases of unexplained changes in zooplankton community structure and in fish population size and juvenile growth rate were linked to overlooked changes in the availability of structured habitats in the littoral zone.

A better understanding of macrophyte importance in relation to lake morphometry may require separate assessment of macrophyte effects on limnological conditions, metabolic processes, and biotic interactions. Due to the experimental limitations of ecosystem manipulation, particularly of large lakes, this will probably be achieved by long-term and comparative studies and possibly by more extensive use of artificial structures in lakes of various sizes.

Acknowledgments. The assistance of Merav Bing and Susan Gilman of the Institute for Nature Conservation Research, of Naomi Paz of the Zoology Department, Tel-Aviv University, and of the staff of the Department of Fisheries and Aquatic Sciences, University of Florida, Gainesville, is gratefully acknowledged. We thank John Barko, Sarig Gafny, and Daniel E. Canfield for constructive remarks.

References

Barko, J.W.; Smart, R.M. Effects of organic matter additions to sediments on the growth of aquatic plants. J. Ecol. 71:161–175; 1983.

Beauchamp, D.A.; Allen, B.B.; Richards R.C.; Wurtsbauch, W.A.; Goldman, C.R. Lake trout spawning in deepwater macrophyte beds. North Am. J. Fish. Manage. 12:442–449; 1992.

Beauchamp, D.A.; Byron, E.R.; Wurtsbauch, W.A. Summer habitat use by littoral-zone fishes in Lake Tahoe and effects of shoreline structures. North Am. J. Fish. Manage. 14:385–394; 1994.

Canfield, D.E., Jr.; Bachmann, R.W. Predictions of total phosphorus concentrations, chlorophyll-a and Secchi depth in natural and artificial lakes. Can. J. Fish. Aquat. Sci. 38:414–423; 1981.

Canfield, D.E., Jr.; Hoyer, M.V. Aquatic macrophytes and their relation to the limnology of Florida lakes. Final Report. Bureau of Aquatic Plants Management, Florida Department of Natural Resources, Tallahassee, FL; 1992.

Canfield, D.E., Jr.; Jones, R.J. Assessing the trophic status of lakes with aquatic macrophytes. Lake and reservoir management. EPA 440/5-84-001. Proceedings of the 3rd Annual Conference, Knoxville, TN, 1984.

Canfield, D.E., Jr.; Langeland, K.A.; Linda, S.B.; Haller, W.T. Relations between water transparency and maximum depth of macrophyte colonization in lakes. J. Aquat. Plant Manage. 23:25–28; 1985.

Carpenter, S.R. Enrichment of Lake Wingra, Wisconsin, by submerged macrophyte decay. Ecology 61:1145–1155; 1980.

Carpenter, S.R. Submersed vegetation: an internal factor in lake ecosystem succession. Am. Nat. 118:372–389; 1981.

Carpenter, S.R. Submersed macrophyte community structure and internal loading: relationship to lake ecosystem productivity and succession. In: Taggart, J., ed. Lake restoration, protection and management. Washington, DC:U.S.E.P.A.; 1983:105–111.

Carpenter, S.R.; Gasith, A. Mechanical cutting of submersed macrophytes: immediate effects on littoral water chemistry and metabolism. Wat. Res. 12:55–57; 1978.

Carpenter, S.R.; Greenlee, J.K. Lake deoxygenation after herbicide use: a simulation model analysis. Aquat. Bot. 11:173–186; 1981.

Carpenter, S.R.; Lodge, D.M. Effects of submersed macrophytes on ecosystem processes. Aquat. Bot. 26:341–370; 1986.

Casterlin, M.E.; Reynolds, W.W. Habitat selection by juvenile bluegill sunfish, *Lepomis macrochirus.* Hydrobiologia 59:75–79; 1978.

Cattaneo, L.B. Grazing on epiphytes. Limnol Oceanogr. 28:124–132; 1983.

Cattaneo, L.B.; Kalff, J. The relative contribution of aquatic macrophytes and their epiphytes to the production of macrophyte beds. Limnol. Oceanogr. 25:280–289; 1980.

Chambers, P.A. Nearshore occurrence of submerged aquatic macrophytes in relation to wave action. Can. J. Fish. Aquat. Sci. 44:1666–1669; 1987.

Chambers, P.A.; Kalff, J. Depth distribution and biomass of submerged macrophyte communities in relation to Secchi depth. Can. J. Fish. Aquat. Sci. 42:701–709; 1985.

Chick, J.H.; Mclvor, C.C. Patterns in the abundance and composition of fishes among beds of different macrophytes: viewing a littoral as a landscape. Can. J. Fish. Aquat. Sci. 51:2873–2882; 1994.

Conrow, R.; Zale, A.V.; Gregory, R.W. Distribution and abundance of early life stages of fishes in a Florida lake dominated by macrophytes. Trans. Am. Fish. Soc. 119:521–528; 1990.

Coops, H.; Boeters, R.; Smit, H. Direct and indirect effects of wave attack on helophytes. Aquat. Bot. 41:333–352; 1991.

Crowder, L.B.; Cooper, W.E. Habitat structural complexity and the interaction between bluegills and their prey. Ecology 63:1802–1813; 1982.

Danehy, R.J. Comparative ecology of fishes associated with natural cobble shoals and sand substrates in Mexico Bay, Lake Ontario. M.S. thesis, S.U.N.Y. College of Environmental Science and Forestry, Syracuse, NY; 1984.

Danehy, R.J.; Ringler, N.H.; Gannon, J.E. Influence of nearshore structure on growth and diets of yellow perch *(Perca flavesens)* and white perch *(Morone americana)* in Mexico Bay, Lake Ontario. J. Great Lake Res. 17:183–193; 1991.

Davidson-Arnott, R.G.D.; Pollard, W.H. Wave climate and potential longshore sediment transport, Nottawasaga Bay, Ontario. J. Great Lake Res. 6:54–67;1980.

Davies, J. Evidence for diurnal horizontal migration in *Daphnia hyalina lacustris* Sars. Hydrobiologia 120:103–105; 1985.

Diehl, S. Foraging efficiency of three freshwater fishes: effects of structural complexity and light. Oikos 53:207–214; 1988.

Dionne, M.; Folt, C.L. An experimental analysis of macrophyte growth forms as fish foraging habitat. Can. J. Fish. Aquat. Sci. 48:123–131; 1991.

Duarte, C.M.; Kalff, J. Littoral slope as a predictor of the maximum biomass of submerged macrophyte communities. Limnol. Oceanogr. 3:1072–1080; 1986.

Duarte, C.M.; Kalff, J. Latitudinal influence on the depth of maximum colonization and maximum biomass of submerged angiosperms in lakes. Can. J. Fish. Aquat. Sci. 44: 1759–1764; 1987.

Duarte, C.M.; Kalff, J.; Peters, R.H. Patterns in biomass and cover of aquatic macrophytes in lakes. Can. J. Fish. Aquat. Sci. 43:1900–1908; 1986.

Dvorak, J.; Best, E.P.H. Macro-invertebrates communities associated with macrophytes of Lake Vechten: structural and functional relationships. Hydrobiologia 95:115–126; 1982.

Eadie, J.; Keast, A. Resource heterogeneity and fish species diversity in lakes. Can. J. Zool. 62:1689–1695; 1984.

Emery, A.R. The basis of fish community structure: marine and freshwater comparison. Environ. Biol. Fish. 3:33–47; 1978.

Engel, S. Role and interactions of submerged macrophytes in a shallow Wisconsin lake. J. Freshwat. Ecol. 4:329–341; 1988.

Fairchild, G.W. Population responses of plant-associated invertebrates to foraging by largemouth bass fry *(Micrpterus salmonides)*. Hydrobiologia 96:169–176; 1982.

Gasith, A. Can littoral resources influence ecosystem processes in large, deep lakes? Verh. Int. Verein. Limnol. 24:1073–1076; 1991.

Godshalk, G.L.; Wetzel, R.G. Decomposition of aquatic angiosperms. Aquat. Bot. 5:281–354; 1978.

Goodyear, C.D.; Edsall, T.A.; Ormsby Dempsy, D.H.; Moss, G.D.; Polanski, P.E. Atlas of the spawning and nursery areas of Great Lakes fishes. Fish and Wildlife Service, U.S. Department of the Interior, Washington, DC; 1982.

Guillory, V.; Jones, M.D.; Rebel, M. A comparison of fish communities in vegetated and beach habitats. Florida Sci. 42:113–122; 1979.

Hakanson, L. The influence of wind, fetch and water depth on the distribution of sediments in Lake Vanern, Sweden. Can. J. Earth Sci. 14:397–412; 1977.

Hakanson, L. Bottom dynamics in lakes. Hydrobiologia 81:47–57;1981.

Hargeby, A.; Andersson, G.; Blindow, I.; Johansson, S. Trophic web structure in a shallow, eutrophic lake during dominance shift from phytoplankton to submersed macrophytes. Hydrobiologia 279/280:83–90; 1994.

Heck, K.L.; Crowder, L.B. Habitat structure and predator-prey interactions in vegetated aquatic systems. In: Bell, S.S.; McCoy, E.D.; Mushinsky, H.R., eds. Habitat structure, the physical arrangement of objects in space. London: Chapman & Hall; 1991:281–299.

Hough, R.A.; Fornwall, M.D.; Thompson, R.L.; Putt, D.A. Plant community dynamics in a chain of lakes: principle factors in the decline of rooted macrophytes with eutrophication. Hydrobiologia 173:199–217; 1989.

Hutchinson, G.E. A treatise on limnology. Vol. 3. Limnological botany. New York: Wiley; 1975.

Jaynes, M.L.; Carpenter, S.R. Effects of vascular and nonvascular macrophytes on sediment redox and solute dynamics. Ecology 67:875–882; 1986.

Jeppesen, E.; Kristensen, P.; Jensen, J.P.; Søndergaard, M.; Mortensen, E.; Lauridsen, T. Recovery resilience following a reduction in external phosphorus loading of shallow, eutrophic Danish Lakes: duration, regulating factors and methods for overcoming resilience. Mem. Ist. Ital. Idrobiol. 48:127–148; 1991.

Johnson, D.L. Lake Erie wetlands: fisheries considerations. In: Krieger, K.A., ed. Lake Erie estuarine systems: issues, resources, status, and management. NOAA Estuarine Programs Office, Washington, DC, NOAA EMS Series 14; 1989:257–274.

Keddy, P.A. Quantifying within-lake gradient of wave energy: interrelationships of wave energy substrate particle size and shoreline plants in Axe Lake Ontario. Aquat. Bot. 14:41–58; 1982.

Keddy, P.A. Shoreline vegetation in Axe Lake, Ontario: effect of exposure on zonation patterns. Ecology 62:331–344; 1983.

Kristensen, P.; Søndergaard, M.; Jeppesen, E. Resuspension in a shallow eutrophic lake. Hydrobiologia 228:101–109; 1992.

Lauridsen, T.L.; Lodge, D.M. Avoidance by *Daphnia magna* of fish and macrophytes: chemical cues and predator mediated use of macrophyte habitat. Limnol. Oceanogr. 41:794–798; 1996.

Lauridsen, T.L.; Pedersen, L.J.; Jeppesen, E.; Søndergaard, M. The importance of macrophyte bed size for cladoceran composition and horizontal migration in a shallow lake. J. Plankton Res. 18:2283–2294; 1996.

Lillie, R.A.; Budd, J. Habitat architecture of *Myriophyllum spicatum* L. as an index to habitat quality for fish and macroinvertebrates. J. Freshwat. Ecol. 7:113–125; 1992

Lorang, M.S.; Stanford, J.A. Variability of shoreline erosion and accretion within a beach compartment of Flathead, Montana. Limnol. Oceanogr. 38:1783–1795; 1993.

MacCrimmon, H.R. Nutrients and sediment retention in a temperate marsh ecosystem. Int. Rev. Ges. Hydrobiol. 65:719–744; 1980.

Miller, A.C.; Beckett, D.C.; Way, C.M.; Bacon, E.J. The habitat value of aquatic macropytes for macroinvertebrates. Technical Report A-89–3, U.S. Army Corps of Engineers, Washington, DC; 1989.

Mittelbach, G.G. Predation and resource partitioning in two sunfishes (Centrachidae). Ecology 65:499–513; 1984.

Mittelbach, G.G. Competition among refuging sunfishes and effects of fish density on littoral zone invertebrates. Ecology 69:614–623; 1988.

Morin, J.O. Initial colonization of periphyton on natural and artificial apices of *Myriophyllum heterophyllum* Michx. Freshwat. Biol. 16:685–694; 1986.

Moss, B. The microwaterscape—a four dimensional view of the interactions among water chemistry, phytoplankton, periphyton, macrophytes, animals and ourselves. Wat. Sci. Techn. 32:105–116; 1995.

Moss, B.; McGowan, S.; Carvalho, L. Determination of phytoplankton crop by top-down and bottom-up mechanisms in a group of English lakes, the West Midland meres. Limnol. Oceanogr. 39:1020–1029; 1994.

O'Gorman, R. Distribution and abundance of larval fish in the nearshore waters of western Lake Erie. J. Great Lakes Res. 9:14–22; 1983.

Osgood, R.A. Lake mixis and internal phosphorus dynamics. Arch. Hydrobiol. 113:629–638; 1988.

Pearsall, W.H. The aquatic and marsh vegetation of Esthwaite water. J. Ecol. 5:180–201; 1917.

Pearsall, W.H. The aquatic vegetation of the English lakes. J. Ecol. 8:163–201; 1920.

Pearsall, W.H. Dynamic factors affecting aquatic vegetation. Proc. Int. Congr. Plant Sci. 1:667–672; 1929.

Planter, M. Physical and chemical conditions in the helophyte zone of the lake littoral. Pol. Arch. Hydrobiol. 20:1–7; 1973.

Prejs, A. Herbivory by temperate freshwater fishes and its consequences. Environ. Biol. Fish. 10:281–296; 1984.

Prentki, R.T.; Adams, M.S.; Carpenter, S.R.; Gasith, A.; Smith, C.S.; Weiler, P.R. The role of submerged weedbeds in internal loading and interception of allochthonous materials in Lake Wingra, Wisconsin. Arch. Hydrobiol. Suppl. 57(2):221–250; 1979.

Rounsefell, G.A. Fish production in lakes as a guide for estimating production in proposed reservoirs. Copeia 1:29–40; 1946.

Savino, J.F.; Stein, R.A. Predator–prey interaction between largemouth bass and bluegills as influenced by simulated submerged vegetation. Trans. Am. Fish. Soc. 111:255–266; 1982.

Scheffer, M.; Hosper, S.H.; Meijer, M.L.; Moss, B.; Jeppesen, E. Alternative equilibria in shallow lakes. Trends. Ecol. Evol. 8:275–279; 1993.

Schram, H.L., Jr.; Jirka, K.J.; Hoyer, M.V. Epiphytic macroinvertebrates on dominant macrophytes in two central Florida lakes. J. Freshwat. Ecol. 4:151–161; 1987.

Schriver, P.; Bøgestrand, J.; Jeppesen, E.; Søndergaard, M. Impact of submerged macrophytes on the interactions between fish, zooplankton and phytoplankton: large scale enclosure experiments in shallow eutrophic lake. Freshwat. Biol. 33:255–270; 1995.

Sculthrope, C.D. The biology of aquatic vascular plants. London: Edward Arnold; 1967.

Sozska, G.J. Ecological relations between invertebrates and submerged macrophytes in the lake littoral. Ekol. Pol. 23:393–415; 1975.

Spence, D.H.N. The zonation of plants in freshwater lakes. In: MacFadyan, A.; Ford, E.D., eds. Advances in ecological research. London: Academic Press; 1982:37–126.

Tabor, R.A.; Wurtsbaugh, W.A. Predation risk and the importance of cover for juvenile rainbow trout in lentic systems. Trans. Am. Fish. Soc. 120:728–738; 1991.

Tilzer, M.M. Specific properties of large lakes. In: Tilzer, M.M.; Cerruya, C., eds. Large lakes, ecological structure and function. New York: Springer-Verlag; 1990:39–45.

Timms, R.M.; Moss, B. Prevention of growth of potentially dense phytoplankton populations by zooplankton grazing, in the presence of zooplanktivorous fish, in a shallow wetland ecosystem. Limnol. Oceanogr. 29:472–486; 1984.

Vollenweider, R.A. Input-output models with special reference to phosphorus loading concept in limnology. Schweiz. Z. Hydrol. 37:53–84; 1975.

Wetzel, R.G. The role of the littoral zone and detritus in lake metabolism. Arch. Hydrobiol. 13:145–161; 1979.

Wetzel, R.G. Detritus, macrophytes and nutrient cycling in lakes. Mem. Ist. Ital. Idrobiol. 47:233–249; 1990.

Wetzel, R.G.; Hough, R.A. Productivity and the role of aquatic macrophytes in lakes. An assessment. Pol. Arch. Hydrobiol. 20:9–19; 1973.

30. Nutrient-Loading Gradient in Shallow Lakes: Report of the Group Discussion

Stephen R. Carpenter, Ellen van Donk, and Robert G. Wetzel

Nutrient gradients are an important means of organizing limnological information. Nutrients are among the most important controls of lake ecosystem processes. Other key factors such as light, macrophytes, and predation change in systematic ways as nutrient availability increases or decreases. Consequently, the nutrient gradient is a useful tool for explaining our understanding of lake diversity. The nutrient gradient can also be used to make predictions about the state of lakes as nutrient status changes.

Our discussion of the nutrient gradient is based on several assumptions. The nutrient gradient represents a continuum of nutrient input (or loading) rates. We do not consider transient dynamics that result from abrupt changes in nutrient input rate. Rather, we consider average or prevalent conditions in lakes that have experienced a given rate of nutrient input for an extended period of time: long enough for several hydrologic flushings, and at least several generations of the dominant organisms.

The generalizations below are based on additional conditions pertinent to the goals of this book. The lakes are too shallow to be thermally stratified. We assume a temperate climate; that the hydraulic residence time, dissolved organic carbon inputs, and alkalinity remain constant; that there are no important species invasions or extirpations; and that harvest rates of the dominant fishes remain constant.

Given these assumptions, discussants converged on a general view of shallow lake ecosystems across the nutrient loading gradient (Fig. 30.1). The possibility of

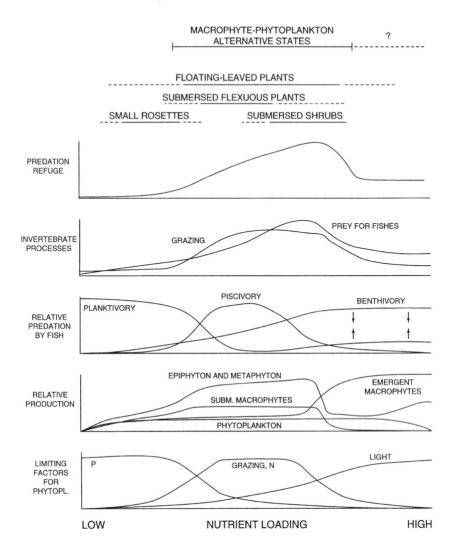

Figure 30.1. Trends across the nutrient-loading gradient in shallow lakes. Solid bars show the range of nutrient input rates where processes or plant types are prevalent; dashed lines show the range where the process or plant type may occur. Curves are plotted for the relative effectiveness of the predation refuge; the relative magnitude of grazing by macroinvertebrates (zooplankton and snails); the relative availability of invertebrate prey to fish; the relative importance of planktivory, piscivory, and benthivory by fish; the relative magnitude of production by phytoplankton, submerged macrophytes, attached algae (epiphyton and metaphyton), and emergent macrophytes; and the prevalent limiting factors for phytoplankton production.

rapid transitions between alternative states of macrophyte or phytoplankton dominance occurs through the middle of the gradient. At low nutrient input rates, nutrient levels are not sufficient to support extensive macrophyte growth. At high nutrient input rates, extreme algal turbidity prevents development of submerged macrophytes.

Patterns described in the remainder of Fig. 30.1 pertain to the macrophyte-dominated state of shallow lakes.

Although floating-leaved macrophytes can grow across most of the nutrient input gradient, the physiognomy of submerged macrophytes shows some trends. Small rosettes (e.g., isoetids or *Lobelia dortmanna*) tend to prevail at the low end of the gradient. Submerged flexuous plants (e.g., *Chara* or *Elodea*) tend to dominate through the middle of the gradient, although they can occur at more extreme nutrient input rates. Large branched plants with multiple shoots resembling "underwater shrubs" (Swindale and Curtis, 1957) include the larger *Myriophyllum* and *Potamogeton* species. These tend to occur at somewhat higher nutrient input rates than the submerged flexuous plants.

Macrophytes provide an important refuge from predation for invertebrates and fish. The effectiveness of this refuge depends on macrophyte physiognomy and abundance. The refuge may be most effective at moderately high nutrient-loading rates.

Invertebrate processes that coincide with macrophyte abundance include grazing by zooplankton and snails and production of prey for fish. Macrophytes shelter herbivorous zooplankton from predation by fish. The flexuous-stemmed macrophytes and underwater shrubs that occur at moderate nutrient loads provide the most effective refuge. The availability of periphyton and epiphyton grazed by snails is also maximal in the middle of the nutrient-loading gradient. Availability of invertebrates to fish predators is the net result of a balance between invertebrate production and the effectiveness of the the shelter from predation that macrophytes provide. Invertebrate prey are most available at moderately high nutrient input rates.

The comparative importance of planktivory, piscivory, and benthivory by fish changes along the nutrient-loading gradient. Planktivory is most important at relatively low nutrient input rates, whereas piscivory is most important at intermediate nutrient input rates. At high nutrient input rates, relative piscivory decreases and fish diets may be dominated by either benthos or plankton.

There are shifts along the nutrient gradient in the relative production by different primary producers. The particular mix of primary producers on a given site will depend on the local topography and hydrology, as well as the ecosystem boundaries fixed by the investigator. For example, in a shallow floodplain lake, emergent macrophytes might account for most of the primary production at all nutrient levels. In sketching Fig. 30.1, we envisioned a well-circumscribed shallow lake in rolling topography and assumed that wetlands were included within the lake boundary. In this situation, the relative contribution of phytoplankton is modest across a broad range of nutrient input rates. However, the presence or absence of buoyant (and sometimes toxic) cyanobacteria can cause significant variations in biological structure. Submerged macrophytes and attached algae attain their highest importance through the middle of the nutrient-loading gradient. Production by emergent macrophytes may predominate when nutrient input rates are relatively high.

The discussants did not agree about the relative productivity of phytoplankton and emergent macrophytes at high nutrient loading. If the lake is defined as open water exclusive of wetland, then phytoplankton production predominates. Conversely, if the lake boundaries include the wetland, then emergent macrophyte production predominates. But the debate is not merely a matter of ecosystem boundaries. It is possible that highly eutrophic shallow lakes are highly variable with respect to the dominant primary producers. For example, year-to-year variation in the timing and duration of spring conditions can affect the relative dominance of different primary producers in summer.

There is substantial change in the factors that limit phytoplankton production across the nutrient gradient. Phosphorus is most limiting when nutrient input rates are relatively low. Nitrogen and/or grazing are most limiting through the middle of the gradient. Limitation by nitrogen appears to be the result of denitrification and the sequestration of nitrogen in macrophyte biomass. At relatively high nutrient-loading rates, phytoplankton become limited by light.

Although this summary represents the consensus of the group discussion, readers should be aware that the patterns were debated, sometimes vigorously. Some of the patterns are well grounded in data, and others represent the collective opinion of the group. The summary is best viewed as a synthesis of current views and a set of hypotheses worthy of further evaluation by careful comparative studies.

Acknowledgments. We thank Kirsten Christofferson, Brian Moss, and Morten Søndergaard for helpful comments on the draft.

Reference

Swindale, D.N.; Curtis, J.T. Phytosociology of the larger submerged plants in Wisconsin lakes. Ecology 38:397–407; 1957.

31. Alternative Stable States

Marten Scheffer and Erik Jeppesen

Introduction

The chapters in this book show how profoundly the presence of abundant submerged macrophytes may change the conditions in a lake. Macrophytes and associated epiphytes compete with phytoplankton by taking up nutrients from the water, by shading, and perhaps also by secreting allelopathic substances (Søndergaard and Moss, this volume, Chapter 6). They reduce wind-induced water movements which results in increased sedimentation and reduced resuspension (Barko and James, this volume, Chapter 10; Van den Berg et al., this volume, Chapter 25). Macrophytes also represent a large surface area and therefore enhance the density of surface-associated organisms (Diehl and Kornijów, this volume, Chapter 2; Jeppesen et al., this volume, Chapter 5). Such a local concentration of organisms together with the reduced water movements often leads to marked gradients in oxygen conditions, which have a significant impact on phosphorus and nitrogen dynamics and pH (Barko and James, this volume, Chapter 10; Søndergaard and Moss, this volume, Chapter 6). The plant-associated organisms comprise animals such as snails that graze periphyton from plant surfaces (Brönmark and Vermaat, this volume, Chapter 3; Crowder et al., this volume, Chapter 14; Diehl and Kornijów, this volume, Chapter 2; Jones et al., this volume, Chapter 4), but plant beds also host macrofiltrators such as mussels (e.g., *Anodonta*) and microfiltrators (e.g., *Sida*), which may occur in such high densities that they help reducing turbidity of the water within the plant beds (Jeppesen et al., this volume, Chapter 5).

Plants are also an important refuge for many animals against predation. Pelagic zooplankton concentrate in plant beds during daytime to reduce predation by fish. At night, they migrate to open water to graze on phytoplankton (Jeppesen et al., this volume, Chapter 5; Lauridsen et al., this volume, Chapter 13). Plant beds also act as a refuge for small fish (Diehl and Kornijów, this volume, Chapter 2; Persson and Crowder, this volume, Chapter 1). In mesotrophic and slightly eutrophic lakes, the presence of vegetation helps, causing a shift in the fish community toward dominance by predatory fish (northern Europe) (Persson and Crowder, this volume, Chapter 1) and molluscivorous fish (Brönmark and Vermaat, this volume, Chapter 3). This has important implications for trophic interactions in the system (Diehl and Kornijów, this volume, Chapter 2; Jeppesen et al., this volume, Chapter 5; Persson and Crowder, this volume, Chapter 1; Søndergaard et al., this volume, Chapter 15). Obviously, plants also promote the abundance of herbivores such as birds and crayfish, which in turn may influence plant biomass and composition (Lodge, et al., this volume, Chapter 8; Mitchell and Perrow, this volume, Chapter 9; Søndergaard et al., this volume, Chapter 20; van Donk, this volume, Chapter 19; Wass and Mitchell, this volume, Chapter 18).

In this chapter, we focus on the idea that several of these vegetation-induced changes in the ecosystem indirectly lead to an improvement in the conditions for growth of submerged plants and that the resulting positive feedback may cause vegetation dominance to be an alternative stable state to a turbid unvegetated situation in shallow freshwater lakes. The remainder of this chapter consists of three sections: a brief review of the work on positive vegetation feedbacks to demonstrate that the phenomenon is by no means restricted to lakes; a summary of the theory on alternative stable states in shallow lakes; and an overview of the regulatory mechanisms considered important for stabilizing the vegetation-dominated state and the factors causing a possible collapse, reflecting the ideas expressed by the participants in the workshop discussion.

Positive Feedbacks in the Vegetation

The idea that the establishment and growth of plants may be stimulated by the presence of other plants was suggested long ago in terrestrial vegetation ecology (Clements et al., 1926; Connell and Slayter, 1977). Although for decades the emphasis in explaining vegetation dynamics has been on competition, there has recently been a renewed interest in the positive effects of plants on other plants (Hunter and Aarssen, 1988; Goldberg, 1990). It has been hypothesized that this so-called facilitation is especially important under harsh conditions (Bertness and Yeh, 1994). Indeed, most terrestrial evidence comes from ecosystems where plants are exposed to severe stress (e.g., as a result of heat and desiccation). In such situations, the establishment of new plants is often restricted to the shade under the canopy of other plants, the so-called nurse plants. Besides ameliorating temperature and moisture conditions, nurse plants have been shown to improve soil properties (accumulation of nutrients and organic matter) and to reduce the probabilities of mechanical or herbivory damage and further colonization by attracting seed dispersers (Hunter et al., 1988).

The discussion on whether facilitation or rather competition is more important in different situations has been confusing at times. The problem is that positive and negative interactions always occur simultaneously. Although improving some environmental conditions, plants will tend to have negative effects on other factors. On the one hand, terrestrial plants may enhance air humidity and prevent extreme temperature fluctuations, but on the other hand, they may limit the potential growth of newly established plants by reducing the availability of light and soil water (Franco and Nobel, 1988, 1989) or by excreting allelopathic substances (Muller, 1953; Muller and Muller, 1956; Callaway et al., 1991). Real facilitation occurs only if the positive effects of plants on other plants outweigh the negative ones (Holmgren et al., 1997).

In dry terrestrial vegetation where water is the main limiting factor, the positive effects of improved moisture and soil conditions in the vicinity of other plants will often dominate over the negative effects of reduced light. Indeed, this positive effect of plants on plants is essential in the understanding of one of the world's most threatening ecological problems: desertification. In very dry areas, savanna vegetation may persist, but once the vegetation disappears due to, for instance, overgrazing, the conditions on the bare soil may be too harsh to allow recolonization. As a result, the change to desert is hard to reverse.

At first glance, the parallel between deserts and lakes may seem quite remote as the factors governing vegetation development of two environments are entirely different. Nonetheless, on a higher abstraction level the disappearance of submerged plants from eutrophic shallow lakes is, in fact, quite similar to desertification. In both cases, the abiotic conditions at the un-vegetated state may be too harsh to allow (re)colonization. Once vegetation is present, however, the plants may ameliorate the conditions sufficiently to ensure vegetation persistence.

In most lakes, light is likely to be a main factor limiting the colonization by submerged plants (Hutchinson, 1975; Chambers and Kalff, 1985; Vant et al., 1986; van Dijk and Van Vierssen, 1991; Van Dijk et al., 1992; Skubinna et al., 1995). A positive feedback is caused by the fact that water clarity tends to increase in the presence of plants (Schreiter, 1928; Canfield et al., 1984; Pokorny et al., 1984; Jeppesen et al., 1990; Faafeng and Mjelde, this volume, Chapter 27; van den Berg et al., this volume, Chapter 25). This may allow a vegetated state to be one of two alternative stable states. The explanation in a nutshell is that in very turbid water, light conditions are insufficient for vegetation development, but once vegetation is present the water clears up and the improved light conditions allow the persistence of a lush vegetation (Scheffer, 1990; Scheffer et al., 1993).

The theory of how the vegetation–turbidity interactions and other mechanisms may lead to alternative stable states in shallow lakes and the evidence for this phenomenon are treated extensively elsewhere (Scheffer, 1997). The next section gives a brief general explanation of the theory of alternative stable states and its application to shallow lakes.

Theory of Alternative Stable States

There is a long tradition of studying the effect of changing conditions on equilibria by using models. Obviously, models are usually rather extreme simplifications of what happens in nature. However, some general observations apply to all the models studied so far and to all "dynamic systems" as well, including ecosystems. Usually the response to changing conditions is gradual, such as the increase in phytoplankton biomass and the decrease in submerged plant abundance with increasing nutrient loading in deep lakes (Fig. 31.1A). However, equilibria may also split, merge, or become unstable or negative. Such qualitative changes are called "bifurcations" in dynamic systems theory and correspond to more remarkable system responses to changing conditions.

Alternative equilibria, the phenomenon of interest here, occur when the equilibrium line is "folded." This is thought to be the case with phytoplankton biomass and vegetation abundance in most shallow lakes (Fig. 31.1B). In this case, three equilibria exist over a certain range of conditions ($n_1 < n < n_2$). Two of these equilibria are stable (the upper and lower branch of the folded curve), whereas the intermediate one (the dashed middle branch) is unstable. The unstable equilibrium marks the "breakpoint" of the system. If the initial state is above this breakpoint, the system state will tend to move to the upper stable equilibrium branch, whereas below the breakpoint it moves toward the lower branch. The unstable and the stable branches merge in the inflection points (f_1 and f_2), called "fold bifurcations."

The system may move from one stable state to another in two distinct ways. First, environmental conditions such as nutrient loading may change. This results merely in gradual changes as long as the system is at one of the stable branches. However, when the inflection point of the current branch is reached, the system will jump to the other branch. In the vicinity of the bifurcation points (f_1 and f_2), a minute change in conditions may result in a major "catastrophic" jump, and this is the reason why the characteristic fold of the equilibrium line is often called a "catastrophe fold." Note that reversal of the change in environmental conditions after a catastrophic transition does not simply lead to a reversed response. Instead, the system stays at the newly reached stable branch until the other bifurcation point is reached and a jump back to the first branch occurs. This tendency to remain at the same state despite changing conditions over a certain range ($n_1 < n < n_2$) is called "hysteresis."

Disturbance is the second way in which the system may be caused to switch from one stable branch of the hysteresis to the other one. In situations with only one stable state, the system will always return to the same state after disturbance (e.g., fish kill or destruction of vegetation). However, in the range of conditions in which two stable states exist, catastrophic systems may be forced from one state to another by disturbance, provided that such disturbance is sufficiently strong to bring the system state over the breakpoint, marking the beginning of the attraction area of the alternative stable state.

Catastrophic systems may lose their catastrophic properties when certain conditions change. In the case of submerged vegetation, depth is considered to be such

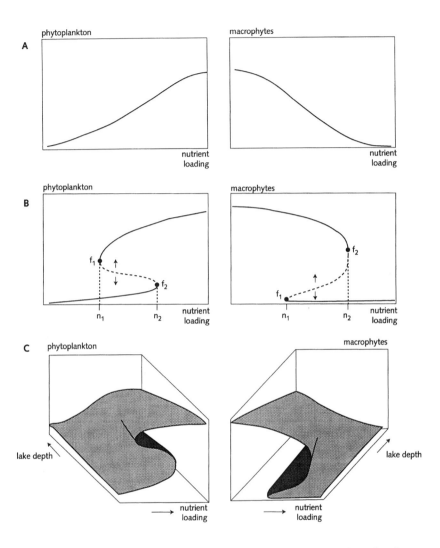

Figure 31.1. In deep lakes, the response of phytoplankton and macrophyte abundance to nutrient loading is usually smooth. (B) In shallow lakes, equilibrium abundance of phytoplankton and macrophytes plotted aginst the nutrient loading (n) has a folded shape, implying that over a certain range of conditions ($n_1 < n < n_2$), two alternative stable states exist (see text for further discussion). (C) In practice, all situations between the extremes represented in the upper two panels (for deep and shallow lakes) may exist.

a condition. As argued, the response to nutrient loading is thought to be smooth rather than catastrophic in deep lakes (Fig. 31.1A). Because lakes range gradually from deep to shallow ones, the corresponding catastrophe folds will range gradually from pronounced to nonexistent ones (Fig. 31.1C). Note that a change in such a second environmental factor (in this case, lake depth) may also be sufficient to let the system meet a fold bifurcation and jump to the alternative stable state.

Clearly, the concept of stable states, or equilibria, is of limited use when studying real ecosystems. Natural communities are never simply in equilibrium. This is not only due to stochasticity of weather and hydrology but also to intrinsic population cycles and the seasonal cycle in light and temperature conditions. In fact, instead of using the terms *equilibrium* or *stable state,* it would be more appropriate to use a more neutral term such as *asymptotic regime* to refer to the dynamic pattern to which a community tends to settle. Despite these complications, the general properties of catastrophic systems described here remain equally relevant. Indeed, studies with more realistic and complicated models reveal that it is perfectly possible to have multiple complex "asymptotic regimes" instead of multiple "stable states" and that these complex structures largely follow the rules explained here for their simpler counterparts.

The two main take-home messages from dynamic systems theory are that (1) the existence of alternative equilibria is always restricted to a limited range of environmental conditions (e.g., nutrients) and that (2) changes in other conditions (e.g., depth) may lead to the disappearance of the hysteresis to the first condition. Because the existence of alternative stable states allows restoration of a clear vegetated state by means of disturbance-oriented methods such as biomanipulation or temporary adjustment of the water level, it would be of great interest to know the conditions under which alternative stable states may exist. So far, the available data suggest that a stable clear vegetated state is unlikely to occur when the total P level of the lake water exceeds 0.05–0.15 mg/L (Hosper and Jagtman, 1990; Jeppesen, 1990) unless the N input is low (Jeppesen et al., 1990). However, the critical nutrient level for the clear state is likely to depend on other factors as well. For instance, it is likely to be inversely related to the size and depth of a lake (Scheffer, 1997).

How It Works in Lakes; Group Discussion

The workshop discussion focused on two questions: (1) the mechanisms thought to be most important for stabilizing the vegetation-dominated state, and (2) the factors that may cause a switch from a vegetation-dominated state to a turbid state with few submerged plants. The opinions expressed are summarized below. They should not be regarded as established facts as they cover much of the range from "fact" to "speculation." In our formulation of this section, we have tried to reflect the degree of consensus among the participants on the various ideas.

What Stabilizes the Vegetated State?

The participants agreed that the low phytoplankton biomass in many vegetation-dominated lakes is largely due to two different mechanisms: reduction of nutrient availability to phytoplankton and enhanced potential for top-down control of algae.

Nitrogen is the nutrient most often found to be limiting in P-rich vegetated lakes, but it was argued that in situations with a moderately dense vegetation and moderate to low total P, phosphorus may become the limiting nutrient. Nutrient uptake by the vegetation contributes to the low nutrient availability in the water column, but it was stressed that periphyton growing on plants may also take up considerable amounts of nutrients from the water column. As for nitrogen, increased denitrification in the presence of aquatic vegetation is thought to be an important explanation of the low water column concentrations.

Enhanced top-down control of phytoplankton is considered to be the result of improved chances of large herbivorous cladocerans to survive predation by planktivorous fish when vegetation is available as a (daytime) refuge. Not all participants were, however, convinced of the effectiveness of this refuge when fish are abundant. Another relevant trend is that the share of piscivorous fish is usually higher in vegetated lakes. Predation pressure by piscivores may reduce planktivore densities and thus release planktivorous predation on cladoceran grazers. Conversely, plants may also act as a refuge for 0^+ fish, increasing the predation pressure on cladocerans. The resultant effects are unclear; they seem to vary from lake to lake and may also depend on the nutrient level. Although free-living and plant-associated cladocerans were mostly mentioned with respect to phytoplankton grazing, it was argued that the grazing pressure by mussel populations may be high in some vegetated systems, and some examples were given.

Suppression of phytoplankton by allelopathic substances excreted by macrophytes was not excluded by the discussants, but this mechanism was regarded as less important than nutrient limitation and top-down control.

It was agreed that prevention of sediment resuspension may be an important reason for enhanced water clarity in vegetated systems, and also sediment stabilization by plants was believed to facilitate establishment of new plants in lakes otherwise having very loose and floculent sediments. Some argued that the more stagnant water in the plant beds facilitates phytoplankton sedimentation and thus high transparency. However, some phytoplankton species may, via their ability to move (e.g., flagellates) or by buoyancy (cyanobacteria), compensate for this and replace the more sedimentation-sensitive species. This is supported by observations made in eutrophic brackish lakes with low zooplankton grazing pressure showing high densities of phytoplankton, even in dense plant beds.

The role of snails in removing periphyton from plants was considered an important mechanism for maintaining plant dominance, whereas the importance of small invertebrates such as chironomids and periphyton-eating microcrustaceans is less clear. Several participants had observed that plants overgrown with periphyton showed earlier senescence than plants kept clean by grazers.

What May Cause the Vegetated State to Disappear?

Increased nutrient loading is considered to be an important reason for the switch from a macrophyte-dominated to an unvegetated turbid state in many shallow lakes. However, it was argued that some lakes may resist extremely high nutrient loading and remain very clear and densely vegetated. Increased growth of phytoplankton and periphyton due to better nutrient availability together with a relatively strong increase in plankti-benthivorous fish eating snails and cladocerans were mentioned as mechanisms through which nutrient loading affects the clear vegetated state. The participants were positive toward the hypothesis that nitrogen loading is a serious threat to vegetated lakes, nitrogen being considered the limiting nutrient to phytoplankton in the many eutrophic vegetated lakes. Several participants have observed high phosphorus concentrations in densely vegetated lakes, probably due to release from anoxic sediments. Provided that nitrogen availability is low, this was, however, considered no great threat to the clearwater state. It was further argued that insecticides may occasionally contribute to a collapse of the vegetated clear state by affecting, for example, the cladoceran population and snails.

Many of the mechanisms suggested as being responsible for a shift to the turbid state involve direct damage to the vegetation. Herbivory by invertebrates was thought to be of minor importance relative to production, and it was suggested that the high C/N ratio of the plant material under such conditions may explain low invertebrate herbivory losses as it makes the plants relatively unattractive to consumers. Introduction of carp or grass-carp, however, may provoke a shift to the unvegetated state in many lakes. The opinions about the potential effect of herbivorous birds were divided, but most evidence indicates that the effects of bird grazing were minor during summer unless plant density is low. However, herbivory by autumn foraging and overwintering bird populations may be severe and a potential threat to the survival of plants such as *Elodea* that overwinter as shoots and do not have hibernation structures enabling them to escape consumption. Likewise, in the recovery phase when plants are about to colonize, bird grazing may delay plant appearance. Treatment of the plants with herbicides may also induce a switch to a stable unvegetated state. In some cases, extreme meteorological conditions are thought to have damaged the vegetation sufficiently to cause a switch to the turbid state. Violent storms are among the more spectacular events reported to have resulted in a permanent loss of vegetation. In Swedish lakes, freezing of the soil during winters with very low water levels is thought to have been the cause of vegetation loss and a switch to a turbid state. In some lakes, a water level increase, either due to meteorological conditions or human interference, has also led to the disappearance of submerged macrophytes.

The participants agreed that there is still a great need for research into the causes of the shift between the turbid and the clearwater state, including especially large-scale and whole-lake experiments, long-term monitoring (comprising open water as well as vegetated areas), paleoecological investigations, and modeling.

Acknowledgments. We thank Milena Holmgren for helping us present an overview of positive feedbacks in terrestrial plant communities. Egbert van Nes and Rita van Leeuwen have helped in gathering and organizing the references.

References

Bertness, M.D.; Yeh, S.M. Cooperative and competitive interactions in the recruitment of marsh elders. Ecology 75:2416–2429; 1994.

Callaway, R.M.; Nadkarni, N.M.; Mahall, B.E. Facilitation and interference of *Quercus douglasii* on understory productivity in central California. Ecology 72:1484–1499; 1991.

Canfield, D.E.J.; Shireman, J.V.; Colle, D.E.; Haller, W.T. Prediction of chlorophyll *a* concentrations in Florida lakes: importance of aquatic macrophytes. Can. J. Fish. Aquat. Sci. 41:497–501; 1984.

Chambers, P.A.; Kalff, J. Depth distribution and biomass of submersed aquatic macrophyte communities in relation to Secchi depth. Can. J. Fish. Aquat. Sci. 42:701–709; 1985.

Clements, F.E.; Weaver, J.; Hanson, H. Plant competition: an analysis of the development of vegetation. Washington, DC: Carnegie Institute; 1926.

Connell, J.H.; Slatyer, R.O. Mechanisms of succession in natural communities and their role in community stability and organization. Am. Nat. 111:1119–1144; 1977.

Franco, A.C.; Nobel, P.S. Interactions between seedlings of *Agave deserti* and the nurse plant *Hilaria rigida*. Ecology 69: 1731–1740; 1988.

Franco, A.C.; Nobel, P.S. Effect of nurse plants on the microhabitat and growth of cacti. J. Ecol. 77:870–886; 1989.

Goldberg, D.E. Components of Resource Competition in Plant Communities. In: Grace, J.B.; Tilman, D., ed. Perspectives on Plant Competition. New York: Academic Press, Inc., 1990: 27–49.

Holmgren, M.; Scheffer, M.; Huston, M.A. . The balance of facilitation and competition in plant communities. Ecology 7017:1966–1975; 1997.

Hosper, S.H.; Jagtman, E. Biomanipulation additional to nutrient control for restoration of shallow lakes in the Netherlands. Hydrobiologia 200–201:523–534; 1990.

Hunter, A.F.; Aarssen, L.W. Plants helping plants. BioScience 38:34–40; 1988.

Hutchinson, G.E. A Treatise on Limnology. Volume III, Limnological Botany. New York: John Wiley & Sons; 1975.

Jeppesen, E.; Jensen, J.P; Kristensen, P.; Søndergaard, M.; Mortensen, E.; Sortkjær, O.; Olrik, K. Fish manipulation as a lake restoration tool in shallow, eutrophic, temperate lakes 2: threshold levels, long-term stability and conclusions. Hydrobiologia 200/201:219–228; 1990.

Muller, C.H. The association of desert annuals with shrubs. Am. J. Bot. 40:53–60; 1953.

Muller, W.H.; Muller, C.H. Association patterns involving desert plants that contain toxic products. Am. J. Bot. 43:354–361; 1956.

Pokorny, J.; Kvet, J.; Ondok, J.P.; Toul, Z.; Ostry, I. Production-ecological analysis of a plant community dominated by *Elodea canadensis*. Aquat. Bot. 19:263–292; 1984.

Scheffer, M. Multiplicity of stable states in freshwater systems. Hydrobiologia 200/201: 475–486; 1990.

Scheffer, M. Ecology of shallow lakes. New York: Chapman & Hall; 1997.

Scheffer, M.; Hosper, S.H.; Meijer, M.-L.; Moss, B.; Jeppesen, E. Alternative equilibria in shallow lakes. Trends Ecol. Evol. 8:275–279; 1993.

Schreiter, T. Untersuchungen über den Einfluss einer Helodeawucherung auf das Netzplankton des Hirschberger Grossteiches in Bohmer in den Jahren 1921 bis 1925 incl. V. Praze. Prague 98; 1928.

Skubinna, J.P.; Coon, T.G.; Batterson, T.R. Increased abundance and depth of submersed macrophytes in response to decreased turbidity in Saginaw Bay, Lake Huron. J. Great Lakes Res. 21:476–488; 1995.

Van Dijk, G.M.; Van Vierssen, V. Survival of a *Potamogeton pectinatus* L. population under various light conditions in a shallow eutrophic lake Lake Veluwe in The Netherlands. Aquat. Bot. 39:121–130; 1991.

Van Dijk, G.M.; Breukelaar, A.W.; Gijlstra, R. Impact of light climate history on seasonal dynamics of a field population of *Potamogeton pectinatus* L. during a three year period 1986–1988. Aquat. Bot. 43:17–41; 1992.

Vant, W.N.; Davies-Colley, R.J.; Clayton, J.S.; Coffey, B.T. Macrophyte depth limits in north island New-Zealand lakes of differing clarity. Hydrobiologia 137:55–60; 1986.

Index

Entries occurring in figures are followed by an f; those occurring in tables, by a t.

Ecological Studies

Volumes published since 1992